Molecular Genetics of Pancreatic Cancer

Diane M. Simeone • Anirban Maitra
Editors

Molecular Genetics of Pancreatic Cancer

Springer

Editors
Diane M. Simeone
University of Michigan
Ann Arbor, MI, USA

Anirban Maitra
Johns Hopkins University
Baltimore, MD, USA

ISBN 978-1-4614-6548-5 ISBN 978-1-4614-6549-2 (eBook)
DOI 10.1007/978-1-4614-6549-2
Springer New York Heidelberg Dordrecht London

Library of Congress Control Number: 2013934964

Printed on acid-free paper

Springer is part of Springer Science+Business Media (www.springer.com)

Contents

Genomic Alterations in Sporadic Pancreatic Cancer

Marco Dal Molin and Anirban Maitra

Abstract The prognosis for most patients afflicted by pancreatic cancer still remains dismal. With the majority of cases being diagnosed at advanced stages, only minimal improvements in survival rates have been achieved using current therapeutic approaches. Nonetheless, remarkable research efforts over the past decade have enabled a detailed understanding of the molecular mechanisms underlying the pathogenesis of pancreatic cancer. According to the current state of knowledge, pancreatic carcinogenesis is a multistep process that requires alterations in a compendium of oncogenes, tumor-suppressor genes and genome-maintenance genes. The most frequent aberrations (somatic point mutations and allelic losses) affect oncogenes (*KRAS2*) and tumor-suppressor genes (*CDKN2A/p16, TP53, SMAD4/DPC4*) that have a key role in transcription, proliferation and regulation of the cell cycle, amongst others. In addition to these known mutational "mountains," a wide number of less frequently altered genes ("hills") have been discovered, which play an important part in defining the unique biology and behavior of each individual pancreatic cancer. A deeper understanding of the genetic landscape of pancreatic cancer, enhanced by "next-generation" high-throughput technologies will hopefully promote the development of new methods for early diagnosis and facilitate improvements in current therapeutic approaches.

Introduction

Extensive clinical and research efforts have been conducted over the last few decades to improve the prognosis of patients with cancer. In some tumor types, such as breast and colorectal cancer, early detection and better therapeutic agents have

M.D. Molin, M.D. • A. Maitra, M.B.B.S. (✉)
The Sol Goldman Pancreatic Cancer Research Center, Johns Hopkins University School of Medicine, CRB II Room 345, 1550 Orleans Street, Baltimore, MD 21231, USA
e-mail: amaitra1@jhmi.edu

D.M. Simeone and A. Maitra (eds.), *Molecular Genetics of Pancreatic Cancer*,
DOI 10.1007/978-1-4614-6549-2_1,

led to a significant decline in mortality rates, even for advanced disease (Berry et al. 2005; Kopetz et al. 2009). Conversely, patients afflicted by pancreatic cancer still harbor a dismal prognosis, with mortality rates that approximate incidence rates (Siegel et al. 2012). Especially at advanced stages, prolonged survival is anecdotal, and although therapeutic regimens have recently shown promising results (Conroy et al. 2011), the overall prognosis remains dismal, underscoring our need for a more detailed molecular knowledge of this disease.

Genomic alterations that translate into gain or loss of function of critical genes represent a hallmark of cancer (Hanahan and Weinberg 2011), and pancreatic cancer is no exception. Molecular and epidemiological data support the importance of key genetic alterations in the pathogenesis of pancreatic cancer. For example, several "driver" genes are mutated at a high frequency in pancreatic cancer, and the altered physiology consequent to these mutations allows the tumor initiating clone to escape the regulatory controls ("niche"), leading to tumor formation (Jones et al. 2008; Yachida et al. 2010). Second, extensive histopathological analyses have led to the recognition of tangible noninvasive precursor lesions that exhibit, with variable frequency, the entire range of genomic alterations that characterize pancreatic cancer (*see Chapter by Offerhaus*) (Kanda et al. 2012; Maitra et al. 2003). Third, genetically engineered mouse models, in which one or more key-mutated genes are expressed in the pancreas, recapitulate the full spectrum of phenotypic alterations of the cognate human disease, from noninvasive precursor lesions (pancreatic intraepithelial neoplasia or PanINs) to metastatic pancreatic cancer (*see Chapter by Pasca di Magliano*) (Hingorani et al. 2003, 2005; Perez-Mancera et al. 2012). Fourth, an increased risk for developing pancreatic cancer has been shown in members of families affected by rare cancer predisposition syndromes (*see Chapter by Petersen*) (Jacobs et al. 2010; Canto et al. 2012). Affected individuals from such high-risk families often harbor germ line mutations that permit the emergence of pancreatic cancer over the lifetime of these patients (Couch et al. 2007; Jones et al. 2009).

The identification of genes involved in pancreatic cancer development was historically obtained through a candidate gene approach. With some notable exceptions (Hahn et al. 1996), the candidate approach was able to establish the role of frequently mutated genes or to identify critical pathways already described in other tumor types, but is inadequate in discovering unexpected molecular alterations or pathways. Recently, the advent of massively parallel high-throughput technologies, such as next-generation sequencing (NGS), has provided the possibility of interrogating cancer genomes at an unprecedented resolution (Wu et al. 2011a; Jiao et al. 2011; Stransky et al. 2011; Parsons et al. 2011; Bettegowda et al. 2011) (*and see chapter by Wei and Kumar*). The information provided by such sensitive methods is expected not only to increase our knowledge of the genetic landscape of human cancers but also, more importantly, to usher in an era of personalized medicine based on tumor-specific genetic aberrations. In the context of pancreatic cancer, there is considerable hope that the translation of new molecular targets into the clinical setting is likely to improve risk assessment, early diagnosis, and the identification of the best possible treatment for each individual patient. In this chapter we describe the spectrum of the most common genetic alterations ("mountains") that drive the development of sporadic pancreatic ductal adenocarcinomas as well as less frequent alterations ("hills") (Vogelstein and Kinzler 2004a). Furthermore, new insights provided by novel high-throughput technologies and their translational relevance are also discussed.

The Genomic Landscape of Pancreatic Cancer: An Overview

Chromosomal Aberrations

Genomic instability represents a hallmark of pancreatic cancer, as well as other cancer types (Campbell et al. 2010; Stephens et al. 2011). Numerous alterations at the chromosomal level are seen in pancreatic cancer and, depending upon the underlying genetic mechanism, they can either occur as chromosomal instability (CIN) or microsatellite instability (MIN). This distinction, which appears to be mutually exclusive, is justified by the unique molecular and histological features of each type of alteration (Goggins et al. 1998; Wilentz et al. 2000).

CIN, which is revealed in the vast majority of pancreatic cancers (97 %) by cytogenetic analysis, is expressed through copy-number gains and losses, translocations, inversions, amplifications and homozygous deletions. Although such alterations may appear to be randomly distributed, they reflect a distinctive pattern in which selected genes that play a critical role in carcinogenesis are targeted and disrupted. In fact, a recent study has elucidated the concept of STOP (suppressors of tumorigenesis and proliferation) and GO (growth enhancers and oncogenes) that contribute negatively and positively towards the neoplastic phenotype, respectively (Solimini et al. 2012). In many instances, areas of hemizygous deletions are enriched for "islands" of high-density STOP genes that each contribute, on the basis of their haploinsufficiency, towards the eventual malignant phenotype, even in the absence of mutations on the remaining allele. Most frequently, numerical changes of the chromosomal architecture in pancreatic cancer are characterized by losses, particularly on chromosomes 6p, 9p, 13q, 17p, and 18q, as well as gains on chromosomes 7q and 20 (Mahlamaki et al. 2004; Holzmann et al. 2004). Several techniques have been used to identify regions of copy number alterations at a high resolution, including dense allelotyping and microarray analysis on single nucleotide polymorphism (SNP), bacterial artificial chromosome (BAC), oligonucleotide, or cDNA arrays (Calhoun et al. 2006; Nowak et al. 2005; Gysin et al. 2005; Chen et al. 2008; Bashyam et al. 2005; Shain et al. 2012; Kwei et al. 2008a). For example, Iacobuzio-Donahue et al. investigated chromosomal alterations in 80 pancreatic cancer xeno grafts by genome wide allelotyping, and confirmed losses in chromosomes 9p, 18p and 17p as the most common copy number alterations, with the regions of overlap encompassing three well known tumor suppressor genes in pancreatic cancer (*CDKN2A, SMAD4/DPC4 and TP53*, respectively) (Iacobuzio-Donahue et al. 2004). Of note, allelotyping of PanINs has revealed imbalances in several chromosomal regions also altered in pancreatic cancer, suggesting that CIN occurs early during the progression from noninvasive precursor lesions to invasive adenocarcinoma (Luttges et al. 2001; Yamano et al. 2000). Kern and colleagues have identified two patterns of CIN in pancreatic cancer using high-density SNP arrays, "original" CIN, characterized by an admixture of allelic loss and copy number changes, and "holey" CIN, exemplified by large regions of homozygous deletions ("holes") in the genome (Calhoun et al. 2006).

The use of array-based approaches to study copy number alterations in pancreatic cancer have helped define the regions of amplification and deletion with unprecedented resolution, including at the level of individual or neighboring genes. Notably, there are many instances wherein genes or pathways are altered predominantly by copy number changes rather than mutations at the nucleotide level. For example, *MYC,* the gene encoding the master transcriptional factor C-myc and located on chromosome 8q, is amplified in 10–20 % of pancreatic adenocarcinomas (Nowak et al. 2005; Bashyam et al. 2005), although somatic mutations have not been reported in this cancer type. Transcriptional overexpression is also observed in the majority of cases (Han et al. 2002), further highlighting the importance of altered C-myc signaling in pancreatic cancer. As recent studies have shown, C-myc plays a crucial role in metabolic reprogramming of cancer cells, allowing them to thrive in the hypoxic, nutrient-deprived environs of the tumor microenvironment (Dang 2010, 2012). Another example of a region of recurrent amplification occurs on chromosome 18q, which targets the gene encoding the transcription factor *GATA6*, amplified in approximately a fifth of pancreatic cancers (Fu et al. 2008; Kwei et al. 2008b). As with *MYC*, somatic mutations of the GATA transcription factor family are rare in pancreatic cancer (Jones et al. 2008). Similarly, inactivation of genes whose encoded products are involved in chromatin remodeling (*ARID1A, ARID1B*, *PBRM1, SMARCA2*, and *SMARCA4*) can be seen in up to a third of pancreatic cancers, only a minor fraction of which occurs via somatic mutations and the majority through copy number alterations (Shain et al. 2012).

Telomere Alterations

Telomeres are tandem repeats of specific noncoding nucleotide sequences (TTAAGGG) present at the ends of chromosomes (Blackburn et al. 2006). Telomeres play a fundamental role as guardians of genomic integrity, protecting chromosomal ends from breakage or fusion with neighboring chromosomes. Since cell cycle results in progressive telomere shortening, telomere length can be maintained by activation of the enzyme telomerase, a feature observed in most human cancers (Harley et al. 1990; Martinez and Blasco 2011). Reactivation of telomerase protects cancer cells from critical telomere shortening and resulting DNA damage, thus allowing limitless replication. Telomerase activation is observed fairly late in the multistep progression of pancreatic cancer, however, and is preceded by an abnormal shortening of telomeres that occurs at the stage of noninvasive precursor lesions (van Heek et al. 2002a). Indeed, more than 90 % of low-grade PanIN lesions

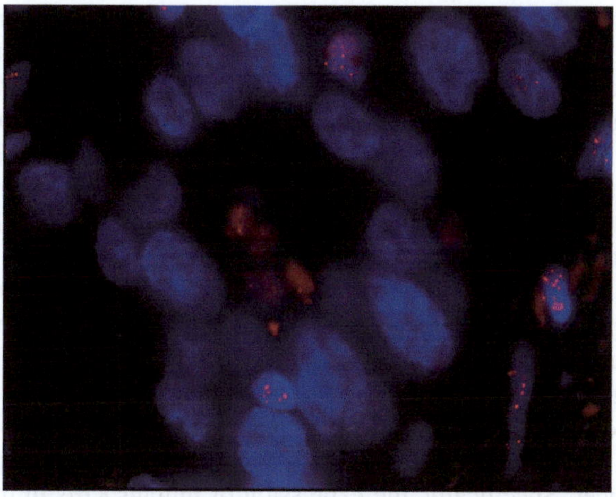

Fig. 1 Attrition in telomere length is one of the earliest detectable molecular alterations in pancreatic cancer, nearly ubiquitously observed at the stage of even low-grade PanIN lesions. A specific fluorescence in situ hybridization probe against telomeric DNA is used for semiquantitative measurement of telomere lengths in archival tissues (TEL-FISH). In this figure, a neoplastic gland from a ductal adenocarcinoma demonstrates near total loss of fluorescence intensity by TEL-FISH. In contrast, bright telomere signals are observed in the adjacent stromal cells, and one infiltrating lymphocyte at the bottom of the gland. Photomicrograph courtesy of Alan Meeker, PhD, Department of Pathology, Johns Hopkins University School of Medicine

demonstrate marked shortening of telomeres, as compared with normal pancreatic ductal epithelium, suggesting that telomere attrition is probably one of the earliest genetic events during pancreatic carcinogenesis (Fig. 1). While the basis for the near uniform telomere dysfunction in precursor lesions is unclear, it is likely that such dysfunction sets the stage for subsequent "breakage-fusion-breakage" cycles, which lead to chromosomal instability and frank neoplasia.

Oncogenes

Somatic activating mutations in the *KRAS2* gene are present in over 90 % of pancreatic adenocarcinomas and PanIN lesions, rendering it the most frequently mutated oncogene in this tumor type (Jones et al. 2008; Kanda et al. 2012). *KRAS2* gene (also known as Kirsten rat sarcoma viral oncogene homolog), located on chromosome 12p, encodes a GTP-binding and hydrolyzing enzyme involved in growth factor signaling pathways (Vigil et al. 2010). The K-ras protein activates multiple downstream effector pathways required for oncogenesis, including cell survival, cell proliferation, cell invasion, and aberrant cellular metabolism (*see chapter by Bar-Sagi*). Principal effectors of K-ras include the mitogen-activated protein kinase (MAPK), phosphoinositide 3-kinase (PI3K)/Akt, and Ral signaling pathways, among others (Young et al. 2009). Under physiological conditions, K-ras is transiently activated by GTP binding, followed rapidly by inactivation due to its intrinsic property of GTP hydrolyzation ("GTPase"). This endogenous GTPase activity is compromised by somatic mutations occurring in the GTP-binding pocket, which causes K-ras to remain constitutively active (DeNicola and Tuveson 2009; Perez-Mancera and Tuveson 2006). Interestingly, the vast majority of *KRAS2* point mutations in human pancreatic cancer are confined to codon 12, and less frequently to codons 13 and 61. In addition to invasive cancer, *KRAS2* mutations are also found in PanINs, including nearly all low-grade PanINs. As recently shown (Kanda et al. 2012), lower grade PanINs represent an admixture of mutant and nonmutant clones of cells, with a progressive increase in the proportion of the mutant clone accompanying histological progression to invasive neoplasia.

Recently developed animals models provide some of the most compelling evidence that K-ras is required for the initiation, maintenance, and progression of pancreatic cancer. Specifically, the expression of mutant *Kras* in the mouse pancreas during development is sufficient to yield the development of murine PanINs (mPanINs), which culminates in invasive adenocarcinoma in a fraction of animals (Hingorani et al. 2003; Aguirre et al. 2003a). More recent studies in transgenic animals have also underscored the importance of *Kras* in the maintenance of pancreatic cancer. This has been accomplished by the use of doxycycline-modulated *Kras* expression in the murine expression, wherein "turning off" mutant protein expression results in regression of established mPanINs and even invasive adenocarcinomas (Collins et al. 2012; Ying et al. 2012). Finally, mouse models of cooperation between mutant *Kras* and *p16* loss have found an intriguing loss of heterozygosity (LOH) of the wild-type *Kras* allele in advanced lesions (metastases), suggesting that the wild-type protein might interfere with the oncogenic function of the mutant K-ras protein (Qiu et al. 2011). In light of the near ubiquitous nature of *KRAS* mutations in pancreatic cancer, and the observed dependence in animal models on sustained Ras signaling, one presumes that pharmacological inhibition of mutant K-ras protein would be a therapy of choice in this malignancy. Unfortunately, clinical trials with inhibitors of farnesyltransferase, a key enzyme in the post-translational processing and membrane targeting of Ras protein, have been disappointing in pancreatic cancer (Kelland 2003; Van Cutsem et al. 2004). Several alternative strategies are currently undergoing evaluation, including targeting of Ras effectors pathways, either singly, or more increasingly, in combination (Feldmann et al. 2011; Collisson et al. 2011).

KRAS2 mutations also represent candidate biomarkers for the diagnosis of pancreatic cancer in biological samples such pancreatic juice, stool, and blood (Goggins 2005). However, in heterogeneous biological samples, the overwhelming presence of wild-type DNA, as opposed to a limited number of mutant molecules, renders *KRAS2* mutations particularly difficult to detect using conventional assays. To overcome these limitations, ultrasensitive assays for the detection of mutant *KRAS2* have been generated in the last few years, which are able to identify low-concentration mutant molecules and estimate differences in the proportion of mutant *KRAS2* molecules between pancreatic cancer and noncancerous conditions.

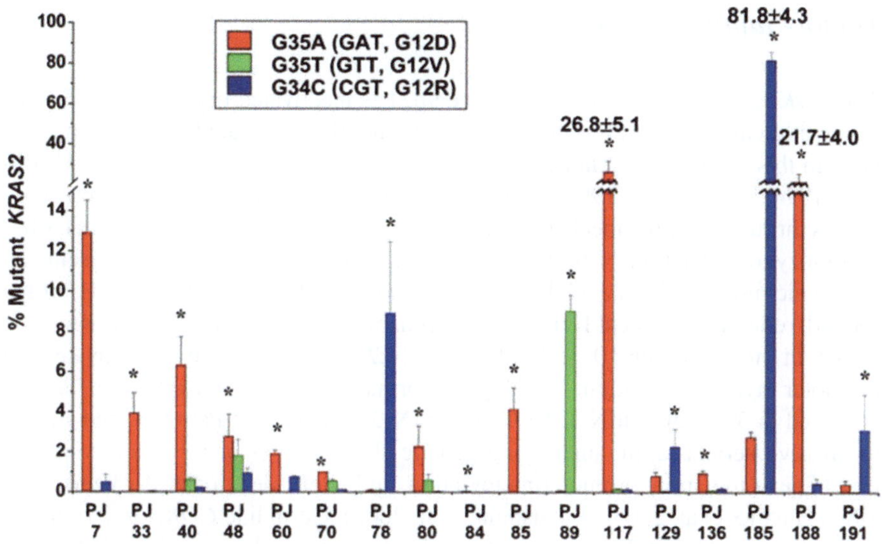

Fig. 2 Quantitative detection of mutant KRAS molecules in pancreatic juice samples obtained from patients with pancreatic adenocarcinoma using ultrasensitive LigAmp technology. Figure reproduced with permission from Shi C et al., Cancer Biol Ther 2008, Landes Bioscience Publishers, Austin TX

For example, a technique known as "LigAmp," which involves sequential DNA ligation and PCR amplification, has been recently developed to detect and quantify *KRAS2* mutant molecules in pancreatic juice samples (Shi et al. 2008) (Fig. 2). In another ultrasensitive approach, known as BEAMing (beads, emulsion, amplification, and magnetics), a single DNA molecule is assigned to a single magnetic bead, PCR-amplified and coupled with specific fluorescent-labeled oligonucleotides (Dressman et al. 2003; Diehl et al. 2008). The percentage of mutant DNA molecules in a mixed population of DNA molecules is then quantified by analyzing fluorescence emission through a flow cytometer. If validated by additional studies, it is expected that these new quantitative assays will greatly improve the diagnostic armamentarium available for the early diagnosis of pancreatic cancer.

In addition to the overwhelming dominance of mutant *KRAS2*, other pathway components can occasionally be altered, and might either be additive, or less frequently substitute for, mutant K-ras function. For example, in rare instances (~1 %), pancreatic cancers may harbor somatic *BRAF* mutations, and some studies have suggested that this preferentially occurs in the setting of *KRAS2*-wild type tumors (Calhoun et al. 2003). In this instance, one envisions that mutant *BRAF* gene product is driving activation of the MAPK signaling pathway. Similarly, amplification of the *AKT2* gene locus on chromosome 19q is observed in ~10 % of pancreatic cancers (Cheng et al. 1996; Ruggeri et al. 1998), and is typically co-existent with a mutant *KRAS2*, likely contributing the abnormal activation of signaling in the Akt oncogenic pathway.

Tumor-Suppressor Genes

The *CDKN2A/p16* gene on chromosome 9p21 is inactivated in more than 95 % of pancreatic cancers, representing the most frequently inactivated tumor suppressor gene in this tumor type (Maitra and Hruban 2008; Rozenblum et al. 1997; Caldas et al. 1994; Schutte et al. 1997). Unlike *KRAS* mutations, *CDKN2A/p16* inactivation occurs through multiple mechanisms: it is estimated that 40 % of the cancers harbor a homozygous deletion of both alleles of the gene, and another 40% presents an intragenic mutation in one allele coupled with loss of heterozygosity (LOH) of the second, reflecting classical Knudsonian mechanisms of gene inactivation (Knudson 1996). In the remaining 10–15 % of cancers, *CDKN2A/p16* gene is inactivated via promoter hypermethylation. Notably, abnormal p16 protein expression is also observed in 30 % of PanIN-1, 55 % of PanIN-2 and 70 % of PanIN-3, and similar to invasive neoplasia, the underlying genetic abnormalities occurs via a combination of gene mutation, promoter methylation, and allelic deletions (Moskaluk et al. 1997; Hustinx et al. 2005a; Fukushima et al. 2002). Germ line *CDKN2A/p16* mutations occur in the familial atypical multiple mole and melanoma (FAMMM) syndrome (*see chapter by Petersen*) (Fusaro and Lynch 2000). Persons affected by this syndrome characteristically present with numerous nevi, including dysplastic nevi characterized by atypical shape, size, and color and a predisposition for developing malignant melanoma. Notably, these patients also harbor nearly a 20-fold lifetime risk of developing pancreatic cancer (Klein et al. 2001), underscoring the importance of *CDKN2A/p16* as a tumor suppressor gene in this cancer type. The gene product of *CDKN2A/p16* regulates cell cycle progression by inhibiting cyclin D1-CDK4/6, a kinase complex that is involved in promoting the G1/S phase transition by inactivating the retinoblastoma protein, Rb (Sherr 2004). The *CDKN2A/p16* locus at chromosome 9p21 has an overlapping reading frame with *Arf*, whose gene product is involved in stabilizing p53 (Kim and Sharpless 2006). In genetically engineered mice, co-deletion of *Cdkn2a/p16* in conjunction with *Arf* plus expression of a mutant *Kras* allele in the pancreas results in rapidly progressive and lethal adenocarcinomas (Aguirre et al. 2003b). Subsequent studies have confirmed that pancreas-specific bi-allelic deletion of *Cdkn2a/p16* alone (with intact *Arf*) in association with mutant *Kras* is sufficient in generating murine pancreatic adenocarcinomas (Bardeesy et al. 2006).

The high frequency of *CDKN2A/p16* abnormalities (especially mutations and promoter methylation) renders this gene as an attractive candidate for biomarker studies. Not surprisingly, both classes of abnormalities of *CDKN2A/p16* can be identified in the pancreatic juice of patients harboring pancreatic cancer, especially using sensitive detection technologies (Bian et al. 2006; Matsubayashi et al. 2006). Interestingly, the gene encoding methylthioadenosine phosphorylase (MTAP), which resides approximately 100 kb telomeric to the *CDKNA2A/p16* gene, is frequently included in the 9p21 homozygous deletions, present in up to 1/3rd of pancreatic cancers overall (Hustinx et al. 2005b). The MTAP enzyme is critical for purine biosynthesis through the salvage pathway, and therefore, pancreatic cancers harboring

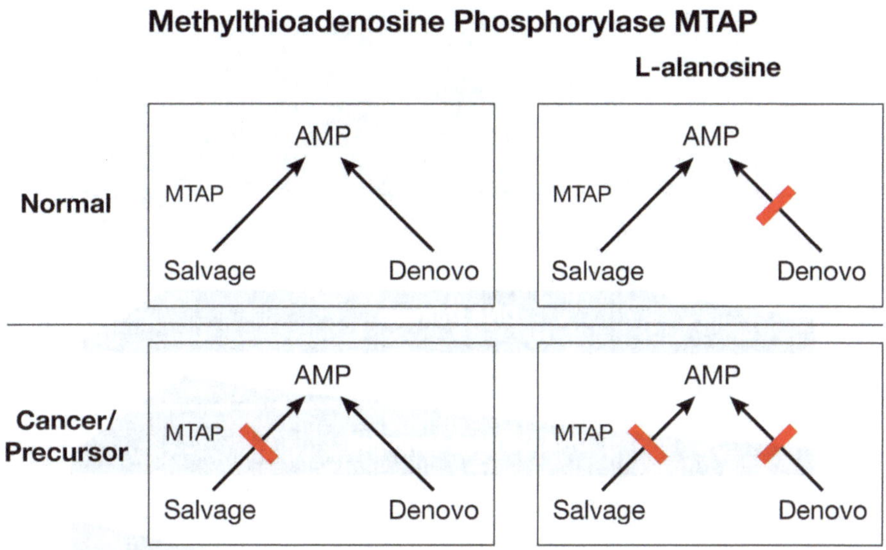

Fig. 3 Purine biosynthesis in cells occurs via either the de novo or the salvage pathways. Methylthioadenosine phosphorylase (MTAP) is the essential enzyme for purine synthesis through the salvage pathway. In pancreatic cancers with homozygous *MTAP* gene deletions, the tumor cells are dependent on de novo purine synthesis. In these cases, blockade with a systemic inhibitor of de novo synthesis like L-alanosine can provide a synthetic lethal effect that is restricted to cancer cells only

MTAP homozygous deletions are potentially susceptible to small molecule inhibitors of de novo purine biosynthesis, providing a great example of a synthetic lethal interaction that is targeted at a passenger, and not a driver alteration (Hustinx et al. 2005b; Karikari et al. 2005; Bertino et al. 2011) (Fig. 3).

The *TP53* gene, located on chromosome 17p, plays a critical role as a "guardian" of the genome. It regulates the G1/S cell cycle phase checkpoint, and induces cell cycle arrest in the setting of DNA damage; the inability to repair damaged DNA then triggers p53-dependent apoptosis (Vazquez et al. 2008). Somatic mutations of *TP53* gene are found in ~50–75 % of invasive pancreatic cancers, which results in the inability of the mutant protein to bind to DNA and activate the p53 transcriptional network (Jones et al. 2008; Hingorani et al. 2005). Several recurrent *TP53* mutations observed in human cancers, such as the R175H mutation, have a dominant-negative "gain-of-function" effect, which attenuates the function of the wild type allele (Jackson et al. 2005; Olive et al. 2004). Thus, loss of the second allele, although generally observed as a chromosome 17p loss of heterozygosity, may not always be necessary to abrogate physiologic p53 protein function. The majority of *TP53* mutations result in stabilization of the encoded protein, and this can be detected as nuclear accumulation of p53 on immunohistochemistry (Baas et al. 1994). In PanINs, nuclear p53 accumulation is typically detected at the stage of PanIN-3 and beyond, suggesting that it is a late anomaly in the multistep progression

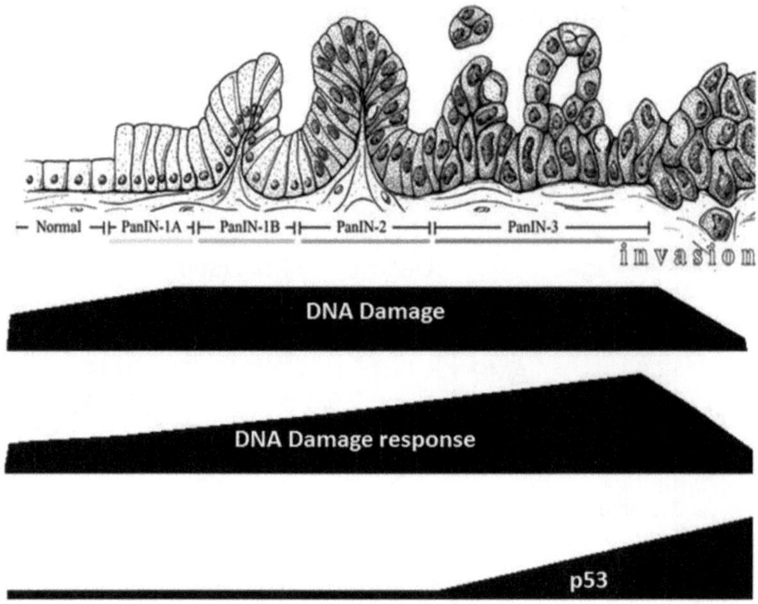

Fig. 4 Retention of p53 function acts as a crucial barrier to cancer progression in the pancreatic epithelium, in response to progressive accumulation of DNA damage and activation of the DNA damage response (DDR). Inactivation of p53 function at the stage of PanIN-3 and beyond is associated with bypass of the DDR checkpoint, and progression to invasive cancer. Figure reproduced with permission from Koorstra et al., Mod Pathol 2009

of pancreatic cancer (Maitra et al. 2003). This is in contrast to markers of DNA damage response (such as phosphorylated ATM and Chk2 proteins), which are observed even in the lowest-grade PanIN lesions (Koorstra et al. 2009). The retention of p53 function in low-grade PanINs (and the resulting checkpoint phenomenon) might explain why pancreatic cancers remain relatively uncommon despite the widespread prevalence of lower grade PanINs in the general population (>50 % harbor such noninvasive lesions above the age of 60 years) (Cubilla and Fitzgerald 1976). Loss of p53 function at the PanIN-3 stage "opens the floodgates" for progression to invasive neoplasia (Fig. 4). The high frequency of *TP53* mutations in pancreatic cancer provides an opportunity for its use as a biomarker in clinical samples, such as pancreatic juice samples (Bian et al. 2006). In addition, the recent development of mutant allele specific p53 targeted small molecule therapeutics (in particular, those that can reactive wild-type function in the R175H allele, the most common mutation in pancreatic cancer) (Yu et al. 2012), provides new therapeutic opportunities against the mutant protein. Another example of selective toxicity against p53-mutant pancreatic cancers has recently been identified in preclinical studies that targeted the Wee1 kinase, which inhibits Cdc2, using a potent and selective small molecule antagonist (Rajeshkumar et al. 2011). Specifically, agents that block Wee1 kinase function, and hence promote Cdc2-mediated G2-M progression result in a phenomenon of so-called "mitotic catastrophe" in the setting of exacerbated DNA damage, such as that induced by concomitant therapy with antineoplastic agents like gemcitabine.

The *DPC4/SMAD4* gene, located on chromosome 18q, encodes for an intracellular protein that transduces growth inhibitory signals upon binding of transforming growth factor β (TGFβ) to its membrane receptors (Siegel and Massague 2003). *DPC4/SMAD4* functions as a key tumor suppressor gene, and homozygous deletion or intragenic inactivating mutation of *DPC4/SMAD4* occur in approximately 55 % of pancreatic adenocarcinomas (Hahn et al. 1996). Of note, loss of *DPC4/SMAD4* is infrequently to rarely seen in other pancreatic neoplasms, such as pancreatic neuroendocrine tumors (PanNETs), or in most extra-pancreatic epithelial neoplasms (Jiao et al. 2011; Schutte et al. 1996). This renders loss of Dpc4/Smad4 protein expression in metastases from occult primaries as a relatively specific, albeit not particularly sensitive, biomarker for pancreatic adenocarcinoma (Tascilar et al. 2001a; van Heek et al. 2002b). Mutations of *DPC4/SMAD4* gene in adenocarcinomas is the only one of the "big four" that has been shown to significantly correlate with decreased survival at both the genetic and protein level (the latter using immunohistochemistry in archival samples) (Blackford et al. 2009; Tascilar et al. 2001b). In addition, mutations of *DPC4/SMAD4* correlate with extensive systemic metastases in terminal pancreatic cancer patients, versus oligo-metastatic or locally advanced disease in those with retained function (*see chapter by Iacobuzio-Donahue*) (Iacobuzio-Donahue et al. 2009). In the multistep progression model, loss of Dpc4/Smad4 protein expression is observed as a relatively "late" event, mostly at the stage of high-grade PanIN lesions (Maitra et al. 2003). Recent chemical genetic approaches have identified compounds that are synthetic lethal to cells with DPC4/SMAD4 mutations, providing an opportunity for molecularly targeted therapies (Wang et al. 2006).

Other tumor suppressor genes have been shown to be inactivated at low frequency in pancreatic cancer (<5 %). Somatic mutations of the *LKB1/STK11* gene, which encodes for a serine threonine kinase, are rarely observed in sporadic pancreatic cancer, but more commonly in the setting of familial pancreatic cancer arising in patients with Peutz-Jeghers syndrome (Su et al. 1999). Individuals affected by this autosomal-dominant syndrome harbor an increased risk of developing colorectal hamartomatous polyps, as well as pancreatic cancer (Giardiello et al. 1987). The *LKB1* gene product is a multifunctional protein involved in metabolic sensing, maintenance of epithelial polarity and in regulating cytoskeletal architecture, amongst others (Hezel and Bardeesy 2008) (Fig. 5). In murine models, intraductal papillary mucinous neoplasms (IPMN) cystic neoplasms develop in the pancreas upon conditional *Lkb1* deletion (Hezel et al. 2008). Notably, loss of Lkb1 protein expression is observed in up to a third of cystic IPMNs of the pancreas (*see chapter by Offerhaus*) (Sahin et al. 2003), although somatic *LKB1* mutations were not seen in the recent sequencing of the IPMN exome (Wu et al. 2011a). Intragenic mutations and homozygous deletions of the *MKK4* gene occur in <5 % of pancreatic cancers (Su et al. 1998). The *MKK4* gene, located on chromosome 17p, encodes for a component of stress-activated protein kinase cascade and plays a role in growth

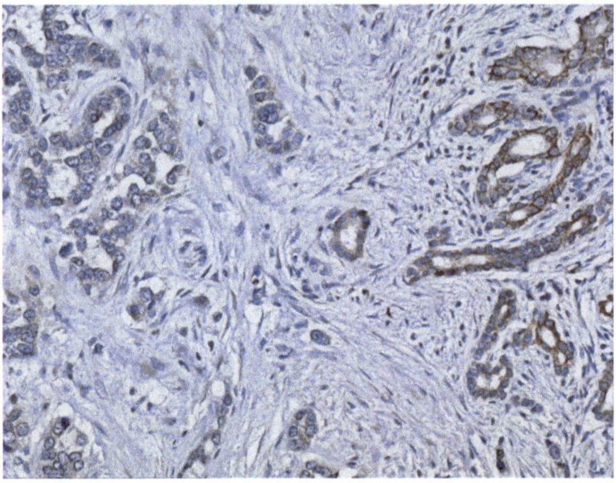

Fig. 5 Loss of Lkb1/Stk11 protein expression by immunohistochemistry in a pancreatic ductal adenocarcinoma. The neoplastic glands (*left half*) are negative for Lkb1 expression, while the intermixed normal ductal epithelium (*right half*) demonstrates robust labelling

control and apoptosis (Robinson et al. 2003; Haeusgen et al. 2011). Furthermore, inactivation of the *MKK4* gene has been documented in subsets of metastatic pancreatic cancer lesions, suggesting that the product of this gene may act as a metastasis suppressor (Xin et al. 2004).

Genome-Maintenance Genes

In addition to oncogenes and tumor-suppressor genes, a third class of genes, collectively defined as genome-maintenance genes, is occasionally inactivated in pancreatic cancer (Vogelstein and Kinzler 2004b). Also known as "caretakers," these genes are involved in the repair of DNA breaks, minimizing errors during DNA replication. One of the most commonly inactivated "caretaker" genes, in approximately 5 % of sporadic pancreatic cancers, is the *BRCA2* gene, located on chromosome 13q (Jones et al. 2008; Naderi and Couch 2002). Germ line mutations of *BRCA2* are observed in 5–10 % of patients with an inherited predisposition to pancreatic cancer, and have a particular propensity to occur in families of Ashkenazi Jewish heritage (*see chapter by Petersen*) (Ozcelik et al. 1997; Goggins et al. 1996; Hahn et al. 2003; Lal et al. 2000). The product of *BRCA2* interacts with proteins encoded by the Fanconi anemia genes (the *FANC* genes) to mediate homologous recombination at sites of DNA double-strand breaks (Gudmundsdottir and Ashworth 2006). Notably, pancreatic cancers that harbor bi-allelic mutations of *BRCA2* are characterized by exquisite sensitivity to DNA cross-linking agents (e.g., mitomycin C, cisplatin) as well as poly (ADP-ribose) polymerase inhibitors (PARP-i), providing an avenue for "personalized" therapy in this malignancy (Gallmeier and Kern 2007; van der Heijden et al. 2005; James et al. 2009). Recently, mutations have also been described in other components of the Fanconi anemia pathway, such as the *Partner and Localizer of BRCA2* (*PALB2*) gene, which encodes for a partner that spatially localizes BRCA1 and BRCA2 proteins at sites of double strand breaks, in order to facilitate repair (Jones et al. 2009). Pancreatic cancers with bi-allelic *PALB2* mutations are similarly sensitive to the effects of cisplatin and mitomycin C (Villarroel et al. 2011). One of the important caveats that have emerged from mouse models of conditional *Brca2* deficiency in the pancreas is that haploinsufficiency for *Brca2*-function might be sufficient for inducing exocrine neoplasia, particularly in combination with mutant *Kras* (Skoulidis et al. 2010). This has therapeutic implications for treating "*BRCA*"-associated human pancreatic adenocarcinomas with PARP-i, since retaining a functional *BRCA2* allele would potentially render the tumors resistant to this class of agents (Fong et al. 2010). The data on somatic loss of the second *BRCA2* allele in pancreatic adenocarcinomas arising in patients with a germ line defect of one allele remains controversial, with at least one study suggesting that it may be retained, rendering such tumors resistant to PARPi-based therapies (Skoulidis et al. 2010).

Other genes involved in DNA repair that have been implicated in pancreatic carcinogenesis include *hMLH1* and *hMSH2*, mostly in the context of familial pancreatic cancers arising on the backdrop of hereditary non-polyposis colorectal cancer (HNPCC) (Lindor et al. 2011; Ghimenti et al. 1999; Yamamoto et al. 2001). Mutations or transcriptional silencing in *hMLH1* and *hMSH2* have been shown to result in replication errors in simple repetitive units known as microsatellites (Parsons et al. 1993; Malkhosyan et al. 1996; Eshleman and Markowitz 1996). As a consequence, microsatellite instability (also known as a defect in mismatch repair or MMR) defines a unique genomic landscape, characterized by very few alterations in chromosome ploidy. Interestingly, pancreatic carcinomas with microsatellite instability exhibit a unique histological pattern, termed as "medullary," comprised of poorly differentiated histology, pushing borders, and large numbers of tumor infiltrating lymphocytes (Wilentz et al. 2000). As additional evidence of the distinct genetic basis for these neoplasms, mutations in the *KRAS2* gene are uncommonly seen in medullary carcinomas (Goggins et al. 1998).

New Perspectives from Exomic and Next-Generation Sequencing Studies

As previously stated, historically, the discovery of molecular alterations in human cancer was based on a candidate gene approach. These methods allowed researchers for the identification of frequently mutated genes (*KRAS*, *CDKN2A/p16*, *SMAD4/DPC4*, *TP53*) in pancreatic adenocarcinoma, although they were often unable to find genes altered at low frequency or in unexpected cancer pathways. The first comprehensive glimpse into the genomic landscape of pancreatic cancer came in 2008, with an exomic sequencing study performed on a series of 24 cancers (Jones et al. 2008). This study utilized automated Sanger sequencing for exome analysis, combined with serial analysis of gene expression (SAGE) for the transcriptome and genome-wide single nucleotide polymorphism (SNP) microrrays for copy number aberrations, in order to generate an integrated assessment of molecular alterations in pancreatic cancer. Using this approach, the sequences of 23,219 transcripts, representing 20,661 protein-coding genes (99.6 % of the coding genome) were determined. Overall, 1,562 somatic mutations were identified, mostly represented by single base substitutions [missense and nonsense mutations, or insertions/deletions (i.e., "indels")]. Pancreatic cancers were found to harbor a median of 66 somatic mutations per tumor. Only a small proportion of the compendium of mutated genes within an individual sample actually contributes to tumorigenesis ("driver genes") and the vast majority simply represent a bystander effect of ongoing genetic instability and clonal evolution ("passenger genes") (Bozic et al. 2010). Genes with a minimum of two genetic alterations (at least one of which was predicted to result in altered function) and a mutation rate > 10 mutations/Mb, calculated by integrating gene size, nucleotide composition and other characteristics, were considered as candidate driver genes ("CAN" genes). Consequently, genes that did not fit these criteria were considered passenger genes. Such an approach led to the identification of 91 CAN genes. Of these, the previously known "big four" (*KRAS2, CDKN2A/p16, TP53, SMAD4/DPC4*) constituted the most obvious "mountains" on the genomic landscape. The rest of the landscape was comprised of low-frequency "hills" and even "private" (unique) mutations, underscoring the considerable genetic heterogeneity amongst the different tumor samples studied. These results might at first appear discouraging to researchers and clinicians in terms of developing targeted therapies. However, such a complexity is significantly reduced if altered genes are considered in the much broader context of biological pathways. In fact, 12 core biological pathways appear to be altered in most cases of pancreatic cancer, many of which are well-established hallmarks of cancer (Hanahan and Weinberg 2011) (Fig. 6). This information may harbor implications for the development of new therapeutic agents that target functional pathways or processes rather than individual products of mutated genes.

Although detailed discussion of the 12 core signaling pathways is beyond the scope of this chapter, one notable theme that has emerged from the pancreatic cancer exome sequencing effort (Jones et al. 2008), as well as other comparable solid tumor studies, has been the emergence of epigenetic modifiers as a major target of genomic alterations (Parsons et al. 2011; Jones et al. 2010, 2012; Varela et al. 2011; Fujimoto et al. 2012). Pancreatic cancers harbor widespread epigenetic alterations, which mimic the multistep genetic progression observed with coding sequences (*see chapter by Goggins*). It is postulated that many of the genomic alterations in chromatin modifying genes represent epigenetic "drivers" of cancer (Elsasser et al. 2011). For example, somatic mutations of the mixed-lineage leukemia 3 (*MLL3*) gene is observed in ~10 % of pancreatic cancers, rendering it as the fourth most commonly mutated tumor suppressor gene in this neoplasm (Jones et al. 2008). The protein encoded by *MLL3* encodes for a histone methyltransferase, which forms part of a multimeric complex involved in regulation of chromatin remodeling (Lee et al. 2009). As previously stated, numerous other chromatin modifying genes are inactivated by copy number alterations in pancreatic cancer (for example, *ARID1A*, *BRG1*, *PRBM1*), with almost a third of tumors demonstrating aberrations in this class of genes (Shain et al. 2012).

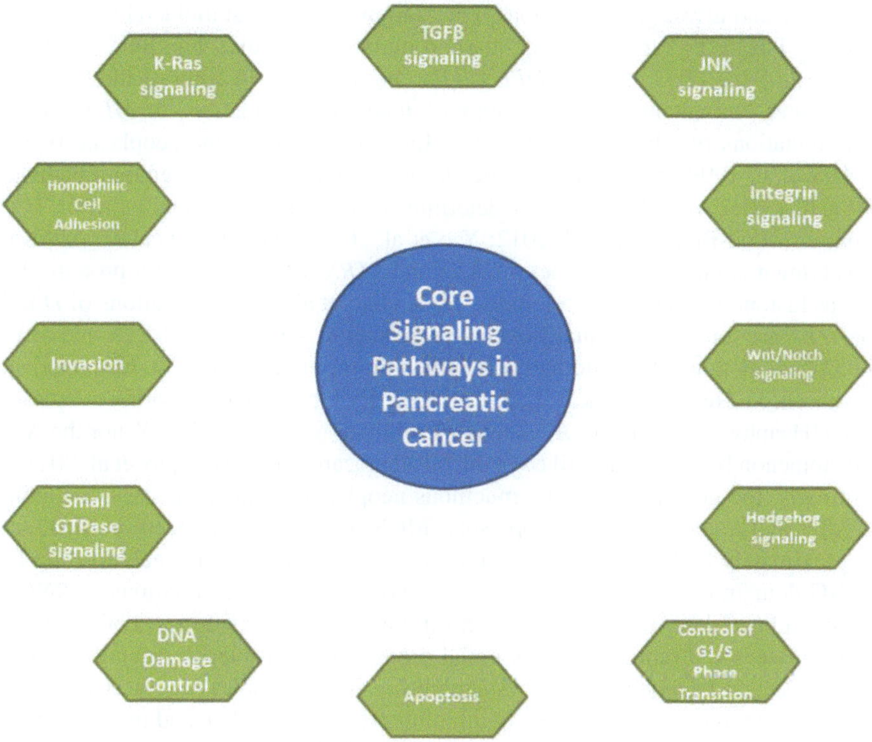

Fig. 6 Core signaling pathways that are altered by somatic mutations in the majority of pancreatic cancers. New data from the ICGC suggests that axonal guidance genes are another important category to be added to this list of core pathways. Figure adapted from Jones et al., Science 2008

The pancreatic adenocarcinoma exome has also been sequenced as part of an international effort known as the International Cancer Genome Consortium (ICGC) (Hudson et al. 2010). In contrast to the Jones et al. study (Jones et al. 2008), the pancreatic cancer ICGC team (led by investigators in Australia and Canada) utilized NGS technology on ~100 primary (Stage I and II) tumors (Biankin et al. 2012). Their data has reaffirmed many of the mutational "mountains" and "hills" uncovered in the Jones study, but also identified novel recurrent mutated pathways in pancreatic cancer. In particular, genes involved in embryonal axonal guidance [members of the *SLIT/ROBO* family of genes (Killeen and Sybingco 2008)] has emerged as recurrently mutated in pancreatic cancer, and appear to impart an adverse prognosis in patients bearing tumors with such somatic alterations.

The pancreas is one of the few organs where not only the most common neoplastic subtype (i.e., ductal adenocarcinoma) has been sequenced at the exome level, but so have nearly all other solid and cystic variant neoplasms as well (Wu et al. 2011a, b; Jiao et al. 2011). These studies, accomplished by harnessing the prowess of NGS have confirmed that "genetics begets morphology"—in that each of the histogenetic subtypes of pancreatic neoplasms is characterized by a unique underlying genomic signature and driver gene mutations. For example, in contrast to ductal adenocarcinomas, PanNETs rarely, if ever, harbor mutations of the "big four" (*KRAS2, CDKN2A/p16, TP53, SMAD4/DPC4*) (Jiao et al. 2011). In contrast these lesions have three "mountains" on their genomic landscape—mutations of *MEN1*, germ line mutations of which are responsible for multiple endocrine neoplasia, type 1 (Marx et al. 1999); mutations of genes in the mammalian TOR signaling pathway (*PIK3CA*, *PTEN*, and *TSC2*) that determines susceptibility to inhibitors of TOR kinase (Meric-Bernstam et al. 2012; Yao et al. 2011); and a novel cancer pathway involving mutations of two genes—*DAXX* and *ATRX*, which encode for proteins that act as histone chaperones at telomeric DNA (Jiao et al. 2011). Mutations of *DAXX* or *ATRX* are found in a mutually exclusive manner in ~50 % of PanNETs, and result in a phenomenon called alternative lengthening of telomeres (ALT), characterized by absence of telomerase activity and abnormally long telomeres within neoplastic cells (Heaphy et al. 2011a). Of note, neither mutations of *ATRX/DAXX*, nor the ALT phenomenon have been described in ductal adenocarcinomas (Heaphy et al. 2011b). Similarly, the genomes of cystic mucinous neoplasms of the pancreas—including IPMNs and mucinous cystic neoplasms (MCNs) have recently been profiled, and approximately half contain inactivating mutations of *RNF43*, a gene encoding for RING domain containing ubiquitin ligase (Wu et al. 2011a). Mutations of *RNF43* have not been described in ductal adenocarcinoma, and the substrates of this ubiquitin ligase could represent the essential proteins responsible for driving exocrine neoplasia along a mucinous and cystic pathway. Recent studies suggest that RNF43 protein functions as a Wnt pathway inhibitor (Hao et al. 2012), and in conjunction with activating *CTNNB1* mutations in a subset of IPMNs (Chetty et al. 2006), aberrant Wnt activation might represent one of the mechanisms by which unique histogenetic differentiation occurs in cystic neoplasms versus "usual" ductal adenocarcinomas.

Conclusion

In conclusion, tremendous advances have been achieved over the last few years in our knowledge of the genomic alterations in spoardic pancreatic cancer. The application of NGS technologies has greatly expanded the scenarios in pancreatic cancer wherein this knowledge can be applied, from developing ultrasensitive early detection assays in biological specimens to more efficacious personalized therapies. In addition, knowledge gleaned from sequencing of the sporadic pancreatic cancer genome has been useful in expanding to the study of genomic alterations in precursor lesions (*see chapter by Offerhaus*) (Wu et al. 2011a, b), discovery of genes involved in familial pancreatic cancer (*see chapter by Petersen*) (Jones et al. 2009; Roberts et al. 2012), to elucidate the genomic complexity of metastases, and construct a timeline for progression to terminal disseminated cancer (*see chapter by Iacobuzio-Donahue*) (Yachida et al. 2010; Campbell et al. 2010). The public dissemination of sequence data using online portals such as the "ICGCMart" (Zhang et al. 2011) is likely to impact research and drug discovery efforts in pancreatic cancer for the next decade.

References

Aguirre AJ, Bardeesy N, Sinha M, Lopez L, Tuveson DA, Horner J, Redston MS, DePinho RA (2003) Activated Kras and Ink4a/Arf deficiency cooperate to produce metastatic pancreatic ductal adenocarcinoma. Genes Dev 17:3112–3126

Baas IO, Mulder JW, Offerhaus GJ, Vogelstein B, Hamilton SR (1994) An evaluation of six antibodies for immunohistochemistry of mutant p53 gene product in archival colorectal neoplasms. J Pathol 172:5–12

Bardeesy N, Aguirre AJ, Chu GC, Cheng KH, Lopez LV, Hezel AF, Feng B, Brennan C, Weissleder R, Mahmood U, Hanahan D, Redston MS, Chin L, Depinho RA (2006) Both p16(Ink4a) and the p19(Arf)-p53 pathway constrain progression of pancreatic adenocarcinoma in the mouse. Proc Natl Acad Sci USA 103:5947–5952

Bashyam MD, Bair R, Kim YH, Wang P, Hernandez-Boussard T, Karikari CA, Tibshirani R, Maitra A, Pollack JR (2005) Array-based comparative genomic hybridization identifies localized DNA amplifications and homozygous deletions in pancreatic cancer. Neoplasia 7:556–562

Berry DA, Cronin KA, Plevritis SK, Fryback DG, Clarke L, Zelen M, Mandelblatt JS, Yakovlev AY, Habbema JD, Feuer EJ (2005) Effect of screening and adjuvant therapy on mortality from breast cancer. N Engl J Med 353:1784–1792

Bertino JR, Waud WR, Parker WB, Lubin M (2011) Targeting tumors that lack methylthioadenosine phosphorylase (MTAP) activity: current strategies. Cancer Biol Ther 11:627–632

Bettegowda C, Agrawal N, Jiao Y, Sausen M, Wood LD, Hruban RH, Rodriguez FJ, Cahill DP, McLendon R, Riggins G, Velculescu VE, Oba-Shinjo SM, Marie SK, Vogelstein B, Bigner D, Yan H, Papadopoulos N, Kinzler KW (2011) Mutations in CIC and FUBP1 contribute to human oligodendroglioma. Science 333:1453–1455

Bian Y, Matsubayashi H, Li CP, Abe T, Canto M, Murphy KM, Goggins M (2006) Detecting low-abundance p16 and p53 mutations in pancreatic juice using a novel assay: heteroduplex analysis of limiting dilution PCRs. Cancer Biol Ther 5:1392–1399

Biankin AV, Waddell N, Kassahn KS, Gingras MC, Muthuswamy LB, Johns AL, Miller DK, Wilson PJ, Patch AM, Wu J, Chang DK, Cowley MJ, Gardiner BB, Song S, Harliwong I, Idrisoglu S, Nourse C, Nourbakhsh E, Manning S, Wani S, Gongora M, Pajic M, Scarlett CJ, Gill AJ, Pinho AV, Rooman I, Anderson M, Holmes O, Leonard C, Taylor D, Wood S, Xu Q, Nones K, Fink JL, Christ A, Bruxner T, Cloonan N, Kolle G, Newell F, Pinese M, Mead RS, Humphris JL, Kaplan W, Jones MD, Colvin EK, Nagrial AM, Humphrey ES, Chou A, Chin VT, Chantrill LA, Mawson A, Samra JS, Kench JG, Lovell JA, Daly RJ, Merrett ND, Toon C, Epari K, Nguyen NQ, Barbour A, Zeps N; Australian Pancreatic Cancer Genome Initiative, Kakkar N, Zhao F, Wu YQ, Wang M, Muzny DM, Fisher WE, Brunicardi FC, Hodges SE, Reid JG, Drummond J, Chang K, Han Y, Lewis LR, Dinh H, Buhay CJ, Beck T, Timms L, Sam M, Begley K, Brown A, Pai D, Panchal A, Buchner N, De Borja R, Denroche RE, Yung CK, Serra S, Onetto N, Mukhopadhyay D, Tsao MS, Shaw PA, Petersen GM, Gallinger S, Hruban RH, Maitra A, Iacobuzio-Donahue CA, Schulick RD, Wolfgang CL, Morgan RA, Lawlor RT, Capelli P, Corbo V, Scardoni M, Tortora G, Tempero MA, Mann KM, Jenkins NA, Perez-Mancera PA, Adams DJ, Largaespada DA, Wessels LF, Rust AG, Stein LD, Tuveson DA, Copeland NG, Musgrove EA, Scarpa A, Eshleman JR, Hudson TJ, Sutherland RL, Wheeler DA, Pearson JV, McPherson JD, Gibbs RA, Grimmond SM (2012) Pancreatic cancer genomes reveal aberrations in axon guidance pathway genes. Nature 491(7424):399–405

Blackburn EH, Greider CW, Szostak JW (2006) Telomeres and telomerase: the path from maize, Tetrahymena and yeast to human cancer and aging. Nat Med 12:1133–1138

Blackford A, Serrano OK, Wolfgang CL, Parmigiani G, Jones S, Zhang X, Parsons DW, Lin JC, Leary RJ, Eshleman JR, Goggins M, Jaffee EM, Iacobuzio-Donahue CA, Maitra A, Cameron JL, Olino K, Schulick R, Winter J, Herman JM, Laheru D, Klein AP, Vogelstein B, Kinzler KW, Velculescu VE, Hruban RH (2009) SMAD4 gene mutations are associated with poor prognosis in pancreatic cancer. Clin Cancer Res 15:4674–4679

Bozic I, Antal T, Ohtsuki H, Carter H, Kim D, Chen S, Karchin R, Kinzler KW, Vogelstein B, Nowak MA (2010) Accumulation of driver and passenger mutations during tumor progression. Proc Natl Acad Sci USA 107:18545–18550

Caldas C, Hahn SA, da Costa LT, Redston MS, Schutte M, Seymour AB, Weinstein CL, Hruban RH, Yeo CJ, Kern SE (1994) Frequent somatic mutations and homozygous deletions of the p16 (MTS1) gene in pancreatic adenocarcinoma. Nat Genet 8:27–32

Calhoun ES, Jones JB, Ashfaq R, Adsay V, Baker SJ, Valentine V, Hempen PM, Hilgers W, Yeo CJ, Hruban RH, Kern SE (2003) BRAF and FBXW7 (CDC4, FBW7, AGO, SEL10) mutations in distinct subsets of pancreatic cancer: potential therapeutic targets. Am J Pathol 163:1255–1260

Calhoun ES, Hucl T, Gallmeier E, West KM, Arking DE, Maitra A, Iacobuzio-Donahue CA, Chakravarti A, Hruban RH, Kern SE (2006) Identifying Allelic Loss and Homozygous Deletions in Pancreatic Cancer without Matched Normals Using High-Density Single-Nucleotide Polymorphism Arrays. Cancer Res 66:7920–7928

Campbell PJ, Yachida S, Mudie LJ, Stephens PJ, Pleasance ED, Stebbings LA, Morsberger LA, Latimer C, McLaren S, Lin ML, McBride DJ, Varela I, Nik-Zainal SA, Leroy C, Jia M, Menzies A, Butler AP, Teague JW, Griffin CA, Burton J, Swerdlow H, Quail MA, Stratton MR, Iacobuzio-Donahue C, Futreal PA (2010) The patterns and dynamics of genomic instability in metastatic pancreatic cancer. Nature 467:1109–1113

Canto MI, Hruban RH, Fishman EK, Kamel IR, Schulick R, Zhang Z, Topazian M, Takahashi N, Fletcher J, Petersen G, Klein AP, Axilbund J, Griffin C, Syngal S, Saltzman JR, Mortele KJ, Lee J, Tamm E, Vikram R, Bhosale P, Margolis D, Farrell J, Goggins M (2012) Frequent detection of pancreatic lesions in asymptomatic high-risk individuals. Gastroenterology 142:796–804

Chen S, Auletta T, Dovirak O, Hutter C, Kuntz K, El-ftesi S, Kendall J, Han H, Von Hoff DD, Ashfaq R, Maitra A, Iacobuzio-Donahue CA, Hruban RH, Lucito R (2008) Copy number alterations in pancreatic cancer identify recurrent PAK4 amplification. Cancer Biol Ther 7:1793–1802

Cheng JQ, Ruggeri B, Klein WM, Sonoda G, Altomare DA, Watson DK, Testa JR (1996) Amplification of AKT2 in human pancreatic cells and inhibition of AKT2 expression and tumorigenicity by antisense RNA. Proc Natl Acad Sci USA 93:3636–3641

Chetty R, Serra S, Salahshor S, Alsaad K, Shih W, Blaszyk H, Woodgett JR, Tsao MS (2006) Expression of Wnt-signaling pathway proteins in intraductal papillary mucinous neoplasms of the pancreas: a tissue microarray analysis. Hum Pathol 37:212–217

Collins MA, Bednar F, Zhang Y, Brisset JC, Galban S, Galban CJ, Rakshit S, Flannagan KS, Adsay NV, Pasca di Magliano M (2012) Oncogenic Kras is required for both the initiation and maintenance of pancreatic cancer in mice. J Clin Invest 122:639–653

Collisson EA, Sadanandam A, Olson P, Gibb WJ, Truitt M, Gu S, Cooc J, Weinkle J, Kim GE, Jakkula L, Feiler HS, Ko AH, Olshen AB, Danenberg KL, Tempero MA, Spellman PT, Hanahan D, Gray JW (2011) Subtypes of pancreatic ductal adenocarcinoma and their differing responses to therapy. Nat Med 17:500–503

Conroy T, Desseigne F, Ychou M, Bouche O, Guimbaud R, Becouarn Y, Adenis A, Raoul JL, Gourgou-Bourgade S, de la Fouchardiere C, Bennouna J, Bachet JB, Khemissa-Akouz F, Pere-Verge D, Delbaldo C, Assenat E, Chauffert B, Michel P, Montoto-Grillot C, Ducreux M (2011) FOLFIRINOX versus gemcitabine for metastatic pancreatic cancer. N Engl J Med 364:1817–1825

Couch FJ, Johnson MR, Rabe KG, Brune K, de Andrade M, Goggins M, Rothenmund H, Gallinger S, Klein A, Petersen GM, Hruban RH (2007) The prevalence of BRCA2 mutations in familial pancreatic cancer. Cancer Epidemiol Biomarkers Prev 16:342–346

Cubilla AL, Fitzgerald PJ (1976) Morphological lesions associated with human primary invasive nonendocrine pancreas cancer. Cancer Res 36:2690–2698

Dang CV (2010) Rethinking the Warburg effect with Myc micromanaging glutamine metabolism. Cancer Res 70:859–862

Dang CV (2012) MYC on the Path to Cancer. Cell 149:22–35

DeNicola GM, Tuveson DA (2009) RAS in cellular transformation and senescence. Eur J Cancer 45(Suppl 1):211–216

Diehl F, Schmidt K, Durkee KH, Moore KJ, Goodman SN, Shuber AP, Kinzler KW, Vogelstein B (2008) Analysis of mutations in DNA isolated from plasma and stool of colorectal cancer patients. Gastroenterology 135:489–498

Dressman D, Yan H, Traverso G, Kinzler KW, Vogelstein B (2003) Transforming single DNA molecules into fluorescent magnetic particles for detection and enumeration of genetic variations. Proc Natl Acad Sci USA 100:8817–8822

Elsasser SJ, Allis CD, Lewis PW (2011) Cancer. New epigenetic drivers of cancers. Science 331:1145–1146

Eshleman JR, Markowitz SD (1996) Mismatch repair defects in human carcinogenesis. Human molecular genetics 5 Spec No:1489–1494

Feldmann G, Mishra A, Bisht S, Karikari C, Garrido-Laguna I, Rasheed Z, Ottenhof N, Dadon T, Alvarez H, Fendrich V, Rajeshkumar NV, Matsui W, Brossart P, Hidalgo M, Bannerji R, Maitra A, Nelkin BD (2011) Cyclin-dependent kinase inhibitor Dinaciclib (SCH727965) inhibits pancreatic cancer growth and progression in murine xenograft models. Cancer Biol Ther 12(7):598–609

Fong PC, Yap TA, Boss DS, Carden CP, Mergui-Roelvink M, Gourley C, De Greve J, Lubinski J, Shanley S, Messiou C, A'Hern R, Tutt A, Ashworth A, Stone J, Carmichael J, Schellens JH, de Bono JS, Kaye SB (2010) Poly(ADP)-ribose polymerase inhibition: frequent durable responses in BRCA carrier ovarian cancer correlating with platinum-free interval. J Clin Oncol 28:2512–2519

Fu B, Luo M, Lakkur S, Lucito R, Iacobuzio-Donahue CA (2008) Frequent genomic copy number gain and overexpression of GATA-6 in pancreatic carcinoma. Cancer Biol Ther 7(10): 1593–1601

Fujimoto A, Totoki Y, Abe T, Boroevich KA, Hosoda F, Nguyen HH, Aoki M, Hosono N, Kubo M, Miya F, Arai Y, Takahashi H, Shirakihara T, Nagasaki M, Shibuya T, Nakano K, Watanabe-Makino K, Tanaka H, Nakamura H, Kusuda J, Ojima H, Shimada K, Okusaka T, Ueno M, Shigekawa Y, Kawakami Y, Arihiro K, Ohdan H, Gotoh K, Ishikawa O, Ariizumi SI, Yamamoto M, Yamada T, Chayama K, Kosuge T, Yamaue H, Kamatani N, Miyano S, Nakagama H, Nakamura Y, Tsunoda T, Shibata T, Nakagawa H (2012) Whole-genome sequencing of liver cancers identifies etiological influences on mutation patterns and recurrent mutations in chromatin regulators. Nat Genet 44(7):760–764

Fukushima N, Sato N, Ueki T, Rosty C, Walter KM, Wilentz RE, Yeo CJ, Hruban RH, Goggins M (2002) Aberrant methylation of preproenkephalin and p16 genes in pancreatic intraepithelial neoplasia and pancreatic ductal adenocarcinoma. Am J Pathol 160:1573–1581

Fusaro RM, Lynch HT (2000) The FAMMM syndrome: epidemiology and surveillance strategies. Cancer Invest 18:670–680

Gallmeier E, Kern SE (2007) Targeting Fanconi anemia/BRCA2 pathway defects in cancer: the significance of preclinical pharmacogenomic models. Clin Cancer Res 13:4–10

Ghimenti C, Tannergard P, Wahlberg S, Liu T, Giulianotti PG, Mosca F, Fornaciari G, Bevilacqua G, Lindblom A, Caligo MA (1999) Microsatellite instability and mismatch repair gene inactivation in sporadic pancreatic and colon tumours. Br J Cancer 80:11–16

Giardiello FM, Welsh SB, Hamilton SR, Offerhaus GJ, Gittelsohn AM, Booker SV, Krush AJ, Yardley JH, Luk GD (1987) Increased risk of cancer in the Peutz-Jeghers syndrome. N Engl J Med 316:1511–1514

Goggins M (2005) Molecular markers of early pancreatic cancer. J Clin Oncol 23:4524–4531

Goggins M, Schutte M, Lu J, Moskaluk CA, Weinstein CL, Petersen GM, Yeo CJ, Jackson CE, Lynch HT, Hruban RH, Kern SE (1996) Germline BRCA2 gene mutations in patients with apparently sporadic pancreatic carcinomas. Cancer Res 56:5360–5364

Goggins M, Offerhaus GJ, Hilgers W, Griffin CA, Shekher M, Tang D, Sohn TA, Yeo CJ, Kern SE, Hruban RH (1998) Pancreatic adenocarcinomas with DNA replication errors (RER+) are associated with wild-type K-ras and characteristic histopathology. Poor differentiation, a syncytial growth pattern, and pushing borders suggest RER+. Am J Pathol 152:1501–1507

Gudmundsdottir K, Ashworth A (2006) The roles of BRCA1 and BRCA2 and associated proteins in the maintenance of genomic stability. Oncogene 25:5864–5874

Gysin S, Rickert P, Kastury K, McMahon M (2005) Analysis of genomic DNA alterations and mRNA expression patterns in a panel of human pancreatic cancer cell lines. Genes Chromosomes Cancer 44:37–51

Haeusgen W, Herdegen T, Waetzig V (2011) The bottleneck of JNK signaling: molecular and functional characteristics of MKK4 and MKK7. Eur J Cell Biol 90:536–544

Hahn SA, Schutte M, Hoque AT, Moskaluk CA, da Costa LT, Rozenblum E, Weinstein CL, Fischer A, Yeo CJ, Hruban RH, Kern SE (1996) DPC4, a candidate tumor suppressor gene at human chromosome 18q21.1. Science 271:350–353

Hahn SA, Greenhalf B, Ellis I, Sina-Frey M, Rieder H, Korte B, Gerdes B, Kress R, Ziegler A, Raeburn JA, Campra D, Grutzmann R, Rehder H, Rothmund M, Schmiegel W, Neoptolemos JP, Bartsch DK (2003) BRCA2 germline mutations in familial pancreatic carcinoma. J Natl Cancer Inst 95:214–221

Han H, Bearss DJ, Browne LW, Calaluce R, Nagle RB, Von Hoff DD (2002) Identification of differentially expressed genes in pancreatic cancer cells using cDNA microarray. Cancer Res 62:2890–2896

Hanahan D, Weinberg RA (2011) Hallmarks of cancer: the next generation. Cell 144:646–674

Hao HX, Xie Y, Zhang Y, Charlat O, Oster E, Avello M, Lei H, Mickanin C, Liu D, Ruffner H, Mao X, Ma Q, Zamponi R, Bouwmeester T, Finan PM, Kirschner MW, Porter JA, Serluca FC, Cong F (2012) ZNRF3 promotes Wnt receptor turnover in an R-spondin-sensitive manner. Nature 485:195–200

Harley CB, Futcher AB, Greider CW (1990) Telomeres shorten during ageing of human fibroblasts. Nature 345:458–460

Heaphy CM, de Wilde RF, Jiao Y, Klein AP, Edil BH, Shi C, Bettegowda C, Rodriguez FJ, Eberhart CG, Hebbar S, Offerhaus GJ, McLendon R, Rasheed BA, He Y, Yan H, Bigner DD, Oba-Shinjo SM, Marie SK, Riggins GJ, Kinzler KW, Vogelstein B, Hruban RH, Maitra A, Papadopoulos N, Meeker AK (2011a) Altered telomeres in tumors with ATRX and DAXX mutations. Science 333:425

Heaphy CM, Subhawong AP, Hong SM, Goggins MG, Montgomery EA, Gabrielson E, Netto GJ, Epstein JI, Lotan TL, Westra WH, Shih Ie M, Iacobuzio-Donahue CA, Maitra A, Li QK, Eberhart CG, Taube JM, Rakheja D, Kurman RJ, Wu TC, Roden RB, Argani P, De Marzo AM, Terracciano L, Torbenson M, Meeker AK (2011b) Prevalence of the alternative lengthening of telomeres telomere maintenance mechanism in human cancer subtypes. Am J Pathol 179:1608–1615

Hezel AF, Bardeesy N (2008) LKB1; linking cell structure and tumor suppression. Oncogene 27:6908–6919

Hezel AF, Gurumurthy S, Granot Z, Swisa A, Chu GC, Bailey G, Dor Y, Bardeesy N, Depinho RA (2008) Pancreatic LKB1 deletion leads to acinar polarity defects and cystic neoplasms. Mol Cell Biol 28:2414–2425

Hingorani SR, Petricoin EF, Maitra A, Rajapakse V, King C, Jacobetz MA, Ross S, Conrads TP, Veenstra TD, Hitt BA, Kawaguchi Y, Johann D, Liotta LA, Crawford HC, Putt ME, Jacks T, Wright CV, Hruban RH, Lowy AM, Tuveson DA (2003) Preinvasive and invasive ductal pancreatic cancer and its early detection in the mouse. Cancer Cell 4:437–450

Hingorani SR, Wang L, Multani AS, Combs C, Deramaudt TB, Hruban RH, Rustgi AK, Chang S, Tuveson DA (2005) Trp53R172H and KrasG12D cooperate to promote chromosomal instability and widely metastatic pancreatic ductal adenocarcinoma in mice. Cancer Cell 7:469–483

Holzmann K, Kohlhammer H, Schwaenen C, Wessendorf S, Kestler HA, Schwoerer A, Rau B, Radlwimmer B, Dohner H, Lichter P, Gress T, Bentz M (2004) Genomic DNA-chip hybridization reveals a higher incidence of genomic amplifications in pancreatic cancer than conventional comparative genomic hybridization and leads to the identification of novel candidate genes. Cancer Res 64:4428–4433

Hudson TJ, Anderson W, Artez A, Barker AD, Bell C, Bernabe RR, Bhan MK, Calvo F, Eerola I, Gerhard DS, Guttmacher A, Guyer M, Hemsley FM, Jennings JL, Kerr D, Klatt P, Kolar P, Kusada J, Lane DP, Laplace F, Youyong L, Nettekoven G, Ozenberger B, Peterson J, Rao TS, Remacle J, Schafer AJ, Shibata T, Stratton MR, Vockley JG, Watanabe K, Yang H, Yuen MM,

Knoppers BM, Bobrow M, Cambon-Thomsen A, Dressler LG, Dyke SO, Joly Y, Kato K, Kennedy KL, Nicolas P, Parker MJ, Rial-Sebbag E, Romeo-Casabona CM, Shaw KM, Wallace S, Wiesner GL, Zeps N, Lichter P, Biankin AV, Chabannon C, Chin L, Clement B, de Alava E, Degos F, Ferguson ML, Geary P, Hayes DN, Johns AL, Kasprzyk A, Nakagawa H, Penny R, Piris MA, Sarin R, Scarpa A, van de Vijver M, Futreal PA, Aburatani H, Bayes M, Botwell DD, Campbell PJ, Estivill X, Grimmond SM, Gut I, Hirst M, Lopez-Otin C, Majumder P, Marra M, McPherson JD, Ning Z, Puente XS, Ruan Y, Stunnenberg HG, Swerdlow H, Velculescu VE, Wilson RK, Xue HH, Yang L, Spellman PT, Bader GD, Boutros PC, Flicek P, Getz G, Guigo R, Guo G, Haussler D, Heath S, Hubbard TJ, Jiang T et al (2010) International network of cancer genome projects. Nature 464:993–998

Hustinx SR, Leoni LM, Yeo CJ, Brown PN, Goggins M, Kern SE, Hruban RH, Maitra A (2005a) Concordant loss of MTAP and p16/CDKN2A expression in pancreatic intraepithelial neoplasia: evidence of homozygous deletion in a noninvasive precursor lesion. Mod Pathol 18:959–963

Hustinx SR, Hruban RH, Leoni LM, Iacobuzio-Donahue C, Cameron JL, Yeo CJ, Brown PN, Argani P, Asfaq R, Fukushima N, Goggins M, Kern SE, Maitra A (2005b) Homozygous deletion of the MTAP gene in invasive adenocarcinoma of the pancreas and in periampullary cancer: a potential new target for therapy. Cancer Biol Ther 4(1):83–86

Iacobuzio-Donahue CA, van der Heijden MS, Baumgartner MR, Troup WJ, Romm JM, Doheny K, Pugh E, Yeo CJ, Goggins MG, Hruban RH, Kern SE (2004) Large-scale allelotype of pancreaticobiliary carcinoma provides quantitative estimates of genome-wide allelic loss. Cancer Res 64:871–875

Iacobuzio-Donahue CA, Fu B, Yachida S, Luo M, Abe H, Henderson CM, Vilardell F, Wang Z, Keller JW, Banerjee P, Herman JM, Cameron JL, Yeo CJ, Halushka MK, Eshleman JR, Raben M, Klein AP, Hruban RH, Hidalgo M, Laheru D (2009) DPC4 gene status of the primary carcinoma correlates with patterns of failure in patients with pancreatic cancer. J Clin Oncol 27(11):1806–1813

Jackson EL, Olive KP, Tuveson DA, Bronson R, Crowley D, Brown M, Jacks T (2005) The differential effects of mutant p53 alleles on advanced murine lung cancer. Cancer Res 65:10280–10288

Jacobs EJ, Chanock SJ, Fuchs CS, Lacroix A, McWilliams RR, Steplowski E, Stolzenberg-Solomon RZ, Arslan AA, Bueno-de-Mesquita HB, Gross M, Helzlsouer K, Petersen G, Zheng W, Agalliu I, Allen NE, Amundadottir L, Boutron-Ruault MC, Buring JE, Canzian F, Clipp S, Dorronsoro M, Gaziano JM, Giovannucci EL, Hankinson SE, Hartge P, Hoover RN, Hunter DJ, Jacobs KB, Jenab M, Kraft P, Kooperberg C, Lynch SM, Sund M, Mendelsohn JB, Mouw T, Newton CC, Overvad K, Palli D, Peeters PH, Rajkovic A, Shu XO, Thomas G, Tobias GS, Trichopoulos D, Virtamo J, Wactawski-Wende J, Wolpin BM, Yu K, Zeleniuch-Jacquotte A (2010) Family history of cancer and risk of pancreatic cancer: a pooled analysis from the Pancreatic Cancer Cohort Consortium (PanScan). Int J Cancer 127:1421–1428

James E, Waldron-Lynch MG, Saif MW (2009) Prolonged survival in a patient with BRCA2 associated metastatic pancreatic cancer after exposure to camptothecin: a case report and review of literature. Anti-cancer drugs 20:634–638

Jiao Y, Shi C, Edil BH, de Wilde RF, Klimstra DS, Maitra A, Schulick RD, Tang LH, Wolfgang CL, Choti MA, Velculescu VE, Diaz LA Jr, Vogelstein B, Kinzler KW, Hruban RH, Papadopoulos N (2011) DAXX/ATRX, MEN1, and mTOR pathway genes are frequently altered in pancreatic neuroendocrine tumors. Science 331:1199–1203

Jones S, Zhang X, Parsons DW, Lin JC, Leary RJ, Angenendt P, Mankoo P, Carter H, Kamiyama H, Jimeno A, Hong SM, Fu B, Lin MT, Calhoun ES, Kamiyama M, Walter K, Nikolskaya T, Nikolsky Y, Hartigan J, Smith DR, Hidalgo M, Leach SD, Klein AP, Jaffee EM, Goggins M, Maitra A, Iacobuzio-Donahue C, Eshleman JR, Kern SE, Hruban RH, Karchin R, Papadopoulos N, Parmigiani G, Vogelstein B, Velculescu VE, Kinzler KW (2008) Core signaling pathways in human pancreatic cancers revealed by global genomic analyses. Science 321:1801–1806

Jones S, Hruban RH, Kamiyama M, Borges M, Zhang X, Parsons DW, Lin JC, Palmisano E, Brune K, Jaffee EM, Iacobuzio-Donahue CA, Maitra A, Parmigiani G, Kern SE, Velculescu VE, Kinzler KW, Vogelstein B, Eshleman JR, Goggins M, Klein AP (2009) Exomic sequencing identifies PALB2 as a pancreatic cancer susceptibility gene. Science 324:217

Jones S, Wang TL, Shih Ie M, Mao TL, Nakayama K, Roden R, Glas R, Slamon D, Diaz LA Jr, Vogelstein B, Kinzler KW, Velculescu VE, Papadopoulos N (2010) Frequent mutations of chromatin remodeling gene ARID1A in ovarian clear cell carcinoma. Science 330:228–231
Jones S, Li M, Parsons DW, Zhang X, Wesseling J, Kristel P, Schmidt MK, Markowitz S, Yan H, Bigner D, Hruban RH, Eshleman JR, Iacobuzio-Donahue CA, Goggins M, Maitra A, Malek SN, Powell S, Vogelstein B, Kinzler KW, Velculescu VE, Papadopoulos N (2012) Somatic mutations in the chromatin remodeling gene ARID1A occur in several tumor types. Hum Mutat 33:100–103
Kanda M, Matthaei H, Wu J, Hong SM, Yu J, Borges M, Hruban RH, Maitra A, Kinzler K, Vogelstein B, Goggins M (2012) Presence of somatic mutations in most early-stage pancreatic intraepithelial neoplasia. Gastroenterology 142:730–733, e9
Karikari CA, Mullendore M, Eshleman JR, Argani P, Leoni LM, Chattopadhyay S, Hidalgo M, Maitra A (2005) Homozygous deletions of methylthioadenosine phosphorylase in human biliary tract cancers. Mol Cancer Ther 4:1860–1866
Kelland LR (2003) Farnesyl transferase inhibitors in the treatment of breast cancer. Expert Opin Investig Drugs 12:413–421
Killeen MT, Sybingco SS (2008) Netrin, Slit and Wnt receptors allow axons to choose the axis of migration. Dev Biol 323:143–151
Kim WY, Sharpless NE (2006) The regulation of INK4/ARF in cancer and aging. Cell 127:265–275
Klein AP, Hruban RH, Brune KA, Petersen GM, Goggins M (2001) Familial pancreatic cancer. Cancer J 7:266–273
Knudson AG (1996) Hereditary cancer: two hits revisited. J Cancer Res Clin Oncol 122:135–140
Koorstra JB, Hong SM, Shi C, Meeker AK, Ryu JK, Offerhaus GJ, Goggins MG, Hruban RH, Maitra A (2009) Widespread activation of the DNA damage response in human pancreatic intraepithelial neoplasia. Mod Pathol 22:1439–1445
Kopetz S, Chang GJ, Overman MJ, Eng C, Sargent DJ, Larson DW, Grothey A, Vauthey JN, Nagorney DM, McWilliams RR (2009) Improved survival in metastatic colorectal cancer is associated with adoption of hepatic resection and improved chemotherapy. J Clin Oncol 27:3677–3683
Kwei KA, Bashyam MD, Kao J, Ratheesh R, Reddy EC, Kim YH, Montgomery K, Giacomini CP, Choi YL, Chatterjee S, Karikari CA, Salari K, Wang P, Hernandez-Boussard T, Swarnalata G, van de Rijn M, Maitra A, Pollack JR (2008) Genomic profiling identifies GATA6 as a candidate oncogene amplified in pancreatobiliary cancer. PLoS Genet 4:e1000081
Lal G, Liu G, Schmocker B, Kaurah P, Ozcelik H, Narod SA, Redston M, Gallinger S (2000) Inherited predisposition to pancreatic adenocarcinoma: role of family history and germ-line p16, BRCA1, and BRCA2 mutations. Cancer Res 60:409–416
Lee S, Roeder RG, Lee JW (2009) Roles of histone H3-lysine 4 methyltransferase complexes in NR-mediated gene transcription. Prog Mol Biol Transl Sci 87:343–382
Lindor NM, Petersen GM, Spurdle AB, Thompson B, Goldgar DE, Thibodeau SN (2011) Pancreatic cancer and a novel MSH2 germline alteration. Pancreas 40:1138–1140
Luttges J, Galehdari H, Brocker V, Schwarte-Waldhoff I, Henne-Bruns D, Kloppel G, Schmiegel W, Hahn SA (2001) Allelic loss is often the first hit in the biallelic inactivation of the p53 and DPC4 genes during pancreatic carcinogenesis. Am J Pathol 158:1677–1683
Mahlamaki EH, Kauraniemi P, Monni O, Wolf M, Hautaniemi S, Kallioniemi A (2004) High-resolution genomic and expression profiling reveals 105 putative amplification target genes in pancreatic cancer. Neoplasia 6:432–439
Maitra A, Hruban RH (2008) Pancreatic cancer. Annu Rev Pathol 3:157–188
Maitra A, Adsay NV, Argani P, Iacobuzio-Donahue C, De Marzo A, Cameron JL, Yeo CJ, Hruban RH (2003) Multicomponent analysis of the pancreatic adenocarcinoma progression model using a pancreatic intraepithelial neoplasia tissue microarray. Mod Pathol 16:902–912
Malkhosyan S, Rampino N, Yamamoto H, Perucho M (1996) Frameshift mutator mutations. Nature 382:499–500

Martinez P, Blasco MA (2011) Telomeric and extra-telomeric roles for telomerase and the telomere-binding proteins. Nat Rev Cancer 11:161–176

Marx SJ, Agarwal SK, Kester MB, Heppner C, Kim YS, Skarulis MC, James LA, Goldsmith PK, Saggar SK, Park SY, Spiegel AM, Burns AL, Debelenko LV, Zhuang Z, Lubensky IA, Liotta LA, Emmert-Buck MR, Guru SC, Manickam P, Crabtree J, Erdos MR, Collins FS, Chandrasekharappa SC (1999) Multiple endocrine neoplasia type 1: clinical and genetic features of the hereditary endocrine neoplasias. Recent Prog Horm Res 54:397–438, discussion 438-9

Matsubayashi H, Canto M, Sato N, Klein A, Abe T, Yamashita K, Yeo CJ, Kalloo A, Hruban R, Goggins M (2006) DNA methylation alterations in the pancreatic juice of patients with suspected pancreatic disease. Cancer Res 66:1208–1217

Meric-Bernstam F, Akcakanat A, Chen H, Do KA, Sangai T, Adkins F, Gonzalez-Angulo AM, Rashid A, Crosby K, Dong M, Phan AT, Wolff RA, Gupta S, Mills GB, Yao J (2012) PIK3CA/PTEN mutations and Akt activation as markers of sensitivity to allosteric mTOR inhibitors. Clin Cancer Res 18:1777–1789

Moskaluk CA, Hruban RH, Kern SE (1997) p16 and K-ras gene mutations in the intraductal precursors of human pancreatic adenocarcinoma. Cancer Res 57:2140–2143

Naderi A, Couch FJ (2002) BRCA2 and pancreatic cancer. Int J Gastrointest Cancer 31:99–106

Nowak NJ, Gaile D, Conroy JM, McQuaid D, Cowell J, Carter R, Goggins MG, Hruban RH, Maitra A (2005) Genome-wide aberrations in pancreatic adenocarcinoma. Cancer Genet Cytogenet 161:36–50

Olive KP, Tuveson DA, Ruhe ZC, Yin B, Willis NA, Bronson RT, Crowley D, Jacks T (2004) Mutant p53 gain of function in two mouse models of Li-Fraumeni syndrome. Cell 119: 847–860

Ozcelik H, Schmocker B, Di Nicola N, Shi XH, Langer B, Moore M, Taylor BR, Narod SA, Darlington G, Andrulis IL, Gallinger S, Redston M (1997) Germline BRCA2 6174delT mutations in Ashkenazi Jewish pancreatic cancer patients. Nat Genet 16:17–18

Parsons R, Li GM, Longley MJ, Fang WH, Papadopoulos N, Jen J, de la Chapelle A, Kinzler KW, Vogelstein B, Modrich P (1993) Hypermutability and mismatch repair deficiency in RER+ tumor cells. Cell 75:1227–1236

Parsons DW, Li M, Zhang X, Jones S, Leary RJ, Lin JC, Boca SM, Carter H, Samayoa J, Bettegowda C, Gallia GL, Jallo GI, Binder ZA, Nikolsky Y, Hartigan J, Smith DR, Gerhard DS, Fults DW, VandenBerg S, Berger MS, Marie SK, Shinjo SM, Clara C, Phillips PC, Minturn JE, Biegel JA, Judkins AR, Resnick AC, Storm PB, Curran T, He Y, Rasheed BA, Friedman HS, Keir ST, McLendon R, Northcott PA, Taylor MD, Burger PC, Riggins GJ, Karchin R, Parmigiani G, Bigner DD, Yan H, Papadopoulos N, Vogelstein B, Kinzler KW, Velculescu VE (2011) The genetic landscape of the childhood cancer medulloblastoma. Science 331: 435–439

Perez-Mancera PA, Tuveson DA (2006) Physiological analysis of oncogenic K-ras. Methods Enzymol 407:676–690

Perez-Mancera PA, Guerra C, Barbacid M, Tuveson DA (2012) What we have learned about pancreatic cancer from mouse models. Gastroenterology 142:1079–1092

Qiu W, Sahin F, Iacobuzio-Donahue CA, Garcia-Carracedo D, Wang WM, Kuo CY, Chen D, Arking DE, Lowy AM, Hruban RH, Remotti HE, Su GH (2011) Disruption of p16 and activation of Kras in pancreas increase ductal adenocarcinoma formation and metastasis in vivo. Oncotarget 2:862–873

Rajeshkumar NV, De Oliveira E, Ottenhof N, Watters J, Brooks D, Demuth T, Shumway SD, Mizuarai S, Hirai H, Maitra A, Hidalgo M (2011) MK-1775, a potent Wee1 inhibitor, synergizes with gemcitabine to achieve tumor regressions, selectively in p53-deficient pancreatic cancer xenografts. Clin Cancer Res 17:2799–2806

Roberts NJ, Jiao Y, Yu J, Kopelovich L, Petersen GM, Bondy ML, Gallinger S, Schwartz AG, Syngal S, Cote ML, Axilbund J, Schulick R, Ali SZ, Eshleman JR, Velculescu VE, Goggins M, Vogelstein B, Papadopoulos N, Hruban RH, Kinzler KW, Klein AP (2012) ATM mutations in patients with hereditary pancreatic cancer. Cancer discovery 2:41–46

Robinson VL, Hickson JA, Vander Griend DJ, Dubauskas Z, Rinker-Schaeffer CW (2003) MKK4 and metastasis suppression: a marriage of signal transduction and metastasis research. Clin Exp Metastasis 20:25–30

Rozenblum E, Schutte M, Goggins M, Hahn SA, Panzer S, Zahurak M, Goodman SN, Sohn TA, Hruban RH, Yeo CJ, Kern SE (1997) Tumor-suppressive pathways in pancreatic carcinoma. Cancer Res 57:1731–1734

Ruggeri BA, Huang L, Wood M, Cheng JQ, Testa JR (1998) Amplification and overexpression of the AKT2 oncogene in a subset of human pancreatic ductal adenocarcinomas. Mol Carcinog 21:81–86

Sahin F, Maitra A, Argani P, Sato N, Maehara N, Montgomery E, Goggins M, Hruban RH, Su GH (2003) Loss of Stk11/Lkb1 expression in pancreatic and biliary neoplasms. Mod Pathol 16:686–691

Schutte M, Hruban RH, Hedrick L, Cho KR, Nadasdy GM, Weinstein CL, Bova GS, Isaacs WB, Cairns P, Nawroz H, Sidransky D, Casero RA Jr, Meltzer PS, Hahn SA, Kern SE (1996) DPC4 gene in various tumor types. Cancer Res 56:2527–2530

Schutte M, Hruban RH, Geradts J, Maynard R, Hilgers W, Rabindran SK, Moskaluk CA, Hahn SA, Schwarte-Waldhoff I, Schmiegel W, Baylin SB, Kern SE, Herman JG (1997) Abrogation of the Rb/p16 tumor-suppressive pathway in virtually all pancreatic carcinomas. Cancer Res 57:3126–3130

Shain AH, Giacomini CP, Matsukuma K, Karikari CA, Bashyam MD, Hidalgo M, Maitra A, Pollack JR (2012) Convergent structural alterations define SWItch/Sucrose NonFermentable (SWI/SNF) chromatin remodeler as a central tumor suppressive complex in pancreatic cancer. Proc Natl Acad Sci USA 109:E252–E259

Sherr CJ (2004) Principles of tumor suppression. Cell 116:235–246

Shi C, Fukushima N, Abe T, Bian Y, Hua L, Wendelburg BJ, Yeo CJ, Hruban RH, Goggins MG, Eshleman JR (2008) Sensitive and quantitative detection of KRAS2 gene mutations in pancreatic duct juice differentiates patients with pancreatic cancer from chronic pancreatitis, potential for early detection. Cancer Biol Ther 7:353–360

Siegel PM, Massague J (2003) Cytostatic and apoptotic actions of TGF-beta in homeostasis and cancer. Nat Rev Cancer 3:807–821

Siegel R, Naishadham D, Jemal A (2012) Cancer statistics, 2012. CA Cancer J Clin 62:10–29

Skoulidis F, Cassidy LD, Pisupati V, Jonasson JG, Bjarnason H, Eyfjord JE, Karreth FA, Lim M, Barber LM, Clatworthy SA, Davies SE, Olive KP, Tuveson DA, Venkitaraman AR (2010) Germline Brca2 heterozygosity promotes Kras(G12D) -driven carcinogenesis in a murine model of familial pancreatic cancer. Cancer Cell 18:499–509

Solimini NL, Xu Q, Mermel CH, Liang AC, Schlabach MR, Luo J, Burrows AE, Anselmo AN, Bredemeyer AL, Li MZ, Beroukhim R, Meyerson M (2012) Elledge SJ. Science, Recurrent Hemizygous Deletions in Cancers May Optimize Proliferative Potential

Stephens PJ, Greenman CD, Fu B, Yang F, Bignell GR, Mudie LJ, Pleasance ED, Lau KW, Beare D, Stebbings LA, McLaren S, Lin ML, McBride DJ, Varela I, Nik-Zainal S, Leroy C, Jia M, Menzies A, Butler AP, Teague JW, Quail MA, Burton J, Swerdlow H, Carter NP, Morsberger LA, Iacobuzio-Donahue C, Follows GA, Green AR, Flanagan AM, Stratton MR, Futreal PA, Campbell PJ (2011) Massive genomic rearrangement acquired in a single catastrophic event during cancer development. Cell 144:27–40

Stransky N, Egloff AM, Tward AD, Kostic AD, Cibulskis K, Sivachenko A, Kryukov GV, Lawrence MS, Sougnez C, McKenna A, Shefler E, Ramos AH, Stojanov P, Carter SL, Voet D, Cortes ML, Auclair D, Berger MF, Saksena G, Guiducci C, Onofrio RC, Parkin M, Romkes M, Weissfeld JL, Seethala RR, Wang L, Rangel-Escareno C, Fernandez-Lopez JC, Hidalgo-Miranda A, Melendez-Zajgla J, Winckler W, Ardlie K, Gabriel SB, Meyerson M, Lander ES, Getz G, Golub TR, Garraway LA, Grandis JR (2011) The mutational landscape of head and neck squamous cell carcinoma. Science 333:1157–1160

Su GH, Hilgers W, Shekher MC, Tang DJ, Yeo CJ, Hruban RH, Kern SE (1998) Alterations in pancreatic, biliary, and breast carcinomas support MKK4 as a genetically targeted tumor suppressor gene. Cancer Res 58:2339–2342

Su GH, Hruban RH, Bansal RK, Bova GS, Tang DJ, Shekher MC, Westerman AM, Entius MM, Goggins M, Yeo CJ, Kern SE (1999) Germline and somatic mutations of the STK11/LKB1 Peutz-Jeghers gene in pancreatic and biliary cancers. Am J Pathol 154:1835–1840

Tascilar M, Offerhaus GJ, Altink R, Argani P, Sohn TA, Yeo CJ, Cameron JL, Goggins M, Hruban RH, Wilentz RE (2001a) Immunohistochemical labeling for the Dpc4 gene product is a specific marker for adenocarcinoma in biopsy specimens of the pancreas and bile duct. Am J Clin Pathol 116:831–837

Tascilar M, Skinner HG, Rosty C, Sohn T, Wilentz RE, Offerhaus GJ, Adsay V, Abrams RA, Cameron JL, Kern SE, Yeo CJ, Hruban RH, Goggins M (2001b) The SMAD4 protein and prognosis of pancreatic ductal adenocarcinoma. Clin Cancer Res 7:4115–4121

Van Cutsem E, van de Velde H, Karasek P, Oettle H, Vervenne WL, Szawlowski A, Schoffski P, Post S, Verslype C, Neumann H, Safran H, Humblet Y, Perez Ruixo J, Ma Y, Von Hoff D (2004) Phase III trial of gemcitabine plus tipifarnib compared with gemcitabine plus placebo in advanced pancreatic cancer. J Clin Oncol 22:1430–1438

van der Heijden MS, Brody JR, Dezentje DA, Gallmeier E, Cunningham SC, Swartz MJ, DeMarzo AM, Offerhaus GJ, Isacoff WH, Hruban RH, Kern SE (2005) In vivo therapeutic responses contingent on Fanconi anemia/BRCA2 status of the tumor. Clin Cancer Res 11: 7508–7515

van Heek NT, Meeker AK, Kern SE, Yeo CJ, Lillemoe KD, Cameron JL, Offerhaus GJ, Hicks JL, Wilentz RE, Goggins MG, De Marzo AM, Hruban RH, Maitra A (2002a) Telomere shortening is nearly universal in pancreatic intraepithelial neoplasia. Am J Pathol 161:1541–1547

van Heek T, Rader AE, Offerhaus GJ, McCarthy DM, Goggins M, Hruban RH, Wilentz RE (2002b) K-ras, p53, and DPC4 (MAD4) alterations in fine-needle aspirates of the pancreas: a molecular panel correlates with and supplements cytologic diagnosis. Am J Clin Pathol 117:755–765

Varela I, Tarpey P, Raine K, Huang D, Ong CK, Stephens P, Davies H, Jones D, Lin ML, Teague J, Bignell G, Butler A, Cho J, Dalgliesh GL, Galappaththige D, Greenman C, Hardy C, Jia M, Latimer C, Lau KW, Marshall J, McLaren S, Menzies A, Mudie L, Stebbings L, Largaespada DA, Wessels LF, Richard S, Kahnoski RJ, Anema J, Tuveson DA, Perez-Mancera PA, Mustonen V, Fischer A, Adams DJ, Rust A, Chan-on W, Subimerb C, Dykema K, Furge K, Campbell PJ, Teh BT, Stratton MR, Futreal PA (2011) Exome sequencing identifies frequent mutation of the SWI/SNF complex gene PBRM1 in renal carcinoma. Nature 469:539–542

Vazquez A, Bond EE, Levine AJ, Bond GL (2008) The genetics of the p53 pathway, apoptosis and cancer therapy. Nat Rev Drug Discov 7:979–987

Vigil D, Cherfils J, Rossman KL, Der CJ (2010) Ras superfamily GEFs and GAPs: validated and tractable targets for cancer therapy? Nat Rev Cancer 10:842–857

Villarroel MC, Rajeshkumar NV, Garrido-Laguna I, De Jesus-Acosta A, Jones S, Maitra A, Hruban RH, Eshleman JR, Klein A, Laheru D, Donehower R, Hidalgo M (2011) Personalizing cancer treatment in the age of global genomic analyses: PALB2 gene mutations and the response to DNA damaging agents in pancreatic cancer. Mol Cancer Ther 10:3–8

Vogelstein B, Kinzler KW (2004) Cancer genes and the pathways they control. Nat Med 10:789–799

Wang H, Han H, Von Hoff DD (2006) Identification of an agent selectively targeting DPC4 (deleted in pancreatic cancer locus 4)-deficient pancreatic cancer cells. Cancer Res 66: 9722–9730

Wilentz RE, Goggins M, Redston M, Marcus VA, Adsay NV, Sohn TA, Kadkol SS, Yeo CJ, Choti M, Zahurak M, Johnson K, Tascilar M, Offerhaus GJ, Hruban RH, Kern SE (2000) Genetic, immunohistochemical, and clinical features of medullary carcinoma of the pancreas: A newly described and characterized entity. Am J Pathol 156:1641–1651

Wu J, Jiao Y, Dal Molin M, Maitra A, de Wilde RF, Wood LD, Eshleman JR, Goggins MG, Wolfgang CL, Canto MI, Schulick RD, Edil BH, Choti MA, Adsay V, Klimstra DS, Offerhaus GJ, Klein AP, Kopelovich L, Carter H, Karchin R, Allen PJ, Schmidt CM, Naito Y, Diaz LA Jr, Kinzler KW, Papadopoulos N, Hruban RH, Vogelstein B (2011a) Whole-exome sequencing of neoplastic cysts of the pancreas reveals recurrent mutations in components of ubiquitin-dependent pathways. Proc Natl Acad Sci USA 108:21188–21193

Wu J, Matthaei H, Maitra A, Dal Molin M, Wood LD, Eshleman JR, Goggins M, Canto MI, Schulick RD, Edil BH, Wolfgang CL, Klein AP, Diaz LA Jr, Allen PJ, Schmidt CM, Kinzler KW, Papadopoulos N, Hruban RH, Vogelstein B (2011b) Recurrent GNAS mutations define an unexpected pathway for pancreatic cyst development. Sci Transl Med 3:92ra66

Xin W, Yun KJ, Ricci F, Zahurak M, Qiu W, Su GH, Yeo CJ, Hruban RH, Kern SE, Iacobuzio-Donahue CA (2004) MAP2K4/MKK4 expression in pancreatic cancer: genetic validation of immunohistochemistry and relationship to disease course. Clin Cancer Res 10:8516–8520

Yachida S, Jones S, Bozic I, Antal T, Leary R, Fu B, Kamiyama M, Hruban RH, Eshleman JR, Nowak MA, Velculescu VE, Kinzler KW, Vogelstein B, Iacobuzio-Donahue CA (2010) Distant metastasis occurs late during the genetic evolution of pancreatic cancer. Nature 467: 1114–1117

Yamamoto H, Itoh F, Nakamura H, Fukushima H, Sasaki S, Perucho M, Imai K (2001) Genetic and clinical features of human pancreatic ductal adenocarcinomas with widespread microsatellite instability. Cancer Res 61:3139–3144

Yamano M, Fujii H, Takagaki T, Kadowaki N, Watanabe H, Shirai T (2000) Genetic progression and divergence in pancreatic carcinoma. Am J Pathol 156:2123–2133

Yao JC, Shah MH, Ito T, Bohas CL, Wolin EM, Van Cutsem E, Hobday TJ, Okusaka T, Capdevila J, de Vries EG, Tomassetti P, Pavel ME, Hoosen S, Haas T, Lincy J, Lebwohl D, Oberg K (2011) Everolimus for advanced pancreatic neuroendocrine tumors. N Engl J Med 364: 514–523

Ying H, Kimmelman AC, Lyssiotis CA, Hua S, Chu GC, Fletcher-Sananikone E, Locasale JW, Son J, Zhang H, Coloff JL, Yan H, Wang W, Chen S, Viale A, Zheng H, Paik JH, Lim C, Guimaraes AR, Martin ES, Chang J, Hezel AF, Perry SR, Hu J, Gan B, Xiao Y, Asara JM, Weissleder R, Wang YA, Chin L, Cantley LC, Depinho RA (2012) Oncogenic Kras Maintains Pancreatic Tumors through Regulation of Anabolic Glucose Metabolism. Cell 149:656–670

Young A, Lyons J, Miller AL, Phan VT, Alarcon IR, McCormick F (2009) Ras signaling and therapies. Adv Cancer Res 102:1–17

Yu X, Vazquez A, Levine AJ, Carpizo DR (2012) Allele-specific p53 mutant reactivation. Cancer Cell 21:614–625

Zhang J, Baran J, Cros A, Guberman JM, Haider S, Hsu J, Liang Y, Rivkin E, Wang J, Whitty B, Wong-Erasmus M, Yao L, Kasprzyk A (2011) International cancer genome consortium data portal—a one-stop shop for cancer genomics data. Database: J Biol Databases Curation 2011:bar026

Molecular Pathology of Pancreatic Cancer Precursor Lesions

Lodewijk A.A. Brosens and G. Johan Offerhaus

Abstract Pancreatic cancer is the fourth leading cause of cancer-related death in the USA. Each year about 44,000 patients are newly diagnosed with pancreatic cancer in the USA. Most of these patients present with advanced disease and have a very poor prognosis.

Given this dismal prognosis, the challenge is to identify pancreatic cancer in an early stage or, better, patients at risk for pancreatic cancer before an incurable invasive carcinoma has developed. Several distinctive precursor lesions of pancreatic cancer are now known, which theoretically allows for detection of patients at risk of developing pancreatic cancer. These precursor lesions are the microscopic pancreatic intraepithelial neoplasia (PanIN) and the macroscopic cystic precursor lesions intraductal papillary mucinous neoplasia (IPMN), intraductal tubulopapillary neoplasm (ITPN), and mucinous cystic neoplasia/mucinous cystadenoma (MCN).

Insight in the molecular biology of pancreatic adenocarcinoma and these precursor lesions has substantially increased during the past decades. Accurate understanding of the successive molecular genetic alterations in these lesions may eventually lead to biomarkers that can predict biological behavior and guide treatment of patients at risk of invasive pancreatic cancer. This chapter reviews the clinical, diagnostic, and molecular genetic aspects of these pancreatic cancer precursor lesions.

Introduction

Pancreatic cancer is the fourth leading cause of cancer-related death in the USA. In 2012, an estimated 44,000 patients are diagnosed with pancreatic cancer and about 37,000 patients will die of this disease (Siegel et al. 2012). Worldwide,

L.A.A. Brosens (✉) • G.J. Offerhaus
Department of Pathology (H04-312), University Medical Center Utrecht,
Heidelberglaan 100, 3584 CX, Utrecht, The Netherlands
e-mail: l.a.a.brosens@umcutrecht.nl; gofferha@umcutrecht.nl

D.M. Simeone and A. Maitra (eds.), *Molecular Genetics of Pancreatic Cancer*,
DOI 10.1007/978-1-4614-6549-2_2, © Springer Science+Business Media New York 2013

approximately 277,000 new patients are diagnosed with pancreatic cancer each year (Maisonneuve and Lowenfels 2010). Depending on stage, the median survival varies from 2.5 to 6.8 months in patients without surgical therapy to 4.5–24.1 months in patients receiving surgery. The overall 5-year survival rate is 3–5 %, whereas the 5-year survival rate is 15–30 % for patients with early-stage disease treated by curative resection. However, more than 70 % of patients present with stage III or IV disease and have a poor prognosis (Bilimoria et al. 2007; Hidalgo 2010; Hruban et al. 2010; Vincent et al. 2011; Siegel et al. 2012). The asymptomatic nature of early pancreatic cancer, the lack of sensitive and specific tools to diagnose early disease, and the lack of response to most forms of treatment all contribute to the high mortality rate of pancreatic cancer. Despite intensive research prognosis of invasive pancreatic cancer has barely improved in the past decades. Postoperative adjuvant chemo- and/or radiation therapies are only marginally effective and there is a high level of chemo- and radioresistance (Hidalgo 2010; Vincent et al. 2011). The most promising way to reduce pancreatic cancer mortality is therefore to identify and treat patients at risk for pancreatic cancer before an incurable invasive carcinoma develops (Maitra et al. 2005; Hruban et al. 2007b).

Although still evolving, knowledge of pancreatic tumorigenesis has significantly improved during the past decades and it is now clear that invasive pancreatic cancer develops from several distinctive precursor lesions. The most common precursor lesion is the microscopic pancreatic intraepithelial neoplasia (PanIN). Less common are the macroscopic cystic precursor lesions intraductal papillary mucinous neoplasm (IPMN), intraductal tubulopapillary neoplasm (ITPN), and mucinous cystic neoplasm/mucinous cystadenoma (MCN) (Maitra et al. 2003; Hruban et al. 2007b, 2010). Detection and treatment of these precursor lesions and thereby preventing development of full-blown invasive pancreatic adenocarcinoma is an important strategy to reduce pancreatic cancer mortality. However, definitive preoperative diagnosis and prediction of biological behavior of these lesions is often difficult but essential for further treatment decisions. Accurate knowledge of molecular genetic alterations in these lesions may lead to biomarkers that can differentiate between and predict biological behavior of these lesions, and thus guide further treatment of patients with these lesions. In this chapter, clinical, histopathological, and molecular aspects of the different pancreatic cancer precursor lesions are discussed.

Pancreatic Intraepithelial Neoplasia

Definition, Clinical Appearance, and Histopathology

PanIN is the most common precursor lesion of conventional pancreatic ductal adenocarcinoma. PanIN is a microscopic precursor lesion arising in small caliber pancreatic ducts and has been recognized for more than a century (Hruban et al. 2004, 2010; Maitra et al. 2005). PanINs occur most frequently in the head of the pancreas and to a lesser extent in the body or tail. The overall prevalence of PanIN is estimated to be about 20 % and the incidence increases with age, present in 6.7 % of people ≤50 years of age, 28 % in people between 50 and 65 years of age, and 37 % of people ≥65 years of age (Kozuka et al. 1979; de Wilde et al. 2012). In addition, PanIN lesions occur more often in pancreata harboring adenocarcinoma (82 %) than pancreata with pancreatitis (60 %) or normal pancreata (16 %) (Andea et al. 2003; Hruban et al. 2008). Moreover, multiple PanINs of all grades are frequently observed in individuals with inherited susceptibility to pancreatic cancer (Shi et al. 2009).

PanINs occur in smaller pancreatic ducts and are less than 5 mm in diameter which is in fact one of the features used to distinguish PanIN from IPMNs which are usually >1 cm diameter. PanINs are microscopic lesions and are not macroscopically detected (Hruban et al. 2004).

Histologically, PanINs are lined by columnar mucinous epithelium instead of the normal cuboidal pancreatic duct epithelium (Hruban et al. 2004). Most PanINs express MUC1, MUC5AC, and MUC6 suggesting gastric foveolar differentiation (Kim et al. 2002). MUC2 expression is not present in PanIN, a distinctive feature to differentiate it from IPMN (Hruban et al. 2004; Maitra et al. 2005).

PanINs are divided in three grades based on the degree of cytonuclear and architectural atypia (Fig. 1a–e) (Hruban et al. 2004, 2010). Low-grade or PanIN-1A lesions typically have flat epithelium consisting of columnar mucinous cells oriented perpendicularly to the basement membrane with basally oriented uniform round to oval nuclei and supranuclear mucin. PanIN-1B lesions have a (micro)papillary architecture, whereas PanIN-2 lesions show even more architectural complexity with pseudostratification, nuclear hyperchromasia, and beginning loss of nuclear polarity consistent with intermediate-grade dysplasia. PanIN-3, or high-grade dysplasia/carcinoma-in situ, is characterized by significant cytological atypia and includes complete loss of nuclear polarity, nuclear hyperchromasia, conspicuous nucleoli, and the presence of (atypical) mitotic figures. In addition, PanIN-3 is characterized by architectural changes including (micro)papillary epithelium and cribriform growth, and there is sometimes luminal necrosis (Hruban et al. 2004; de Wilde et al. 2012). Interestingly, PanINs are often surrounded by lobular parenchymal atrophy which can be detected by imaging techniques (e.g., endoscopic ultrasound) and may be used as a biomarker in a subset of patients with a high-risk pancreatic cancer (Meckler et al. 2001; Detlefsen et al. 2005; Brune et al. 2006).

Molecular Characteristics of PanIN

Molecular genetic alterations in PanIN confirm the stepwise progression from normal epithelium to low-grade, subsequent high-grade dysplasia and invasive carcinoma. A simplified model of this histologic–genetic progression is called the "PanINgram" and shows that accumulation of molecular alterations correlates with increasing grades of dysplasia (Hruban et al. 2000) (Fig. 2). Early genetic alterations that can initiate PanIN development are mainly found in the *KRAS* oncogene and

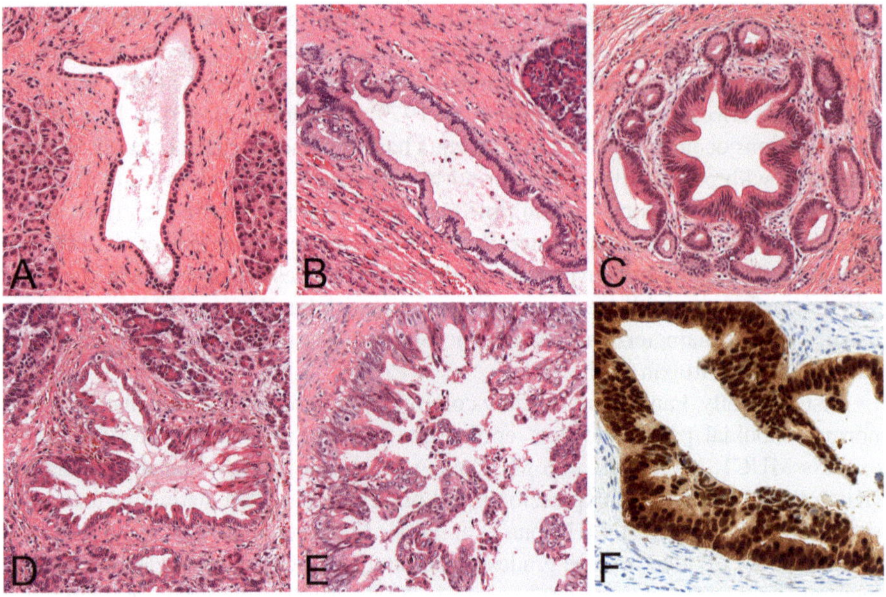

Fig. 1 Histology of PanIN lesions. (**a**) Normal pancreatic duct lined by cuboidal epithelium. (**b**) PanIN-1A. Pancreatic duct lined by flat epithelium consisting of columnar mucinous cells with basally oriented uniform round to oval nuclei and supranuclear mucin. (**c**) PanIN-1B. Pancreatic duct lined by epithelium consisting of columnar mucinous cells and micropapillary architecture. (**d**) PanIN-2. Pancreatic duct lined by columnar cells with nuclear hyperchromasia, pseudostratification, and papillary architecture. (**e**) PanIN-3. Pancreatic duct lined by columnar cells with severe cytonuclear pleiomorphism, loss of nuclear polarity, and complex architecture with (micro)papillary epithelium and cribriform growth pattern. (**f**) p53 immunohistochemistry in a PanIN-3 lesion showing accumulation of the p53 protein consistent with *TP53* mutation

less frequently in *p16/CDKN2A*, *GNAS*, or *BRAF* (Kanda et al. 2012). In addition, telomere shortening is found in >90 % of PanIN lesions of all grades but this may rather be a consequence of activation of oncogene stress-induced senescence programs than an initiator of PanIN (van Heek et al. 2002; Kanda et al. 2012).

Previous studies have shown an increase of *KRAS* mutations correlating with neoplastic progression (i.e., 36 % in PanIN-1A, 44 % in PanIN-1B, 87 % in PanIN-2/3, and >90 % in PDAC), suggesting that *KRAS* mutation is more involved after PanIN initiation than responsible for initiation of tumorigenesis (Moskaluk et al. 1997; Hruban et al. 2000; Lohr et al. 2005). However, a recent study using more sensitive mutation detection methods identified *KRAS* mutations in >90 % of both low- and high-grade PanIN lesions. Interestingly, the average concentration of mutant *KRAS* alleles increased in subsequent PanIN grades, which is consistent with a gradual expansion of the *KRAS*-mutant clone during progression of PanIN. This finding can also explain the lower prevalence of *KRAS* mutations in low-grade lesions found in prior studies that used less sensitive sequence methods (Kanda et al. 2012). *BRAF* mutations were only found in a small subset of *KRAS*-wild-type

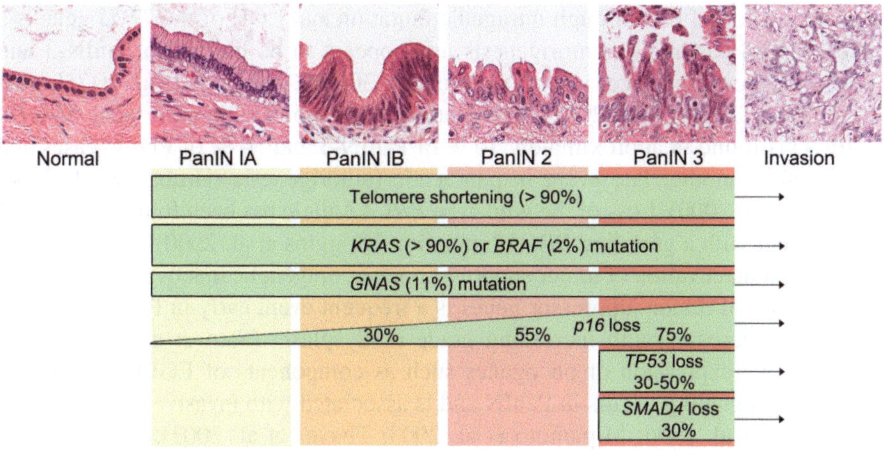

Fig. 2 Progression model of pancreatic cancer. Each step in the progression from normal epithelium to low-grade PanIN, subsequent high-grade PanIN and eventually invasive adenocarcinoma is accompanied by additional genetic alterations. More than 99 % of the earliest stage PanIN-1 lesions contain mutations in *KRAS*, *p16/CDKN2A*, *GNAS*, or *BRAF*

PanINs and pancreatic cancers (Jones et al. 2008; Kanda et al. 2012). A subset of PanINs (~11 %) harbored a *GNAS* mutation, an oncogene that was recently discovered to be mutated in ~60 % of IPMNs (Wu et al. 2011b; Kanda et al. 2012). Interestingly, in some PanINs a *GNAS* mutation was the only identified mutation and in other PanINs the *GNAS* mutation seemed to have occurred earlier than the *KRAS* mutation. In total, >99 % of the earliest stage PanIN-1 lesions contain mutations in *KRAS*, *p16/CDKN2A*, *GNAS*, or *BRAF*, indicating that somatic mutations are required for the early development of all PanIN lesions which can be used as an argument against the hypothesis that PanINs begin as metaplasia. However, it also appears that *KRAS* mutation alone provides only a modest selective advantage over neighboring cells and that additional genetic or epigenetic events are needed for neoplastic progression (Kanda et al. 2012).

p16/CDKN2A mutation is a relatively early event in PanIN and may be the additional genetic event needed for PanINs with *KRAS* mutation to progress (Hruban et al. 2000; Kanda et al. 2012). *p16/CDKN2A* mutations were found in 11 % of low-grade (i.e., PanIN-1/2) lesions and were more often found in PanIN lesions without a *KRAS* mutation (Kanda et al. 2012). Previously, loss of p16/CDKN2 protein expression was already shown to increase with PanIN grade (i.e., p16/CDKN2 protein expression was lost in 30 % of PanIN-1A/B, 55 % of PanIN-2, and 71 % of PanIN-3 lesions) (Wilentz et al. 1998). Loss of function occurs through homozygous deletions, mutation and loss of heterozygosity (LOH) or promotor hypermethylation, each of these mechanisms accounting for approximately one-third of p16 silencing (Schutte et al. 1997). In addition, overexpression of cyclin D1 is noted in 29 % of PanIN-2 and 57 % of PanIN-3 lesions (Maitra et al. 2003).

Inactivation of p53 through intragenic mutation and LOH of the *TP53* gene is a late event in pancreatic tumorigenesis and appears to be limited to PanIN-3 and invasive pancreatic cancer where it is found in 30–50 % of cases (Fig. 1f) (Hruban et al. 2000; Luttges et al. 2001). Inactivation of the tumor suppressor gene *SMAD4* (*DPC4*) is found in approximately 30 % of PanIN-3 and 50 % of PDAC cases and is therefore another late event in pancreatic tumorigenesis (Hruban et al. 2000; Wilentz et al. 2000). Loss of the wild-type *BRCA2* allele has been found in PanIN-3 in a patient with a germline *BRCA2* mutation (Goggins et al. 2000). In addition to mutational inactivation of tumor suppressor genes, epigenetic inactivation by hypermethylation of tumor suppressor genes is a frequent event early in PanIN development and increases with increasing grade of dysplasia (Sato et al. 2008). Also aberrant overexpression of oncogenes such as components of EGFR, Notch and Hedgehog signaling occurs in PanIN and is associated with invasive adenocarcinomas (Day et al. 1996; Miyamoto et al. 2003; Thayer et al. 2003). Lastly, PanIN lesions show aberrant expression of many microRNAs, which is likely to be important in pancreatic carcinogenesis. Interestingly, expression of some microRNAs, such as miR-196b, appears specific for high-grade lesions (PanIN-3 and PDAC) and may therefore be useful as diagnostic markers (Yu et al. 2012).

Intraductal Papillary Mucinous Neoplasm

Definition, Clinical Appearance, and Histopathology

IPMN is a macroscopically visible cystic mucin producing tumor arising in a main pancreatic duct or one of its branches. IPMNs are quite common lesions and account for approximately 3 % of exocrine pancreatic neoplasms and for 20 % of cystic pancreatic neoplasms (Kosmahl et al. 2004; Adsay et al. 2010; Shi and Hruban 2012). Most IPMNs are found in patients between 60 and 70 years of age and the mean age of diagnosis varies from 63 to 66 years (Fukushima et al. 1997; Chari et al. 2002). Patients with an IPMN with an associated invasive carcinoma tend to be 3–5 years older than patients with an IPMN without invasive carcinoma. IPMNs are slightly more common in males (~60 % of cases) than females (Shi and Hruban 2012).

IPMNs have been reported in individuals with a family history of pancreatic cancer and in patients with Peutz-Jeghers syndrome (Sato et al. 2001; Canto et al. 2012).

IPMNs are divided in main duct, branch duct, and combined or mixed type, which is mainly based on its appearance on imaging and to a lesser extent on gross pathologic examination (Crippa et al. 2010; Shi and Hruban 2012). Main-duct IPMNs usually occur in the pancreatic head and often produce copious thick mucin which gives rise to a (diffusely) dilated main pancreatic duct and associated symptoms. These symptoms include abdominal or back pain, nausea, vomiting, weight loss, or recurrent episodes of pancreatitis. Approximately 60 % of main-duct IPMNs harbor high-grade dysplasia and associated invasive carcinoma is found in about

45 % of main-duct IPMNs (Salvia et al. 2004; Kawamoto et al. 2006; Crippa et al. 2010; Shi and Hruban 2012). Branch-duct IPMNs occur mainly in the head and uncinate process and are often multicystic grapelike structures with thin cyst walls involving side branches of the main pancreatic duct. Branch-duct IPMNs are usually asymptomatic and are therefore often incidental findings on imaging studies for other medical reasons. One study found an unsuspected pancreatic cyst (most of which were probably IPMN) in 2.6 % of asymptomatic patients and this number increased with age (Laffan et al. 2008). Most branch-duct IPMNs are low-grade lesions with an indolent behavior, although high-grade dysplasia and invasive carcinoma are found in about 25 and 20 % of branch-duct IPMNs meeting the "Sendai criteria", respectively (Terris et al. 2000; Kawamoto et al. 2006; Rodriguez et al. 2007; Crippa et al. 2010; Shi and Hruban 2012). Mixed-type IPMNs involve both the main and branch ducts. Both main and branch-duct IPMNs can be associated with atrophy of the adjacent pancreatic parenchyma.

Prognosis of IPMN is mainly determined by the presence or absence of associated invasive carcinoma. The 5-year survival rate for patients with an IPMN without an associated invasive carcinoma is 90–100 %, whereas this is about 30–60 % for patients with an IPMN with associated invasive carcinoma (Chari et al. 2002; Maire et al. 2002; Raimondo et al. 2002; D'Angelica et al. 2004; Salvia et al. 2004; Nara et al. 2008; Crippa et al. 2010). Invasive carcinoma in IPMN has a better prognosis than primary PDAC which maybe mainly due to the lower stage at which IPMN-associated adenocarcinoma is usually diagnosed (Poultsides et al. 2010).

The "Sendai criteria" are international consensus guidelines for the management of IPMNs (Tanaka et al. 2006). These criteria advise surgical resection of all main-duct IPMNs and resection of branch-duct IPMNs that are symptomatic, >3 cm, harbor a mural nodule, or are associated with significant dilatation of the pancreatic duct. In addition, lesions should be resected if cytology shows severe cytonuclear atypia (Tanaka et al. 2006; Shi and Hruban 2012).

Grossly, IPMNs can be lined by flat epithelium (ductectatic pattern) or by epithelium with papillary projections (villous growth). By definition, IPMNs are >0.5 cm and most IPMNs are >1 cm, with the size varying from 1 cm to the entire pancreas (Hruban et al. 2007a). Careful gross examination to differentiate between main-duct and branch-duct IPMNs is important in view of the higher risk of high-grade dysplasia and invasive carcinoma in the main-duct type (Crippa et al. 2010). Because invasive carcinoma can be very focal within an IPMN, these lesions should be thoroughly sampled for histological examination. Gross features suggestive of invasive adenocarcinoma are irregular heterogeneous thickening of cyst walls, fibrotic foci, and the presence of solid nodules (Shi and Hruban 2012; de Wilde et al. 2012).

Microscopically, IPMNs are classified according to the degree of dysplasia and the direction of differentiation of the neoplastic epithelium, which can be intestinal-, pancreatobiliary-, gastric-, or oncocytic type. Because multiple histological types of epithelium can often be found in an IPMN, the dominant component defines the subtype (Adsay et al. 2010; Shi and Hruban 2012). It is important to recognize the histological subtype of an IPMN because this appears to be an independent predictor of patient prognosis (Furukawa et al. 2011). Moreover some IPMN subtypes

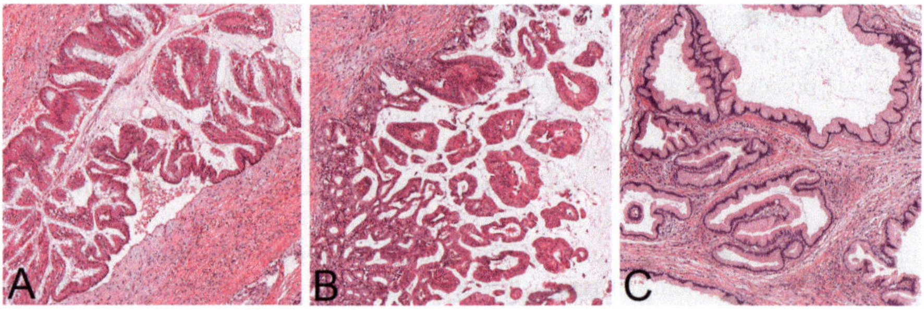

Fig. 3 Histologic subtypes of IPMN. (**a**) Intestinal-type IPMN with intermediate-grade dysplasia lined by columnar mucin-producing cells with cigar-shaped pseudostratified nuclei and scattered goblet-like cells. (**b**) Pancreatobiliary IPMN with high-grade dysplasia lined by cuboidal cells with round hyperchromatic nuclei with prominent nucleoli, cytoplasm containing less mucin than in the intestinal-type IPMN and more complex papillary architecture. (**c**) Gastric-type IPMN with low-grade dysplasia lined by a single layer of cells with basally oriented small nuclei and abundant apical cytoplasmic mucin resembling gastric foveolar epithelium

are associated with distinct types of invasive carcinoma with varying prognosis. For instance, colloid carcinoma (associated with intestinal-type IPMN) and oncocytic carcinoma (associated oncocytic-type IPMN) have better a prognosis than the tubular type carcinoma (associated with gastric-, pancreatobiliary-, or intestinal-type IPMN) which has a course similar as PDAC (Mino-Kenudson et al. 2011).

Main-duct IPMNs are usually lined by intestinal- and pancreatobiliary-type epithelium, whereas branch-duct IPMNs are typically lined by gastric-type epithelium (Adsay et al. 2010). The intestinal-type IPMN (Fig. 3a) shows long papillae lined by columnar mucin-producing cells with cigar-shaped pseudostratified nuclei and basophilic cytoplasm, resembling a villous adenoma of the colon. Often goblet-like cells are encountered. Intermediate to high-grade dysplasia is usually seen in this type (Adsay et al. 2010; Shi and Hruban 2012). The neoplastic cells of intestinal-type IPMN do not express MUC1, weakly express MUC6 and strongly express MUC5A, MUC2, and CDX2 (Adsay et al. 2004; Basturk et al. 2010).

Pancreatobiliary IPMNs (Fig. 3b) are lined by cuboidal cells with round hyperchromatic nuclei with prominent nucleoli and cytoplasm containing less mucin than in the intestinal-type IPMN. These IPMNs are further characterized by more complex thin papillae with branching and cribriform growth and therefore tend to be high-grade lesions (Adsay et al. 2010; Shi and Hruban 2012). Pancreatobiliary IPMNs have an immunohistochemical expression pattern similar to that of PanIN and usually express MUC1 and MUC5A, sometimes MUC6 but not MUC2 (Adsay et al. 2004; Ban et al. 2006; Basturk et al. 2010).

Gastric foveolar-type IPMNs (Fig. 3c) are lined by cells with abundant apical cytoplasmic mucin and basally oriented small nuclei, resembling gastric foveolar epithelium (Furukawa et al. 2005). These IPMNs are usually lined by a single flat layer of epithelium lining dilated ducts. Papillary projections are uncommon in these lesions and there is mostly low-grade dysplasia. The neoplastic cells often extend along the pancreatic ducts into adjacent pancreatic tissue resulting in acinar-ductal metaplasia, acinar atrophy, and fibrosis. Gastric foveolar-type IPMNs strongly express gastric-type mucins MUC5A and MUC6 but not MUC1 and MUC2 (Furukawa et al. 2005; Ban et al. 2006; Basturk et al. 2010).

Oncocytic-type IPMNs, also known as intraductal oncocytic papillary neoplasms (IOPNs), are composed of cells with abundant granular eosinophilic cytoplasm due to accumulation of mitochondria. The architecture of IOPNs is very complex with arborizing papillae, cribriform growth, and solid nests, growing into the lumen of the dilated duct. Intraepithelial and intracellular mucin is frequently present and scattered goblet cells can be observed. The stratified oncocytic neoplastic cells have abundant eosinophilic granular cytoplasm and large round uniform nuclei. Because of the marked cytonuclear and architectural atypia most IOPNs are classified as having high-grade dysplasia (Adsay et al. 2010; Shi and Hruban 2012). Sometimes it can be difficult to appreciate the intraductal nature of this lesion. IOPNs express MUC1 and MUC6, whereas expression of CDX2, MUC2, and MUC5A is restricted to the goblet cells (Basturk et al. 2010; Liszka et al. 2010; Shi and Hruban 2012). Invasive carcinoma arising from IOPN is a relatively well-circumscribed tumor composed of cells with the characteristic oncocytic features growing in the periductal stroma as small solid nests and glands (Patel et al. 2002). Although only few cases have been described, genetic changes seem distinct from typical pancreatic adenocarcinoma which may explain the indolent clinical behavior of IOPN (Patel et al. 2002; Xiao et al. 2011).

In the fourth edition of WHO classification of tumors of the digestive system, ITPN is recognized as a subtype of the intraductal pancreatic neoplasms and is therefore discussed separately (Adsay et al. 2010).

Molecular Characteristics of IPMN

A recent study investigating eight IPMNs by whole-exome sequencing showed that IPMNs contain an average of 26 somatic mutations (Wu et al. 2011a). The most common genetic alteration in IPMN is mutation of codon 12 and to a lesser extent codon 13 of the *KRAS* gene which is found in >80 % of IPMNs (Wu et al. 2011b). Previous studies have shown that the prevalence of *KRAS* mutation increases with increasing grade of dysplasia (Sessa et al. 1994; Satoh et al. 1996; Schonleben et al. 2007). In addition, this study identified mutations in *GNAS*, a well-known oncogene functioning as a signal transducer between hormonal receptors and adenylyl cyclase, to be present in 66 % of IPMNs. Interestingly, it was suggested that *GNAS* mutations are specific for IPMN since mutations in this gene were not found in other types of cystic pancreatic neoplasms (i.e., serous cystadenoma, MCN, and solid pseudopapillary neoplasm) or in invasive adenocarcinomas not associated with IPMNs, whereas *GNAS* mutations were found in adenocarcinomas developing in association with IPMNs (Wu et al. 2011b).

Taken together, about 50 % of IPMNs harbor both a *GNAS* and a *KRAS* mutation, whereas either a *KRAS* or a *GNAS* mutation can be found in 96 % of IPMNs. Because *KRAS* and *GNAS* gene mutations can be detected in cyst fluid, mutation analysis of these genes in cyst fluid aspirates may prove to be a valuable asset for preoperative diagnostic workup of IPMNs (Wu et al. 2011b). Importantly, both *KRAS* and *GNAS* mutations are restricted to specific codons (*GNAS* codon 201 and *KRAS* codon 12 or 13) which makes analysis of these molecular alterations relatively straight forward and suitable for routine diagnostics (Wu et al. 2011b).

Different subtypes of IPMN appear to follow different pathways of neoplastic progression. For instance, gastric- and pancreatobiliary-type IPMNs show higher rates of *KRAS* mutation than intestinal-type IPMNs, whereas *GNAS* mutations are most prevalent in the intestinal-type IPMNs and absent in IOPN (Mohri et al. 2012; Wu et al. 2011b). In addition, *KRAS* mutation and p53 overexpression are less prevalent in IOPN than in pancreatobiliary-type IPMN (17 % vs. 58 % and 11 % vs. 58 %, respectively) (Xiao et al. 2011). Whole-exome sequencing also identified *RNF43*, encoding a protein with intrinsic E3 ubiquitin ligase activity, as a gene that is frequently mutated in IPMN (6 of 8 cases). Although *RNF43* mutations were not specific for IPMN, since mutation of this gene was also found in a subset of MCNs, this finding highlights the importance of inactivation of ubiquitin ligase in cystic pancreatic tumors (Wu et al. 2011a).

The mTOR pathway may be involved in IPMN tumorigenesis via loss of *LKB1/STK11* which is a serine threonine kinase upstream of mTOR. *LKB1/STK11* loss is found in IPMNs arising in patients with Peutz-Jeghers syndrome (caused by germline *LKB1/STK11* mutation) and also in about 25 % of sporadic IPMNs (Su et al. 1999; Sato et al. 2001). In addition, *PIK3CA*, which also encodes a protein upstream of AKT-mTOR, is mutated in a subset of IPMNs (~10 %), but *PIK3CA* mutation may be more specific for ITPNs than for IPMNs (Schonleben et al. 2008b; Yamaguchi et al. 2011).

Other genetic alterations in IMPN are found with variable frequencies. *TP53* mutation represents a late event in neoplastic development of IPMN and is found in 0–50 % of IPMNs (Sessa et al. 1994; Kawahira et al. 2000; Sasaki et al. 2003; Xiao et al. 2011). Loss of *p16/CDKN2A* has been reported in 0–80 % of IPMNs and increases with grade of dysplasia (Biankin et al. 2002; Sasaki et al. 2003). SMAD4 is only rarely inactivated in noninvasive IPMN and protein expression is preserved in most IPMNs regardless of grade of dysplasia (Iacobuzio-Donahue et al. 2000a; Biankin et al. 2002). *APC* and *HER2* mutations are very rare in IPMN (Schonleben et al. 2008a; Schonleben et al. 2008b; Wu et al. 2011a; Xiao et al. 2011). Allelic loss of at least one chromosome region is found in most IPMNs (7 of 8) (Fritz et al. 2009). By array-CGH it has been shown that copy number alterations are frequently found in IPMNs with moderate- and high-grade dysplasia but not in IPMNs with low-grade dysplasia. Commonly lost regions were located on chromosomes 5q, 6q, 10q, 11q, 13q, 18q, and 22q (Fritz et al. 2009).

Gene expression analysis of IPMN has identified a number of genes that are associated with progression to invasive carcinoma, including *claudin 4*, *CXCR4*, *S100A4*, and *mesothelin*, which may serve as biomarkers to identify high-risk IPMNs (Sato et al. 2004; Habbe et al. 2009; Tsutsumi et al. 2011; Jury et al. 2012). Expression of *MSX-2* has been linked to neoplastic progression of branch-duct IPMN (Satoh et al. 2010). Overexpression of Sonic Hedgehog is an early event in the development of IPMN (Ohuchida et al. 2006). In addition, aberrant DNA methylation occurs frequently in IPMNs and contributes to inactivation of tumor suppressor genes and neoplastic progression (Sato et al. 2002; Hong et al. 2008, 2012). Interestingly, methylation of specific genes, including *BNIP3*, *PTCHD2*, *SOX17*, *NXPH1*, and *EBF3*, may predict the presence of high-grade dysplasia in an IPMN (Hong et al. 2012). Also, overexpression of microRNAs, in particular miR-21 and miR-155, has been described in IPMN (Habbe et al. 2009). Lastly, telomere shortening has been shown in IPMN and the average telomere length decreases with tumor progression (Hashimoto et al. 2008).

Intraductal Tubulopapillary Neoplasm

Definition, Clinical Appearance, and Histopathology

ITPN is a recently described rare variant of an intraductal neoplasm of the pancreas accounting for <1 % of all exocrine pancreatic neoplasms and for 3 % of pancreatic intraductal neoplasms (Tajiri et al. 2005; Yamaguchi et al. 2009; Adsay et al. 2010). Limited data is available about prognosis for patients with ITPN, but 5-year survival is likely more than 30 %. No significant correlation between invasive growth and survival has been found which may be due to the microscopic nature of the invasion or because small foci of invasion may have been missed due to inadequate sampling (Adsay et al. 2010).

ITPN is a generally large (average size 6 cm; range 0.8–15.0 cm) macroscopically visible solid nodular tumor filling the dilated pancreatic duct. In contrast to IPMN, these tumors lack overt mucin production and have a predominantly tubular growth pattern although papillae can be found in some lesions (Suda et al. 1996; Yamaguchi et al. 2009, 2011). The tumor consists a proliferation of back-to-back acinar glands lined by cuboidal cells with modest amount of eosinophilic to amphophilic cytoplasm and round to oval moderately to marked atypical nuclei (Fig. 4). Typically ITPNs express cytokeratins 7 and 19 and MUC1. About 60 % of cases also express MUC6, whereas MUC2 and MUC5AC are not expressed, which can be helpful in distinguishing these lesion from IPMNs (Tajiri et al. 2005; Yamaguchi et al. 2009). There is homogenous high-grade dysplasia and complex architecture throughout the lesion and, in contrast to IPMNs, foci of necrosis are frequently encountered. In about 40 % of cases an associated invasive carcinoma is found (Suda et al. 1996; Yamaguchi et al. 2009).

Molecular Characteristics of ITPN

Few studies have investigated the molecular characteristics of ITPN. Abnormal expression of p53 and SMAD4 has been described in 1 case. No aberrant expression of β-catenin or mutations in *KRAS* of *BRAF* have been found (Yamaguchi et al.

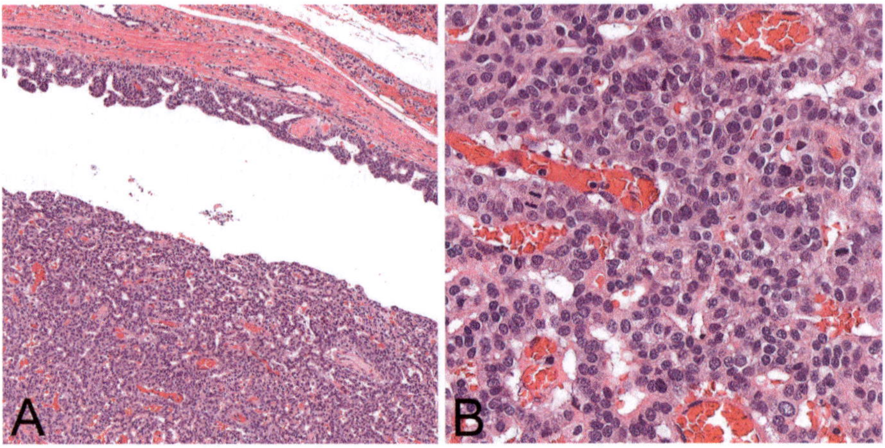

Fig. 4 Histologic appearance of ITPN. (**a**) Intraductal tubulopapillary neoplasm showing an intraductal proliferation of back-to-back acinar glands lined by cuboidal cells with marked cytonuclear pleiomorphism. (**b**) Detail of (**a**) showing proliferation of cuboidal cells with hyperchromatic anisomorphic nuclei and several mitoses

2009, 2011), whereas a *KRAS* mutation is found >80 % of IPMNs (Sarr et al. 2001; Crippa et al. 2008, 2010; Wu et al. 2011a). Interestingly, a recent study investigating molecular alterations in 11 ITPNs and 50 IPMNs found mutations in *PIK3CA* in a subset of ITPNs (3 of 11) but in none of the IPMNs. In addition, *PIK3CA* mutations were associated with strong expression of phosphorylated AKT. As previously reported, no *BRAF* of *KRAS* gene mutations were found in any of the ITPNs. These results suggest a role of the phosphatidylinositol 3-kinase pathway in ITPNs and the activated phosphatidylinositol 3-kinase pathway may therefore be a potential target for molecular diagnosis and therapy of ITPNs.

Mucinous Cystic Neoplasm

Definition, Clinical Appearance, and Histopathology

MCN of the pancreas is a macroscopically visible cystic neoplasm accounting for approximately 8 % of all resected cystic lesions of the pancreas (Kosmahl et al. 2004; Fukushima and Fukayama 2007; Zamboni et al. 2010). These lesions are most often found in the body and tail of the pancreas and, in contrast to IPMNs, usually do not communicate with the pancreatic duct system. Almost all MCNs occur in female patients with a female to male ratio of 20:1. However, male gender cannot be used to rule out the diagnosis since sporadic MCNs have been reported in males (Wouters et al. 1998). The mean age at diagnosis is between 40 and 50 years with a range of 14–95 years (Thompson et al. 1999; Wilentz et al. 1999; Zamboni et al.

1999; Fukushima and Fukayama 2007). On average, patients with an associated invasive carcinoma are 5–10 years older than patients with noninvasive MCN (Zamboni et al. 2010).

Clinical manifestations of MCN depend on the size of the lesion. Lesions smaller than <3 cm are often found incidentally in patients imaged for another indication. Larger lesions often give rise to nonspecific complaints such as abdominal discomfort and the sensation of a mass in the epigastric region. About one-third of resected MCNs have an associated invasive carcinoma, which usually resembles a common pancreatic ductal adenocarcinoma. However, the number of MCNs with associated adenocarcinoma may decrease since more MCNs are being detected incidentally in patients imaged for another reason (Wilentz et al. 1999; Zamboni et al. 1999, 2010; Tanaka et al. 2006; Fukushima and Fukayama 2007; Crippa et al. 2008; Yamao et al. 2011). Patients with a surgically resected noninvasive MCN have an excellent prognosis, but the 5-year survival rate for patients with an MCN with an associated invasive carcinoma is about 50–60 %. Since the invasive component can be very focal MCNs should undergo extensive histological examination before invasion is excluded (Wilentz et al. 1999; Zamboni et al. 1999; Fukushima and Fukayama 2007). In contrast to IPMNs, MCNs are almost always unifocal and after surgery for an MCN there is minimal risk of metachronous disease (de Wilde et al. 2012).

Macroscopically, MCNs are single spherical lesions with a mean diameter of 6–10 cm (range 2–35 cm) and a fibrous pseudocapsule. The tumor can be unilocular or multilocular with cysts varying from millimeters to several centimeters containing thick mucinous and/or hemorrhagic or necrotic material. Low-grade lesions usually have a smooth and glistering internal surface, whereas high-grade lesions often show papillary projections. MCNs with an associated invasive carcinoma are often large and multilocular and contain papillary projections or mural nodules (Zamboni et al. 1999; Fukushima and Fukayama 2007).

Histologically, MCNs are defined by the presence of distinctive ovarian-type stroma consisting of densely packed spindle cells with round to elongated nuclei and a small amount of cytoplasm expressing inhibin, estrogen and progesterone receptors, as well as vimentin, smooth-muscle actin, and desmin (Fig. 5) (Fukushima and Mukai 1997; Ridder et al. 1998; Thompson et al. 1999; Zamboni et al. 1999; Tanaka et al. 2006). In some lesions it may be difficult to identify the ovarian-type stroma since the stroma may become fibrotic and hypocellular and some areas can resemble corpora albicantia (Fukushima and Fukayama 2007; Zamboni et al. 2010).

The epithelium overlying the ovarian-type stroma and lining the cyst consists of mucin-producing tall columnar epithelial cells that can have pseudopyloric, gastric-foveolar, small- or large-intestinal differentiation. Rarely squamous differentiation is noted (Zamboni et al. 2010). The columnar epithelial cells express cytokeratins 7, 8, 18, and 19, the gastric-type mucin MUC5A, and pancreatic-type mucin DUPAN-2 and CA19-9. Scattered goblet-like cells express the intestinal mucin MUC2. MUC1 expression is observed in most MCNs with invasive ductal adenocarcinoma (Luttges et al. 2002). Within a single MCN the degree of epithelial atypia can vary greatly and change abruptly from minimal to severe dysplasia or even focal invasive growth. MCNs should therefore be extensively sampled for histologic examination before

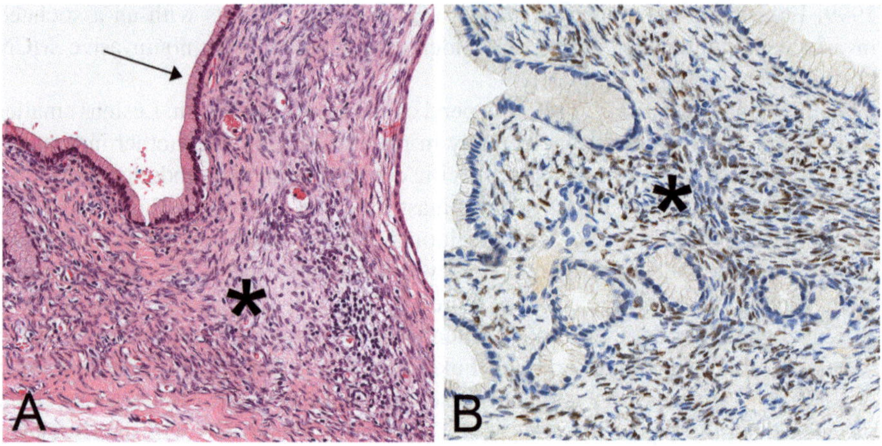

Fig. 5 Histologic appearance of MCN. (**a**) Mucinous cystic neoplasm showing ovarian type stroma (*asterisk*) and lining by mucin-producing tall columnar epithelial with low-grade dysplasia (*arrow*). (**b**) Estrogen receptor expression in stromal cells (*brown staining*) (*asterisk*)

excluding invasive growth. MCNs are categorized based on the highest degree of architectural and cytonuclear atypia present, as MCN with either low-grade, intermediate-grade, or high-grade dysplasia (Zamboni et al. 2010).

Two theories about pathogenesis of MCN prevail in the literature. One hypothesis argues that MCNs are a result of ectopic gonadal mesenchyme that may be incorporated in the pancreas during the fourth and fifth weeks of embryogenesis as a result of the close proximity of the left primordial gonad to the dorsal pancreatic anlage which gives rise to the body and tail of the pancreas (Zamboni et al. 1999; Erdogan et al. 2006). However, this theory does not explain the rare occurrence of MCNs in male patients. An alternative theory suggests that neoplastic epithelial cells of MCNs induce ovarian stromal differentiation in cells that are normally present in the pancreas (Zamboni et al. 2010).

Molecular Characteristics of MCN

A recent study investigated genetic alterations in MCN by whole-exome sequencing and found that MCNs contain an average of 16 somatic mutations and relatively few allelic losses (Wu et al. 2011a). *KRAS* is the most frequently mutated gene in MCN and correlates with the degree of neoplastic progression. *KRAS* mutations have been found in 26 % (7/27) of MCNs with low-grade dysplasia, 38 % (5/13) of MCNs with intermediate-grade dysplasia, and 89 % (8/9) of MCNs with high-grade dysplasia or carcinoma (Jimenez et al. 1999). *p53* mutation appears to be a relatively late event occurring only in areas with severe dysplasia or carcinoma (Jimenez et al. 1999).

A newly discovered and relatively frequently mutated gene in MCN is *RNF43* which was mutated in three of eight MCNs and encodes a protein with intrinsic E3 ubiquitin ligase activity (Wu et al. 2011a). Allelic loss at 3p25, the chromosomal location of *VHL* gene, has been reported in 17 % (2/12) of MCNs (Kim et al. 2003). In addition, loss of SMAD4 and p16/CDKN2A expressions is found in lesions with associated invasive carcinoma (Iacobuzio-Donahue et al. 2000b). Hypermethylation of *p14* and *p16* has been reported in about 15 % of benign or borderline MCNs (Kim et al. 2003).

Global gene expression profiling identified a number of genes that are upregulated in the epithelium of MCNs, including *S100P*, *PSCA*, *c-myc*, *STK6/STK15*, *cathepsin E*, *TCF4*, and *pepsinogen C*. In addition, activation of the Notch pathway was shown in the epithelial component by the demonstration of overexpression of Jagged1 and the downstream Notch pathway member Hes1. Overexpression of steroidogenic acute regulatory protein (*STAR*) and estrogen receptor 1 (*ESR1*) occurs in the stroma (Fukushima et al. 2004).

Conclusions

Molecular genetic alterations in pancreatic cancer have largely been unraveled in the past decade and knowledge about pancreatic cancer precursor lesions has substantially grown. Recently, some important steps have been made in the molecular characterization of pancreatic cancer precursor lesions which may ultimately prove to be useful in diagnostic workup of patients with these lesions and may lead to new targets for therapy.

All pancreatic cancer precursor lesions share a high frequency of somatic mutation of the *KRAS* oncogene. In PanIN, it was recently shown that somatic mutations in *KRAS* or *GNAS* are already present in virtually all of the earliest PanIN lesions. In addition, *GNAS* mutations are found in the majority of IPMNs but not in other cystic pancreatic tumors such as MCN or serous cystic adenoma. Furthermore, cystic pancreatic tumors appear to share defects in genes that play a role in the ubiquitin ligase complex. *RNF43* mutations were identified in IPMNs and MCNs but not in serous cystic adenomas or solid pseudopapillary neoplasms. Serous cystic adenomas are characterized by mutations in *VHL* and solid pseudopapillary neoplasms by mutations in *CTNNB1*.

Testing for these and other molecular genetic alterations in pancreatic cyst fluid can potentially be used to distinguish different cyst types on a molecular level and may lead to more accurate diagnosis (Wu et al. 2011a). However, further studies are needed to validate these findings and to test the potential of these genetic alterations for diagnostic use. In addition, it is important to develop biomarkers that can distinguish between high-grade or low-grade lesions and predict biological behavior.

Acknowledgments We thank Folkert Morsink for help with the figures.

References

Adsay NV, Merati K, Basturk O et al (2004) Pathologically and biologically distinct types of epithelium in intraductal papillary mucinous neoplasms: delineation of an "intestinal" pathway of carcinogenesis in the pancreas. Am J Surg Pathol 28(7):839–848

Adsay NV, Fukushima N, Furukawa H et al (2010) Intraductal neoplasms of the pancreas. In: Bosman FT, Carneiro F, Hruban RH, Theise N (eds) WHO Classification of Tumours of the Digestive System World Health Organisation Classification of Tumors. IARC, Lyon, pp 304–313

Andea A, Sarkar F, Adsay VN (2003) Clinicopathological correlates of pancreatic intraepithelial neoplasia: a comparative analysis of 82 cases with and 152 cases without pancreatic ductal adenocarcinoma. Mod Pathol 16(10):996–1006

Ban S, Naitoh Y, Mino-Kenudson M et al (2006) Intraductal papillary mucinous neoplasm (IPMN) of the pancreas: its histopathologic difference between 2 major types. Am J Surg Pathol 30(12):1561–1569

Basturk O, Khayyata S, Klimstra DS et al (2010) Preferential expression of MUC6 in oncocytic and pancreatobiliary types of intraductal papillary neoplasms highlights a pyloropancreatic pathway, distinct from the intestinal pathway, in pancreatic carcinogenesis. Am J Surg Pathol 34(3):364–370

Biankin AV, Biankin SA, Kench JG et al (2002) Aberrant p16(INK4A) and DPC4/Smad4 expression in intraductal papillary mucinous tumours of the pancreas is associated with invasive ductal adenocarcinoma. Gut 50(6):861–868

Bilimoria KY, Bentrem DJ, Ko CY et al (2007) Validation of the 6th edition AJCC Pancreatic Cancer Staging System: report from the National Cancer Database. Cancer 110(4):738–744

Brune K, Abe T, Canto M et al (2006) Multifocal neoplastic precursor lesions associated with lobular atrophy of the pancreas in patients having a strong family history of pancreatic cancer. Am J Surg Pathol 30(9):1067–1076

Canto MI, Hruban RH, Fishman EK et al (2012) Frequent detection of pancreatic lesions in asymptomatic high-risk individuals. Gastroenterology 142(4):796–804

Chari ST, Yadav D, Smyrk TC et al (2002) Study of recurrence after surgical resection of intraductal papillary mucinous neoplasm of the pancreas. Gastroenterology 123(5):1500–1507

Crippa S, Salvia R, Warshaw AL et al (2008) Mucinous cystic neoplasm of the pancreas is not an aggressive entity: lessons from 163 resected patients. Ann Surg 247(4):571–579

Crippa S, Fernandez-Del Castillo C, Salvia R et al (2010) Mucin-producing neoplasms of the pancreas: an analysis of distinguishing clinical and epidemiologic characteristics. Clin Gastroenterol Hepatol 8(2):213–219

D'Angelica M, Brennan MF, Suriawinata AA et al (2004) Intraductal papillary mucinous neoplasms of the pancreas: an analysis of clinicopathologic features and outcome. Ann Surg 239(3):400–408

Day JD, Digiuseppe JA, Yeo C et al (1996) Immunohistochemical evaluation of HER-2/neu expression in pancreatic adenocarcinoma and pancreatic intraepithelial neoplasms. Hum Pathol 27(2):119–124

de Wilde RF, Hruban RH, Maitra A et al (2012) Reporting precursors to invasive pancreatic cancer: pancreatic intraepithelial neoplasia, intraductal neoplasms and mucinous cystic neoplasm. Diagn Histopathol 18(1):17–30

Detlefsen S, Sipos B, Feyerabend B et al (2005) Pancreatic fibrosis associated with age and ductal papillary hyperplasia. Virchows Arch 447(5):800–805

Erdogan D, Lamers WH, Offerhaus GJ et al (2006) Cystadenomas with ovarian stroma in liver and pancreas: an evolving concept. Dig Surg 23(3):186–191

Fritz S, Fernandez-del Castillo C, Mino-Kenudson M et al (2009) Global genomic analysis of intraductal papillary mucinous neoplasms of the pancreas reveals significant molecular differences compared to ductal adenocarcinoma. Ann Surg 249(3):440–447

Fukushima N, Fukayama M (2007) Mucinous cystic neoplasms of the pancreas: pathology and molecular genetics. J Hepatobiliary Pancreat Surg 14(3):238–242

Fukushima N, Mukai K (1997) "Ovarian-type" stroma of pancreatic mucinous cystic tumor expresses smooth muscle phenotype. Pathol Int 47(11):806–808

Fukushima N, Mukai K, Kanai Y et al (1997) Intraductal papillary tumors and mucinous cystic tumors of the pancreas: clinicopathologic study of 38 cases. Hum Pathol 28(9):1010–1017

Fukushima N, Sato N, Prasad N et al (2004) Characterization of gene expression in mucinous cystic neoplasms of the pancreas using oligonucleotide microarrays. Oncogene 23(56): 9042–9051

Furukawa T, Kloppel G, Volkan Adsay N et al (2005) Classification of types of intraductal papillary-mucinous neoplasm of the pancreas: a consensus study. Virchows Arch 447(5):794–799

Furukawa T, Hatori T, Fujita I et al (2011) Prognostic relevance of morphological types of intraductal papillary mucinous neoplasms of the pancreas. Gut 60(4):509–516

Goggins M, Hruban RH, Kern SE (2000) BRCA2 is inactivated late in the development of pancreatic intraepithelial neoplasia: evidence and implications. Am J Pathol 156(5):1767–1771

Habbe N, Koorstra JB, Mendell JT et al (2009) MicroRNA miR-155 is a biomarker of early pancreatic neoplasia. Cancer Biol Ther 8(4):340–346

Hashimoto Y, Murakami Y, Uemura K et al (2008) Telomere shortening and telomerase expression during multistage carcinogenesis of intraductal papillary mucinous neoplasms of the pancreas. J Gastrointest Surg 12(1):17–28, discussion 28-9

Hidalgo M (2010) Pancreatic cancer. N Engl J Med 362(17):1605–1617

Hong SM, Kelly D, Griffith M et al (2008) Multiple genes are hypermethylated in intraductal papillary mucinous neoplasms of the pancreas. Mod Pathol 21(12):1499–1507

Hong SM, Omura N, Vincent A et al (2012) Genome-wide CpG island profiling of intraductal papillary mucinous neoplasms of the pancreas. Clin Cancer Res 18(3):700–712

Hruban RH, Goggins M, Parsons J et al (2000) Progression model for pancreatic cancer. Clin Cancer Res 6(8):2969–2972

Hruban RH, Takaori K, Klimstra DS et al (2004) An illustrated consensus on the classification of pancreatic intraepithelial neoplasia and intraductal papillary mucinous neoplasms. Am J Surg Pathol 28(8):977–987

Hruban RH, Pitman MB, Klimstra DS (2007a) Intraductal neoplasms. In: Hruban RH, Pitman MB, Klimstra DS (eds) Tumors of the pancreas: AFIP atlas of tumor pathology. AFIP atlas of tumor pathology, vol 6. American Registry of Pathology in collaboration with Armed Forces Institute of Pathology, Washington, DC, pp 75–110

Hruban RH, Takaori K, Canto M et al (2007b) Clinical importance of precursor lesions in the pancreas. J Hepatobiliary Pancreat Surg 14(3):255–263

Hruban RH, Maitra A, Goggins M (2008) Update on pancreatic intraepithelial neoplasia. Int J Clin Exp Pathol 1(4):306–316

Hruban RH, Boffetta P, Hiraoka N et al (2010) Ductal adenocarcinoma of the pancreas. In: Bosman FT, Carneiro F, Hruban RH, Theise ND (eds) WHO Classification of tumors of the digestive system World Health Organisation Classification of Tumors. IARC, Lyon, pp 281–291

Iacobuzio-Donahue CA, Klimstra DS, Adsay NV et al (2000a) Dpc-4 protein is expressed in virtually all human intraductal papillary mucinous neoplasms of the pancreas: comparison with conventional ductal adenocarcinomas. Am J Pathol 157(3):755–761

Iacobuzio-Donahue CA, Wilentz RE, Argani P et al (2000b) Dpc4 protein in mucinous cystic neoplasms of the pancreas: frequent loss of expression in invasive carcinomas suggests a role in genetic progression. Am J Surg Pathol 24(11):1544–1548

Jimenez RE, Warshaw AL, Z'Graggen K et al (1999) Sequential accumulation of K-ras mutations and p53 overexpression in the progression of pancreatic mucinous cystic neoplasms to malignancy. Ann Surg 230(4):501–509, discussion 509-11

Jones S, Zhang X, Parsons DW et al (2008) Core signaling pathways in human pancreatic cancers revealed by global genomic analyses. Science 321(5897):1801–1806

Jury RP, Thibodeau BJ, Fortier LE et al (2012) Gene expression changes associated with the progression of intraductal papillary mucinous neoplasms. Pancreas 41(4):611–618

Kanda M, Matthaei H, Wu J et al (2012) Presence of somatic mutations in most early-stage pancreatic intraepithelial neoplasia. Gastroenterology 142(4):730–733

Kawahira H, Kobayashi S, Kaneko K et al (2000) p53 protein expression in intraductal papillary mucinous tumors (IPMT) of the pancreas as an indicator of tumor malignancy. Hepatogastroenterology 47(34):973–977

Kawamoto S, Lawler LP, Horton KM et al (2006) MDCT of intraductal papillary mucinous neoplasm of the pancreas: evaluation of features predictive of invasive carcinoma. AJR Am J Roentgenol 186(3):687–695

Kim GE, Bae HI, Park HU et al (2002) Aberrant expression of MUC5AC and MUC6 gastric mucins and sialyl Tn antigen in intraepithelial neoplasms of the pancreas. Gastroenterology 123(4):1052–1060

Kim SG, Wu TT, Lee JH et al (2003) Comparison of epigenetic and genetic alterations in mucinous cystic neoplasm and serous microcystic adenoma of pancreas. Mod Pathol 16(11):1086–1094

Kosmahl M, Pauser U, Peters K et al (2004) Cystic neoplasms of the pancreas and tumor-like lesions with cystic features: a review of 418 cases and a classification proposal. Virchows Arch 445(2):168–178

Kozuka S, Sassa R, Taki T et al (1979) Relation of pancreatic duct hyperplasia to carcinoma. Cancer 43(4):1418–1428

Laffan TA, Horton KM, Klein AP et al (2008) Prevalence of unsuspected pancreatic cysts on MDCT. AJR Am J Roentgenol 191(3):802–807

Liszka L, Pajak J, Zielinska-Pajak E et al (2010) Intraductal oncocytic papillary neoplasms of the pancreas and bile ducts: a description of five new cases and review based on a systematic survey of the literature. J Hepatobiliary Pancreat Sci 17(3):246–261

Lohr M, Kloppel G, Maisonneuve P et al (2005) Frequency of K-ras mutations in pancreatic intraductal neoplasias associated with pancreatic ductal adenocarcinoma and chronic pancreatitis: a meta-analysis. Neoplasia 7(1):17–23

Luttges J, Galehdari H, Brocker V et al (2001) Allelic loss is often the first hit in the biallelic inactivation of the p53 and DPC4 genes during pancreatic carcinogenesis. Am J Pathol 158(5):1677–1683

Luttges J, Feyerabend B, Buchelt T et al (2002) The mucin profile of noninvasive and invasive mucinous cystic neoplasms of the pancreas. Am J Surg Pathol 26(4):466–471

Maire F, Hammel P, Terris B et al (2002) Prognosis of malignant intraductal papillary mucinous tumours of the pancreas after surgical resection. Comparison with pancreatic ductal adenocarcinoma. Gut 51(5):717–722

Maisonneuve P, Lowenfels AB (2010) Epidemiology of pancreatic cancer: an update. Dig Dis 28(4–5):645–656

Maitra A, Adsay NV, Argani P et al (2003) Multicomponent analysis of the pancreatic adenocarcinoma progression model using a pancreatic intraepithelial neoplasia tissue microarray. Mod Pathol 16(9):902–912

Maitra A, Fukushima N, Takaori K et al (2005) Precursors to invasive pancreatic cancer. Adv Anat Pathol 12(2):81–91

Meckler KA, Brentnall TA, Haggitt RC et al (2001) Familial fibrocystic pancreatic atrophy with endocrine cell hyperplasia and pancreatic carcinoma. Am J Surg Pathol 25(8):1047–1053

Mino-Kenudson M, Fernandez-Del Castillo C, Baba Y et al (2011) Prognosis of invasive intraductal papillary mucinous neoplasm depends on histological and precursor epithelial subtypes. Gut 60(12):1712–1720

Miyamoto Y, Maitra A, Ghosh B et al (2003) Notch mediates TGF alpha-induced changes in epithelial differentiation during pancreatic tumorigenesis. Cancer Cell 3(6):565–576

Mohri D, Asaoka Y, Ijichi H et al (2012) Different subtypes of intraductal papillary mucinous neoplasm in the pancreas have distinct pathways to pancreatic cancer progression. J Gastroenterol 47(2):203–213

Moskaluk CA, Hruban RH, Kern SE (1997) p16 and K-ras gene mutations in the intraductal precursors of human pancreatic adenocarcinoma. Cancer Res 57(11):2140–2143

Nara S, Shimada K, Kosuge T et al (2008) Minimally invasive intraductal papillary-mucinous carcinoma of the pancreas: clinicopathologic study of 104 intraductal papillary-mucinous neoplasms. Am J Surg Pathol 32(2):243–255

Ohuchida K, Mizumoto K, Fujita H et al (2006) Sonic hedgehog is an early developmental marker of intraductal papillary mucinous neoplasms: clinical implications of mRNA levels in pancreatic juice. J Pathol 210(1):42–48

Patel SA, Adams R, Goldstein M et al (2002) Genetic analysis of invasive carcinoma arising in intraductal oncocytic papillary neoplasm of the pancreas. Am J Surg Pathol 26(8):1071–1077

Poultsides GA, Reddy S, Cameron JL et al (2010) Histopathologic basis for the favorable survival after resection of intraductal papillary mucinous neoplasm-associated invasive adenocarcinoma of the pancreas. Ann Surg 251(3):470–476

Raimondo M, Tachibana I, Urrutia R et al (2002) Invasive cancer and survival of intraductal papillary mucinous tumors of the pancreas. Am J Gastroenterol 97(10):2553–2558

Ridder GJ, Maschek H, Flemming P et al (1998) Ovarian-like stroma in an invasive mucinous cystadenocarcinoma of the pancreas positive for inhibin. A hint concerning its possible histogenesis. Virchows Arch 432(5):451–454

Rodriguez JR, Salvia R, Crippa S et al (2007) Branch-duct intraductal papillary mucinous neoplasms: observations in 145 patients who underwent resection. Gastroenterology 133(1): 72–79, quiz 309-10

Salvia R, Fernandez-del Castillo C, Bassi C et al (2004) Main-duct intraductal papillary mucinous neoplasms of the pancreas: clinical predictors of malignancy and long-term survival following resection. Ann Surg 239(5):678–685, discussion 685-7

Sarr MG, Kendrick ML, Nagorney DM et al (2001) Cystic neoplasms of the pancreas: benign to malignant epithelial neoplasms. Surg Clin North Am 81(3):497–509

Sasaki S, Yamamoto H, Kaneto H et al (2003) Differential roles of alterations of p53, p16, and SMAD4 expression in the progression of intraductal papillary-mucinous tumors of the pancreas. Oncol Rep 10(1):21–25

Sato N, Rosty C, Jansen M et al (2001) STK11/LKB1 Peutz-Jeghers gene inactivation in intraductal papillary-mucinous neoplasms of the pancreas. Am J Pathol 159(6):2017–2022

Sato N, Ueki T, Fukushima N et al (2002) Aberrant methylation of CpG islands in intraductal papillary mucinous neoplasms of the pancreas. Gastroenterology 123(1):365–372

Sato N, Fukushima N, Maitra A et al (2004) Gene expression profiling identifies genes associated with invasive intraductal papillary mucinous neoplasms of the pancreas. Am J Pathol 164(3):903–914

Sato N, Fukushima N, Hruban RH et al (2008) CpG island methylation profile of pancreatic intraepithelial neoplasia. Mod Pathol 21(3):238–244

Satoh K, Shimosegawa T, Moriizumi S et al (1996) K-ras mutation and p53 protein accumulation in intraductal mucin-hypersecreting neoplasms of the pancreas. Pancreas 12(4):362–368

Satoh K, Hamada S, Kanno A et al (2010) Expression of MSX2 predicts malignancy of branch duct intraductal papillary mucinous neoplasm of the pancreas. J Gastroenterol 45(7):763–770

Schonleben F, Qiu W, Bruckman KC et al (2007) BRAF and KRAS gene mutations in intraductal papillary mucinous neoplasm/carcinoma (IPMN/IPMC) of the pancreas. Cancer Lett 249(2):242–248

Schonleben F, Allendorf JD, Qiu W et al (2008a) Mutational analyses of multiple oncogenic pathways in intraductal papillary mucinous neoplasms of the pancreas. Pancreas 36(2):168–172

Schonleben F, Qiu W, Remotti HE et al (2008b) PIK3CA, KRAS, and BRAF mutations in intraductal papillary mucinous neoplasm/carcinoma (IPMN/C) of the pancreas. Langenbecks Arch Surg 393(3):289–296

Schutte M, Hruban RH, Geradts J et al (1997) Abrogation of the Rb/p16 tumor-suppressive pathway in virtually all pancreatic carcinomas. Cancer Res 57(15):3126–3130

Sessa F, Solcia E, Capella C et al (1994) Intraductal papillary-mucinous tumours represent a distinct group of pancreatic neoplasms: an investigation of tumour cell differentiation and K-ras, p53 and c-erbB-2 abnormalities in 26 patients. Virchows Arch 425(4):357–367

Shi C, Hruban RH (2012) Intraductal papillary mucinous neoplasm. Hum Pathol 43(1):1–16
Shi C, Klein AP, Goggins M et al (2009) Increased prevalence of precursor lesions in familial pancreatic cancer patients. Clin Cancer Res 15(24):7737–7743
Siegel R, Naishadham D, Jemal A (2012) Cancer statistics, 2012. CA Cancer J Clin 62(1):10–29
Su GH, Hruban RH, Bansal RK et al (1999) Germline and somatic mutations of the STK11/LKB1 Peutz-Jeghers gene in pancreatic and biliary cancers. Am J Pathol 154(6):1835–1840
Suda K, Hirai S, Matsumoto Y et al (1996) Variant of intraductal carcinoma (with scant mucin production) is of main pancreatic duct origin: a clinicopathological study of four patients. Am J Gastroenterol 91(4):798–800
Tajiri T, Tate G, Inagaki T et al (2005) Intraductal tubular neoplasms of the pancreas: histogenesis and differentiation. Pancreas 30(2):115–121
Tanaka M, Chari S, Adsay V et al (2006) International consensus guidelines for management of intraductal papillary mucinous neoplasms and mucinous cystic neoplasms of the pancreas. Pancreatology 6(1–2):17–32
Terris B, Ponsot P, Paye F et al (2000) Intraductal papillary mucinous tumors of the pancreas confined to secondary ducts show less aggressive pathologic features as compared with those involving the main pancreatic duct. Am J Surg Pathol 24(10):1372–1377
Thayer SP, di Magliano MP, Heiser PW et al (2003) Hedgehog is an early and late mediator of pancreatic cancer tumorigenesis. Nature 425(6960):851–856
Thompson LD, Becker RC, Przygodzki RM et al (1999) Mucinous cystic neoplasm (mucinous cystadenocarcinoma of low-grade malignant potential) of the pancreas: a clinicopathologic study of 130 cases. Am J Surg Pathol 23(1):1–16
Tsutsumi K, Sato N, Cui L et al (2011) Expression of claudin-4 (CLDN4) mRNA in intraductal papillary mucinous neoplasms of the pancreas. Mod Pathol 24(4):533–541
van Heek NT, Meeker AK, Kern SE et al (2002) Telomere shortening is nearly universal in pancreatic intraepithelial neoplasia. Am J Pathol 161(5):1541–1547
Vincent A, Herman J, Schulick R et al (2011) Pancreatic cancer. Lancet 378(9791):607–620
Wilentz RE, Geradts J, Maynard R et al (1998) Inactivation of the p16 (INK4A) tumor-suppressor gene in pancreatic duct lesions: loss of intranuclear expression. Cancer Res 58(20): 4740–4744
Wilentz RE, Albores-Saavedra J, Zahurak M et al (1999) Pathologic examination accurately predicts prognosis in mucinous cystic neoplasms of the pancreas. Am J Surg Pathol 23(11):1320–1327
Wilentz RE, Iacobuzio-Donahue CA, Argani P et al (2000) Loss of expression of Dpc4 in pancreatic intraepithelial neoplasia: evidence that DPC4 inactivation occurs late in neoplastic progression. Cancer Res 60(7):2002–2006
Wouters K, Ectors N, Van Steenbergen W et al (1998) A pancreatic mucinous cystadenoma in a man with mesenchymal stroma, expressing oestrogen and progesterone receptors. Virchows Arch 432(2):187–189
Wu J, Jiao Y, Dal Molin M et al (2011a) Whole-exome sequencing of neoplastic cysts of the pancreas reveals recurrent mutations in components of ubiquitin-dependent pathways. Proc Natl Acad Sci USA 108(52):21188–21193
Wu J, Matthaei H, Maitra A et al (2011b) Recurrent GNAS mutations define an unexpected pathway for pancreatic cyst development. Sci Transl Med 3(92):92ra66
Xiao HD, Yamaguchi H, Dias-Santagata D et al (2011) Molecular characteristics and biological behaviours of the oncocytic and pancreatobiliary subtypes of intraductal papillary mucinous neoplasms. J Pathol 224(4):508–516
Yamaguchi H, Shimizu M, Ban S et al (2009) Intraductal tubulopapillary neoplasms of the pancreas distinct from pancreatic intraepithelial neoplasia and intraductal papillary mucinous neoplasms. Am J Surg Pathol 33(8):1164–1172
Yamaguchi H, Kuboki Y, Hatori T et al (2011) Somatic Mutations in PIK3CA and Activation of AKT in Intraductal Tubulopapillary Neoplasms of the Pancreas. Am J Surg Pathol 35(12): 1812–1817

Yamao K, Yanagisawa A, Takahashi K et al (2011) Clinicopathological features and prognosis of mucinous cystic neoplasm with ovarian-type stroma: a multi-institutional study of the Japan pancreas society. Pancreas 40(1):67–71

Yu J, Li A, Hong SM et al (2012) MicroRNA alterations of pancreatic intraepithelial neoplasias. Clin cancer res 18(4):981–992

Zamboni G, Scarpa A, Bogina G et al (1999) Mucinous cystic tumors of the pancreas: clinicopathological features, prognosis, and relationship to other mucinous cystic tumors. Am J Surg Pathol 23(4):410–422

Zamboni G, Fukushima N, Hruban RH et al (2010) Mucinous cystic neoplasms of the pancreas. In: Bosman FT, Carneiro F, Hruban RH, Theise ND (eds) WHO Classification of tumors of the digestive system World Health Organisation Classification of Tumors. IARC, Lyon, pp 300–303

Genetic Epidemiology and Pancreatic Cancer

Li Jiao and Donghui Li

Abstract Gene mutations that are associated with cancer syndromes explain a small portion of pancreatic cancer cases. The majority of the sporadic pancreatic cancer cases are perhaps the consequence of a joint effect of genetic factors and environmental or lifestyle risk factors. Studies on common genetic variants via the candidate gene approach have observed risk modifications by genes involved in various biological process and signaling pathways. However, most of these findings were made in studies that lacked adequate statistical power or replication effort. Recent genome-wide association studies (GWAS) have identified several genes and loci associated with the risk of pancreatic cancer: *ABO*, *NR5A2*, and *TERT1* in individuals with European ancestry, *FOXQ1*, *BICD1*, and *DPP6* in the Japanese population, and *BACH1*, *DAB2*, *PRLHR*, *TFF1*, and *FAM19A5* in the Chinese population. Future completion of larger scale GWAS in pancreatic cancer, mining of GWAS data using novel statistical approaches, and functional studies on the mechanistic links between identified genes and the disease will provide new insights into genetic susceptibility to and the molecular mechanisms of pancreatic cancer.

L. Jiao, M.D., Ph.D.
Section of Gastroenterology and Hepatology, Section of Health Services Research, Department of Medicine, Dan L. Duncan Cancer Center, Baylor College of Medicine, 2002 Holcombe Blvd, 152, Houston, TX 77030, USA
e-mail: jiao@bcm.edu

D. Li, Ph.D. (✉)
Department of Gastrointestinal Medical Oncology, The University of Texas MD Anderson Cancer Center, 1515 Holcombe Boulevard, Unit 426, Houston, TX 77030, USA
e-mail: dli@mdanderson.org

D.M. Simeone and A. Maitra (eds.), *Molecular Genetics of Pancreatic Cancer*,
DOI 10.1007/978-1-4614-6549-2_3,

Introduction

Familial pancreatic cancer accounts for approximately 10 % of pancreatic cancer cases in the general population. Mutations in genes that are associated with cancer syndromes also explain a small portion of pancreatic cancer cases. The majority of the sporadic cases are perhaps the consequence of a joint effect of genetic factors and environmental or lifestyle risk factors. Cigarette smoking, high body mass index (BMI), long-term type 2 diabetes, and possibly higher intake of red meats or fat are major nongenetic modifiable risk factors for this disease. Because only a portion of individuals with these modifiable risk factors ever develop pancreatic cancer, genetic susceptibility factors alone or in combination with epidemiological factors may play a major role in pancreatic carcinogenesis. Research on common genetic variants via the candidate gene approach and via genome-wide association studies (GWAS) has generated a large amount of information on potential genetic susceptibility genes for this disease. In this chapter, we summarize recent information and discuss future directions in this research field.

The Candidate Gene Approach

Since 1994, several large, retrospective case–control studies in the USA (Duell et al. 2002a; Gross et al. 1999; Li 2001; McWilliams et al. 2008; Prizment et al. 2012; Asomaning et al. 2008), China (Li et al. 2011; Zhao et al. 2009), the Czech Republic (Vrana et al. 2009), and Japan (Suzuki et al. 2008a) have achieved adequate sample size to address the main effect of common single-nucleotide polymorphisms (SNPs) on the risk of sporadic pancreatic cancer. The genes and SNPs selected in those studies included those involved in carcinogen or nutrient metabolism (Vrana et al. 2009; Suzuki et al. 2008a; Ayaz et al. 2008; Bartsch et al. 1998; Duell et al. 2002b; 2010; Jiao et al. 2007a, b; Kanda et al. 2009; Li et al. 2005, 2006; Liu et al. 2000; Miyasaka et al. 2005, 2010; Mohelnikova-Duchonova et al. 2010; Ockenga et al. 2003; Ohnami et al. 2008; Piepoli et al. 2006; Suzuki et al. 2008b; Verlaan et al. 2005; Wang et al. 2005; Vrana et al. 2010), DNA repair (Duell et al. 2002a; McWilliams et al. 2008; Dong et al. 2011a; Gargiulo et al. 2009; Jiao et al. 2006, 2007c, 2008; Li et al. 2009; McWilliams et al. 2009a; Zhang et al. 2011a), cell cycle regulation and apoptosis (Asomaning et al. 2008; Li et al. 2011; Chen et al. 2007, 2008, 2010; Couch et al. 2009, 2010; Grochola et al. 2010; Naccarati et al. 2010; Reid-Lombardo et al. 2011; Sonoyama et al. 2011; Theodoropoulos et al. 2010a; Wang et al. 2007; Yang et al. 2008), antioxidant defense (Lyn-Cook et al. 2006; Mohelnikova-Duchonova et al. 2011; Tang et al. 2010), inflammation and the immune system (Zhao et al. 2009; Reid-Lombardo et al. 2011; Duell et al. 2006; Hamacher et al. 2009; Lang et al. 2012; Olson et al. 2007; Ozhan et al. 2011;

Sun et al. 2008; Talar-Wojnarowska et al. 2011; Yang et al. 2012), and mitochondrial function (Wang et al. 2007; Lynch et al. 2011). Other genes and SNPs include those related to familial pancreatic cancer (McWilliams et al. 2009b), other cancers (Couch et al. 2009; Lang et al. 2012; Chen et al. 2011), or medical conditions such as insulin resistance (Suzuki et al. 2008c; Dong et al. 2011b) or obesity, and diabetes (Prizment et al. 2012; Wang et al. 2007; Tang et al. 2010, 2011 Fong et al. 2010; Pierce et al. 2011). Researcher's selection of candidate genes is largely based on existing knowledge of risk factors for pancreatic cancer and hallmarks of cancer. With the evolution of genotyping technology, PCR-RFLP, Taqman, mass spectrometry, Sequenom, Illumina GoldenGate, and other methods have been used in different studies. In most of these studies, weak main effects of the genes were observed occasionally; interactions with known nongenetic risk factors were reported more frequently. In this section, we briefly summarize the major findings from the existing research. Findings on genes involved in xenobiotic metabolism, oxidative stress, and cell cycle control published after 2009 are summarized in Table 1. Prior studies were summarized in a recent review (Lin et al. 2011) and a meta-analysis (Mazaki et al. 2011). Studies that included fewer than 100 cases (Ayaz et al. 2008; Bartsch et al. 1998; Piepoli et al. 2006; Hamacher et al. 2009; Fong et al. 2010; Krechler et al. 2009; Lukic et al. 2011; Scola et al. 2009; Theodoropoulos et al. 2010b) are not reviewed in this section.

Xenobiotic Metabolizing Genes

Because cigarette smoking is a major risk factor for pancreatic cancer, carcinogen metabolic genes and DNA repair genes were among the first genes studied in a wave of research on genetic variants in pancreatic cancer. Studies conducted at the University of Texas MD Anderson Cancer Center reported positive associations between the *CYP1A2*, *NAT1*, and *NAT2* genotypes and risk of pancreatic cancer independently or jointly with exposure to tobacco carcinogens (Jiao et al. 2007a; Li et al. 2006; Suzuki et al. 2008b). None of four studies on glutathione S-transferase (*GST*) genes found a significant main effect on risk of pancreatic cancer (Vrana et al. 2009; Duell et al. 2002b; Jiao et al. 2007b; Liu et al. 2000). Of those four studies, one observed a possible interaction between *GSTT1* gene deletion and heavy smoking among Caucasians, in particular among women (Duell et al. 2002b), and two reported an age-related effect of the *GSTP1*-codon 105 SNP on risk of pancreatic cancer (Vrana et al. 2009; Jiao et al. 2007b). The rs743572 SNP of *CYP17A1*, a gene encoding an enzyme involved in estrogen and testosterone biosynthesis, was associated with risk of pancreatic cancer in Caucasians (Duell et al. 2010). One SNP of *CYP1B1* (rs1056836) was associated with pancreatic cancer in a Czech Republic population. However, the confounding factors were not evaluated in this study (Vrana et al. 2010) (Table 1).

Table 1 Selected candidate gene studies on pancreatic cancer (published mostly since 2009)

Reference	Study location (ethnicity)	No. of cases/controls	Genes and SNPs assessed	OR (95 % CI) with $P<0.05$	Covariates	Joint effect of risk factors and SNP	Test for joint effect
Duell et al. (2010)	USA-San Francisco (77 % Caucasian)	308/964	*CYP17A1*: rs743572 34 T/C(A1/A2)	A1/A2 vs. A1A1: 0.77 (0.58–1.00) A2/A2 vs. A1/A1: 0.63 (0.42–0.93)	Age, sex, race, smoking	Not detected	NA
Vrana et al. (2010)	Czech Republic (100 % Caucasian)	156/337	*CYP1B1*: rs1056836 Leu432Val rs1800400 Asn453Ser	rs1056836: Val/Val vs. Leu/Leu: 0.59 (0.36–0.96)	No	Not examined	NA
Wheatley-Price et al. (2008)	USA-Massachusetts General Hospital (100 % Caucasian)	122/331	*SOD2*: rs4880 Ala16Val *MPO*: –G463A	rs4880: Val/Val vs. any Ala: 2.05 (1.2–3.6) –G463A: any A vs. GG: 0.57 (0.4–0.9)	Age, sex smoking, alcohol use	Not examined	NA
Tang et al. (2010)	USA-MDACC (91 % Caucasian)	575/648	*SOD2*: rs4880, rs2758346 *GSTA4*: rs1802061, rs182623, rs316141 *CAT*: rs1001179 *GPX1*: rs1050450	No	Age, sex, race, smoking, alcohol, diabetes and family history of cancer	SOD2 rs4880 stronger for any Ala and diabetes	$P=0.05$
Mohelnikova-Duchonova et al. (2011)	Czech Republic (100 % Caucasian)	235/265	*SOD2*: rs4880 *SOD3*: rs1799895 *NQO1*: rs1800566 *NQO2*: rs1143684	No	Age, sex, weight, diabetes, pancreatitis, smoking, alcohol, coffee, and tea consumption	Not detected	NA

Zhang et al. (2011a)	USA-Minnesota (93.5 % Caucasian)	189/486	*SOD2*: rs4880 *CAT*: rs1001179 *hOGG1*: rs1052133 *XRCC1*: rs25487	rs4880: any Val vs. Ala/Ala: 0.57 (0.37–0.89)	Age, sex, race, education, smoking, alcohol use, physical activity, and energy intake	Stronger for Val allele and lower intake of antioxidants	NR
Sonoyama et al. (2011)	Japan (Japanese)	226/448	*TP53*: rs1042522 Arg72/Pro	Pro/Pro vs. Arg/Arg: 1.70 (1.01–2.88)	Age, sex, smoking and alcohol use	Stronger for male heavy smoking, or heavy drinking	NR
Naccarati et al. (2010)	Czech Republic (100 % Caucasian)	240/743	*TP53*: rs1042522 G>C Arg72Pro, rs17878362 A(1)>A(2), rs12947788 C>T, rs17884306 G>A	rs1042522: any Pro vs. Arg/Arg: 1.73 (1.26–2.39)	Age, sex, BMI, smoking	Not examined	NA
Grochola et al. (2010)	UK (100 % Caucasian)	103/499	*MDM2*: 309 T/G	G-allele vs. T-allele was associated with earlier onset of pancreatic cancer	Not examined	Stronger for male	NR
Asomaning et al. (2008)	USA- Massachusetts General Hospital (100 % Caucasian)	123/372	*MDM2*: 309 T/G	T/G vs. TT: 1.89 (1.20–2.99) G/G vs. T/T: 2.07 (1.03–4.16)	Age, sex, smoking status, pack-years of smoking	Not examined	NA

BMI body mass index, *MDACC* The University of Texas MD Anderson Cancer Center, *NA* not applicable, *NR* not reported, *OR* (95 % CI) odds ratio (95 % confidence interval) of pancreatic cancer, *SNPs* single-nucleotide polymorphisms

DNA Repair Genes

Studies on various DNA repair pathways—such as base excision repair, nucleotide excision repair, homologous recombination repair and non-homologous end joining, and mismatch repair—have observed some weak main effects of variants of DNA repair genes on the risk of pancreatic cancer, such as *LIG3* and *ATM* (Li et al. 2009), *MGMT* and *PMS2* (Dong et al. 2011a). Some joint effects of *XRCC1*, *APE1*, *MGMT*, *XRCC2*, and *XPD* variants with smoking (Jiao et al. 2006, 2007c; 2008) and *ATM* and *LIG4* variants with diabetes (Li et al. 2009) were also reported. However, two studies on the interaction between the *XPD* D312N SNP (rs1799793) and heavy smoking showed opposite directions: the minor allele was associated with increased risk in one study (McWilliams et al. 2008) and decreased risk in the other (Jiao et al. 2007c). Three studies in the USA consistently found a null association of the *XRCC1* rs25487 with risk of pancreatic cancer (Duell et al. 2002a; McWilliams et al. 2008; Jiao et al. 2006). Using a tagging SNP approach, a Mayo Clinic study examined 236 tag-SNPs of 26 DNA repair genes and identified that the genotype and haplotype of the *MMS19L* gene, which is involved in nucleotide excision repair, were associated with risk of pancreatic cancer (McWilliams et al. 2009a). Three studies have investigated *hOGG1* SNPs (McWilliams et al. 2008; Zhang et al. 2011a; Li et al. 2002), but only one found an association between the variant allele of rs1052133, and the risk of pancreatic cancer (OR: 1.57, 95 % CI: 1.04–2.39, any 326Cys compared with Ser326Ser) (Zhang et al. 2011a).

Oxidative Stress-Associated Genes

Oxidative stress is one of the mechanisms whereby cigarette smoking can contribute to pancreatic cancer development. A number of studies have investigated the association between SNP rs4880 of *SOD2* and the risk of pancreatic cancer (Zhang et al. 2011a; Mohelnikova-Duchonova et al. 2011; Tang et al. 2010; Wheatley-Price et al. 2008). A study with a Czech population showed neither main effects nor interactions with smoking and alcohol, coffee, or tea consumption (Mohelnikova-Duchonova et al. 2011). A U.S. study showed that the valine allele of *SOD2* rs4880 interacted with diabetes and antioxidant use in modifying the risk of pancreatic cancer (Zhang et al. 2011a; Tang et al. 2010). No association was reported for other genes involved in oxidative stress, including *SOD3*, *CAT*, *NQO1*, and *NQO2*, in pancreatic cancer (Table 1). Mitochondria play a key role in the production of reactive oxygen species. Oxidative stress could cause mitochondrial damage and affect mitochondrial DNA copy numbers. A Mayo Clinic study found no association between 24 mitochondrial SNPs or haplogroup and risk of pancreatic cancer (Wang et al. 2007). In a nested case–control study within a Finnish male smoker cohort, a significantly higher copy number of mitochondrial DNA was detected (Lynch et al. 2011).

Inflammation and Immunity Genes

Accumulating evidence suggests that chronic inflammation may be one of the underlying mechanisms that contribute to pancreatic cancer development (Farrow et al. 2004). Several studies have evaluated the polymorphisms of selected inflammatory genes in association with pancreatic cancer. A Mayo Clinic study examined 1,538 SNPs of 102 genes involved in nuclear factor κB–mediated inflammatory pathways and found significant associations between the *CD101* rs10923193 or four SNPs of *NOS1* (rs3782203, rs9658350, rs532967, and rs547954) and the risk of pancreatic cancer. However, the significant associations could not be validated in a PanScan cohort and case–control consortium study (Reid-Lombardo et al. 2011). Two other studies found possible interactions of *TNFα* –308 G/A and *RANTES* –403 G/A with pancreatitis, *CCR5* –Δ32 with smoking (Duell et al. 2006), and *IL-4R* G3017T with allergic response (Olson et al. 2007) in modifying risk of pancreatic cancer.

Cytotoxic T-lymphocyte antigen-4 (CTLA-4) plays important roles in down-regulating T-cell activation, thereby attenuating antitumor responses and increasing cancer susceptibility. The *CTLA-4* 49 G>A SNP (rs231775) weakens the binding affinity of CTLA-4 to B7.1, leading to attenuated CTLA-4-triggered inhibition of T-cell activation and proliferation (Sun et al. 2008). Two independent studies in China showed that the *CTLA-4* 49A allele was significantly associated with a higher risk of pancreatic cancer (Yang et al. 2008; Lang et al. 2012).

Cyclooxygenase-2 (COX-2) is a key enzyme in the arachidonic acid pathway. A Chinese study and a Polish study both showed a positive association of the –1195AA *COX-2* genotype with risk of pancreatic cancer (Zhao et al. 2009; Talar-Wojnarowska et al. 2011). The Chinese study also revealed that the –765GC genotype increased the risk of pancreatic cancer both independently and jointly with cigarette smoking (Zhao et al. 2009). However, the Polish study did not find such an association (Talar-Wojnarowska et al. 2011). A small hospital-based study in Turkey found that two haplotypes of *COX2* were more frequent in patients than in control subjects (Ozhan et al. 2011).

Folate- and Alcohol-Metabolizing Genes

Observations have been inconsistent on the 5,10-methylenetetrahydrofolate reductase (*MTHFR*) C677T SNP (rs1801133) in pancreatic cancer (Suzuki et al. 2008a; Li et al. 2005; Ohnami et al. 2008; Wang et al. 2005; Matsubayashi et al. 2005) and null for the *MTHFR* A1298C SNP (Li et al. 2005; Wang et al. 2005; Matsubayashi et al. 2005). Those studies were summarized in two previous review articles (Lin et al. 2011; Mazaki et al. 2011). The latter article, a meta-analysis, concluded that the *MTHFR* 677TT genotype in Caucasian smokers conferred a 1.66- and 2.52-fold higher risk of pancreatic cancer compared with the CC and CT genotypes, respectively (Mazaki et al. 2011). In a step-wise genotyping study, a Japanese study investigated 227 SNPs of 46 selected genes that are involved in folate metabolism.

The variant alleles of the methionine synthase reductase (*MTRR*) gene SNPs rs162049 and rs10380 were associated with increased risk of pancreatic cancer (Ohnami et al. 2008), but the results from other previously reported SNPs of *MTHFR* and *NAT1* were not replicated in this study. Another Japanese study did not find any main effect of the folate metabolic genes, but a potential interaction of some SNPs of *MTHFR* and *MTRR* with heavy alcohol consumption was suggested (Suzuki et al. 2008a).

One of two studies on the thymidylate synthase (*TS*) variable number of tandem repeat variants found no association with pancreatic cancer in a Japanese population (Suzuki et al. 2008a). The second study found an increased risk of pancreatic cancer for the *TS* 5'-untranslated region 3Rc/3Rc genotype in a Chinese population (Dong et al. 2011a).

Heavy alcohol consumption (>4 drinks per day) has been associated with an increased risk of pancreatic cancer (Jiao et al. 2009). However, a case–control study in the Czech Republic did not find an association between the alcohol dehydrogenase *ADH1B* and *ADH1C* variants and pancreatic cancer risk (Mohelnikova-Duchonova et al. 2010). A meta-analysis of studies on the aldehyde dehydrogenase (*ALDH)2* gene found a marginally significant effect of alcohol intake on the risk of pancreatic cancer among the heterozygous *1*2 genotype carriers but not among the *2*2 homozygous genotype carriers (Mazaki et al. 2011).

Cell Cycle Regulation- and Apoptosis-Related Genes

Two studies found that the *P53* Arg72Pro minor allele conferred a higher risk of pancreatic cancer (Naccarati et al. 2010; Sonoyama et al. 2011). Mouse double minute 2 homologue (MDM2) is an E3 ubiquitin ligase that blocks the transcriptional activation of p53 and is overexpressed in human pancreatic cancer (Dong et al. 2005). Two small studies provided evidence that a common *MDM2* T309G SNP was associated with a higher risk of pancreatic cancer (Asomaning et al. 2008; Grochola et al. 2010). A U.S. study of 509 cases and 462 controls reported a main effect of *P21* SNP rs1801270 but not *P27* SNP rs2066827 in pancreatic cancer (Chen et al. 2010).

The FAS/FASL system plays a crucial role in modulating apoptosis and maintaining homeostasis. A study of Chinese Han subjects found that the functional SNPs of *FasL* (−844 T-C) and caspase-8 (*CASP8*) (−652 6N ins→del) were both independently and jointly associated with risk of pancreatic cancer. Furthermore, these two genetic variants interacted with smoking and diabetes to modify this risk (Yang et al. 2008).

Other Cancer-Related Genes

Hypothesis-driven analyses of existing GWAS data can be a cost-efficient approach to investigating genetic susceptibility to pancreatic cancer. A series of studies investigated SNPs that predispose individuals to other forms of cancers. SNPs of *CASP8* (rs1045485) and *MAP3K1* (rs889312), *APC* (rs2431238) and *NIN* (rs10145182), which have been implicated in breast cancer, were shown to be associated with pancreatic cancer in the same Caucasian population (Couch et al. 2009, 2010). However, in an MD Anderson Cancer Center study, two SNPs that have been implicated in lung cancer, rs8034191 and rs1051730, which are located in the 15q24-25.1 region, were not associated with risk of pancreatic cancer (Chen et al. 2011). Genetic variations that contribute to hereditary pancreatic cancer do not seem to contribute to sporadic pancreatic cancer: polymorphisms of *PRSS1*, *PRSS2*, *CDKN2A* and 28 genes directly and indirectly involved in the Fanconi/BRCA pathway had no effect on pancreatic cancer risk (McWilliams et al. 2009b).

Diabetes and Obesity-Related Genes

Type 2 diabetes and obesity have been consistently associated with increased risk of pancreatic cancer. Therefore, an association between diabetes or obesity-associated SNPs and pancreatic cancer is biologically plausible. SNPs of the genes for *GCKR*, *FTO*, *PPARγ*, *MTNR1B*, *MADD*, and *BCL11A* have all been associated with risk of pancreatic cancer (Prizment et al. 2012; Pierce et al. 2011). An interaction of the *FTO* and *ADIPOQ* SNPs and BMI was detected (Tang et al. 2011).

Strong experimental evidence supports the role of insulin-like growth factor (IGF) in pancreatic carcinogenesis. Thus far, three studies have investigated the IGF axis genes in association with pancreatic cancer (Suzuki et al. 2008c; Dong et al. 2010; Nakao et al. 2011a). An MD Anderson study observed that genotypes of the *IGF1*, *IGF1R*, and *IGFBP1* genes and haplotypes of the *IGF2R* and *IGFBP3* genes were significantly associated with pancreatic cancer risk (Dong et al. 2010). These studies also showed that genetic variants of IGF axis genes act jointly with diabetes, BMI, and alcohol consumption to affect susceptibility to pancreatic cancer. Notably, a 3′-untranslated region variant of the *IGF1* gene (rs5742714) was implicated in two independent studies (Suzuki et al. 2008c; Nakao et al. 2011a). The other study also found genetic variations of somatostatin receptor (*SSTR5*) and glucose metabolizing enzyme that modified, independently or jointly with smoking or diabetes, the risk of pancreatic cancer (Li et al. 2011; Dong et al. 2010).

Copy Number Variation

Structural variations of the human genome, including copy number variation (CNV), have been recognized as a common type of genetic variation that predisposes individuals to sporadic cancer (Ionita-Laza et al. 2009; Kuiper et al. 2010). Loss of chromosome 6q13 is a frequent event in pancreatic cancer (Harada et al. 2007). CNVR2966.1 is a common CNV in a gene desert region on 6q13. A Chinese study revealed that individuals carrying one copy of CNVR2966.1 had a significantly higher risk of pancreatic cancer compared with those carrying two copies (adjusted OR: 1.31, 95 % CI: 1.08–1.60) (Huang et al. 2012). Moreover, this study found that CNVR2966.1 functions as a potential *trans*-acting regulator of the *CDKN2B* gene that is a cell growth regulator controlling cell cycle G1 progression.

Summary

In summary, efforts using the candidate gene approach to identify low-penetrating and common gene traits (minor allele frequency >5 %) that modify the risk of sporadic pancreatic cancer have been largely unsuccessful. Some weak main effects of *NAT*, *SOD2*, *TP53*, *COX2*, *IGF1* and *MTHFR* variants were reported while findings on other genes have not been independently validated in different study populations. Therefore, additional genetic epidemiologic studies of pancreatic cancer are needed to establish the relevance of the intriguing findings on genes involved in DNA repair, inflammatory response, IGF signaling, as well as obesity and diabetes. Further examination of possible gene–environment interactions are required in adequately powered studies to resolve the problem of imprecise risk estimates. Such studies will rely on accurate assessment of the major risk factors such as smoking, alcohol use, diet, BMI, and diabetes. Findings from such studies need to be replicated in racial and ethnic groups other than non-Hispanic Caucasians. Despite limited success in the past, retrospective case–control studies will likely continue to contribute to the genetic association study of pancreatic cancer in the format of consortium studies. To date, a number of consortia of preexisting studies exist, and they may facilitate the identification of additional low-penetrating variants, gene–environment and gene–gene interactions using the high throughput technology. Large consortium studies are needed to have the requisite power to examine genetic variants in minority populations, CNV, and common and rare SNPs in various pathways. However, the consortium studies should not prevent the generation of additional well-designed, sufficiently powered studies that apply uniform criteria for case selection, acquisition of environmental exposure information, and biological sample collection.

Genome-Wide Association Studies

GWAS have identified numerous gene traits that predispose individuals to cancer. The comprehensive coverage of a large number of gene variants in this approach has uncovered novel gene variants that had previously not been considered in relation to cancer. Stringent criteria are applied in the statistical analysis of GWAS data to minimize the false-positive discoveries associated with multiple testing.

GWAS Publications

To date, four GWAS have been conducted in association with risk of pancreatic cancer (Table 2). Two of those studies were conducted with people mostly with European ancestry (Amundadottir et al. 2009; Petersen et al. 2010), one study with a Japanese population (Low et al. 2010), and one with a Chinese population (Wu et al. 2012).

PanScan I and PanScan II

The first GWAS for pancreatic cancer (PanScan I) was conducted by the National Cancer Institute using 1,896 cases and 1,939 controls pooled from 12 cohort studies and one case–control study by the Pancreatic Cancer Cohort Consortium (Amundadottir et al. 2009). Approximately 550,000 SNPs were genotyped, and the most significant ones (top 100 hits with small P values) were tested in the replication stage using 2,457 cases and 2,654 controls from eight case–control studies of the Pancreatic Cancer Case–Control Consortium (Petersen and Boffetta 2012). The initial scan identified a significant association of an *ABO* gene variant (rs505922) with risk of pancreatic cancer, and this observation was confirmed in the replication study. A significant association was also detected for some *sonic hedgehog* (*SHH*) gene variants (rs167020 and rs172310) in the initial scan, but that finding was not replicated.

The second GWAS for pancreatic cancer (PanScan II) was performed with 1,955 cases and 1,995 controls drawn from the same eight case–control studies used in the replication stage of PanScan I (Petersen et al. 2010). Approximately 620,000 SNPs were genotyped, and the combined dataset of PanScans I and II revealed three additional loci in association with the risk of pancreatic cancer. Two SNPs (rs9543325 and rs9564966) identified on the chromosome 13q22.1 region map to a non-genic region between *KLF5* and *KLF12* genes, which code for the kruppel-like transcription factors that regulate cell growth and transformation. This chromosome segment is frequently deleted in many cancers, including pancreatic cancer, and thus an unidentified tumor suppressor gene may be harbored in this region. Five SNPs on the chromosome 1q32.1 region map to the nuclear receptor subfamily 5, group A, member 2 (*NR5A2*) gene (also known as *liver receptor homologue 1*, *LRH1*); the strongest signal was rs3790844. A single SNP (rs401681) resides on the chromosome 5p15.33 region, which contains the cleft lip and palate transmembrane 1-like gene (*CLPTM1L*) and the telomerase reverse transcriptase gene (*TERT*), has been associated with multiple cancers.

GWAS with a Japanese Population

The third GWAS was conducted with a Japanese population involving 991 cases and 5,209 controls without a replication step (Low et al. 2010). Three genes were

Table 2 Genome-wide association studies of pancreatic cancer

Study cohort (reference)	No. of cases/controls	Chromosome region	Top hits	Gene	*P* value	OR (95 % CI)
PanScan I (Huang et al. 2012)	1,896/1,939 Replication in 2,457/2,654	9q34	rs505922	*ABO*	5.37×10^{-8}	1.20 (1.12–1.28)[a]
PanScans I and II (Amundadottir et al. 2009)	3,851/3,934	13q22.1	rs9543325	NA	3.27×10^{-11}	1.26 (1.18–1.35)[b]
		13q22.1	rs9564966	NA	5.86×10^{-8}	1.21 (1.13–1.30)
		1q32.1	rs3790844	*NR5A2*	2.45×10^{-10}	0.77 (0.71–0.84)
		5p15.33	rs401681	*CLPTM1L-TERT*	3.66×10^{-7}	1.19 (1.11–1.27)
Japanese (Petersen et al. 2010)	991/5,209	6p25.3	rs9502893	*FOXQ1*	3.30×10^{-7}	1.29 (1.17–1.43)[b]
		12p11.21	rs708224	*BICD1*	3.30×10^{-7}	1.32 (1.19–1.47)[b]
		7q36.2	rs6464375	*DPP6*	4.41×10^{-7}	3.73 (2.24–6.21)[c]
Han Chinese (Low et al. 2010)	981/1,191 replication in 2,603/2,877	21q21.3	rs372883	*BACH1*	2.24×10^{-13}	0.79 (0.75–0.84)[b]
		5p13.1	rs2255280	*DAB2*	4.18×10^{-10}	0.81 (0.76–0.87)
		10q26.11	rs12413624	*PRLHR*	5.12×10^{-11}	1.23 (1.16–1.31)
		21q22.3	rs1547374	*TFF1*	3.71×10^{-13}	0.79 (0.74–0.84)
		13q22.1	rs4885093	NA	1.57×10^{-12}	1.25 (1.18–1.33)
		13q22.1	rs9573163	NA	5.14×10^{-13}	1.26 (1.18–1.34)
		22q13.32	rs5768709	*FAM19A5*	1.41×10^{-10}	1.25 (1.17–1.34)

NA non-genic region, *OR* (95 % CI) odds ratio (95 % confidence interval)
[a]Heterozygous OR
[b]Allelic OR
[c]Homozygous OR

significantly associated with the risk of pancreatic cancer: *FOXQ1* SNP rs9502893, located on chromosome 6p25.3, *BICD1* SNP rs708224 on chromosome 12p11.21, and *DPP6* SNP rs6464374 on chromosome 7q36.2. None of the GWAS top hits reported in PanScan I or PanScan II were confirmed in this Japanese study.

GWAS with a Han Chinese Population

The most recent GWAS on pancreatic cancer was conducted with a Han Chinese population. This two-stage study involved 981 cases and 1,191 controls in the initial scan and 2,603 cases and 2,877 controls in the replication phase (Wu et al. 2012). Five genes were found to be highly significantly associated with pancreatic cancer: *BACH1*, *DAB2*, *PRLHR*, *TFF1*, and *FAM19A5*, which are located on chromosomes 21q21.3, 5p13.1, 10q26.11, 21q22.3, and 22q13.32, respectively. Furthermore, two of the top hits of PanScans I and II, one located on the non-genic region of chromosome 13q22.1 and one on chromosome 5p15.33, were replicated in this population.

Validation and Functional Characterization of Genes Identified in GWAS

Understanding the biological mechanisms that link the GWAS top hits with the phenotype is crucial to the application of these findings in disease intervention. Among the genes/SNPs identified in pancreatic cancer GWAS, few have been validated in different populations or functionally characterized in experimental models (Table 3).

ABO Genotype

The association between *ABO* genotypes and risk of pancreatic cancer has been validated in several studies. In two large prospective cohort studies (the Nurses' Health Study and the Health Professionals Follow-up Study), individuals with non-O serotypes had a 1.32- to 1.72-fold higher risk of pancreatic cancer than those with the O blood type; as much as 17 % of the cases could be explained by the non-O blood types (Wolpin et al. 2009). Similar findings were reported when the *ABO* genotype was imputed using SNPs examined in the PanScan I GWAS: the non-O genotypes contributed to 19.5 % of the pancreatic cancer cases (Wolpin et al. 2010a). Furthermore, the *ABO* A1 allele, which is associated with higher glycosyltransferase activity, was responsible for the increased risk of pancreatic cancer (Wolpin et al. 2010b). Although the GWAS conducted in the Japanese and Han Chinese populations did not confirm the association between *ABO* genotype and risk of pancreatic cancer (Low et al. 2010; Wu et al. 2012), this association was reported by another Japanese study of 185 pancreatic cancer cases and 1,465 controls (Nakao et al. 2011b). The mechanisms underlying the association between *ABO* and pancreatic

Table 3 GWAS top hits and possible links with pancreatic cancer

Gene symbol	Full gene name	Known protein function	Potential mechanism
ABO	ABO blood group (transferase A, α-1-3-*N*-acetylgalactosaminy l-transferase; transferase B, α-1-3-galactosyltransferase)	Glycosyltransferase	Inflammation, cell adhesion
BACH1	BTB and CNC homology 1 (basic leucine zipper transcription factor 1)	Transcription factor	Antioxidant-response-element-mediated gene regulation?
BICD1	Bicaudal D homolog 1	Mediator of dynein function	Telomere length, G protein signaling
CLPTM1L-TERT1	Cleft lip and palate transmembrane 1-like–telomerase reverse transcriptase	Telomerase reverse transcriptase	Genomic stability
DAB2	Disabled homolog 2	Mitogen-responsive phosphoprotein	Growth factor or Ras pathway modulation
DPP6	Dipeptidyl-peptidase 6	Bind specific voltage-gated potassium channels	Electrophysiological properties?
FAM19A5	Family with sequence similarity 19 (chemokine [C-C motif]-like), member A5	Secreted protein	Immune and nervous cell regulation
FOXQ1	Forkhead box Q1	Transcription factor	Embryonic development, cell cycle, epithelial–mesenchymal transition
NR5A2	Nuclear receptor subfamily 5, group A member 2	Nuclear receptor	Pancreas development and differentiation, steroidogenesis, cholesterol and bile acid homeostasis, cell proliferation
PRLHR	Prolactin-releasing hormone receptor	G protein–coupled receptor	?
TFF1	Trefoil factor 1	Secretory protein	Activation of NF-κB-mediated inflammation

cancer risk are not understood. Several studies have shown a significant association between *ABO* genotype and the plasma level of proteins involved in inflammatory response, cell adhesion, and vascular functions, such as tumor necrosis factor α, intercellular adhesion molecule 1, E-selectin, and P-selectin (Melzer et al. 2008; Barbalic et al. 2010; Paterson et al. 2009). Whether ABO plays a regulatory role in inflammatory response, which in turn contributes to pancreatic carcinogenesis, requires further investigation (Lennon et al. 2010).

NR5A2 Gene

Among the genes identified in the PanScans I and II GWAS, some have known functional significance in regulating biological processes, such as organ development and cell differentiation, cell cycle, and genomic stability, all of which have important roles in tumorigenesis. For example, NR5A2 plays a role in controlling pancreas differentiation during embryonic development and in regulating cholesterol and bile acid homeostasis, steroidogenesis, and cell proliferation (Fayard et al. 2004). A recent study reported a critical role for *NR5A2* in the phosphatidylcholine signaling pathway regulating fatty acid and glucose homeostasis (Lee et al. 2011). *NR5A2* was overexpressed in pancreatic cancer, and its knockdown by small interfering RNA significantly inhibited pancreatic cancer cell proliferation in vitro (Benod et al. 2011), suggesting an oncogenic property of this gene in pancreatic cancer.

BACH1 Gene

BACH1 (BTB and CNC homology 1) is a basic leucine zipper transcription factor and an *Nrf2* target gene. Induction of *BACH1* by *Nrf2* serves as a feedback-inhibitory mechanism for antioxidant-response-element–mediated gene regulation (Jyrkkanen et al. 2011). BACH1 effects DNA helicase activities and physically interacts with BRCA1 and MLH1 (mutL homologue 1), which differentially control DNA double-stranded break repair processes. Because *BRCA1* and *BACH1* mutations targeting the BRCA1-BACH1 interaction have been associated with breast cancer susceptibility, *BACH1* has been suggested as a tumor suppresser gene (Dohrn et al. 2012).

DAB2 Gene

DAB2 encodes a mitogen-responsive phosphoprotein. This protein binds to the SH3 domains of GRB2, an adaptor protein that couples tyrosine kinase receptors to SOS (a guanine nucleotide exchange factor for Ras). Thus, this protein may modulate growth factor/Ras signaling pathways by competing with SOS for binding to GRB2 (Wang et al. 2002). Knockdown of *DAB2* in human mammary epithelial cells leads to increased Ras/MAPK signaling and promotes epithelial-to-mesenchymal transition (Martin et al. 2010).

TTF1 Gene

TFF1 (trefoil factor 1) is a stable secretory protein expressed in gastrointestinal mucosa. The function of this gene is ill defined, but it may protect the mucosa from insults, stabilize the mucus layer, and affect healing of the epithelium. Overexpression of *TFF1* has been reported in many types of human cancers and preneoplastic lesions. In one study, recombinant *TFF1* stimulated the motility of both human pancreatic cancer cells and human pancreas stellate cells in vitro, and overexpression of *TFF1* in pancreatic cancer cells greatly increased metastasis in vivo (Arumugam et al. 2011). Loss of TFF1 is associated with activation of nuclear factor κB–mediated inflammation and gastric neoplasia in mice and humans (Soutto et al. 2011).

DPP6 and *PRLHR* Genes

The functional significance of *DPP6* (dipeptidyl-peptidase 6) and *PRLHR* (prolactin-releasing hormone receptor and their potential roles in the development of pancreatic cancer are intriguing. *DPP6* encodes a single-pass type II membrane protein that is a member of the S9B family in clan SC of the serine proteases. This protein has no detectable protease activity but binds specific voltage-gated potassium channels and alters their expression and biophysical properties. Genetic variation in *DPP6* has been associated with susceptibility to amyotrophic lateral sclerosis (van Es et al. 2008). PRLHR is a seven-transmembrane domain receptor for prolactin-releasing hormone and is a G protein-coupled receptor. Physical activity and a genetic variant of PRLHR have been associated with hypertension (Franks et al. 2004; Bhattacharyya et al. 2003).

Other Genes

Still other genes have been identified by GWAS for pancreatic cancer. *FOXQ1* is a member of the *FOX* gene family, which are involved in embryonic development, cell cycle regulation, tissue-specific gene expression, cell signaling, and tumorigenesis (Bieller et al. 2001). Recent studies showed that FOXQ1 regulates epithelial cell differentiation and epithelial–mesenchymal transition in human cancers (Qiao et al. 2011; Zhang et al. 2011b; Feuerborn et al. 2011). TERT1 plays an essential role in maintaining telomere length and preventing fusion of chromosome ends. In addition to its role in regulating G protein signaling and internalization (Swift et al. 2010), BICD1 has been associated with telomere length (Mangino et al. 2008). *BICD1* gene variants have been associated with risk of aggressive but not indolent prostate cancer (Xu et al. 2010). The *FAM19A5* gene codes for a small secreted protein. These proteins contain conserved cysteine residues at fixed positions and are distantly related to MIP-1α, a member of the CC-chemokine family (Tom Tang et al. 2004). TAFA proteins are predominantly expressed in specific regions of the brain and are postulated to function as brain-specific chemokines or neurokines that regulate immune and nervous cells.

Summary

Overall, genes identified by GWAS seem to have diverse functions that might contribute to cancer development. Fine mapping to identify the responsible variants and mechanistic studies on the biological and functional significance of these genes in relation to pancreatic cancer are required before the value of these GWAS findings can be appreciated.

Post-GWAS Data Analysis

Candidate Pathway Analysis

Single-locus analysis of GWAS data may miss some markers and genes that are related to a phenotype but do not pass the stringent statistical threshold. Furthermore, most genes work as a network or via a signaling transduction pathway; thus, moderate changes in the expression or function of genes involved in the same biological pathways may alter phenotypic outcomes. To further explore other genetic susceptibility factors in pancreatic cancer, a pathway-based analysis was conducted using PanScan data (Li et al. 2012). A total of 577 genes belong to 23 pathways or groups of genes known or hypothesized to be important in pancreatic carcinogenesis were analyzed using the adaptive rank truncated product method and the logic regression method.

Among the pathways, the pancreatic development pathway showed the most statistically significant association with risk of pancreatic cancer ($P=2.0\times10^{-6}$) (Li et al. 2012). The major contributing genes to this pathway included *NR5A2*, *HNF1A*, *HNF4G*, *PDX1*, and *HNF1B*. These genes are important components of the transcriptional networks that govern embryonic pancreatic development and differentiation and maintain pancreatic homeostasis in adults (Maestro et al. 2007; Martin et al. 2007). Mutations in *HNF1A*, *PDX1*, and *HNF1B* are responsible for maturity-onset diabetes of young (MODY) types 3, 4, and 5, respectively (Glucksmann et al. 1997; Carette et al. 2007). Mutations in and common variants of *HNF1A* and *HNF1B* have also been associated with risk of type 2 diabetes (Voight et al. 2010; Furuta et al. 2002; Holmkvist et al. 2006). Notably, *HNF1A* was the top hit for pancreatic cancer in a separate analysis of PanScan data as identified by assessing markers previously identified in a GWAS of phenotypes other than pancreatic cancer (Pierce and Ahsan 2011). *HNF1A* gene mutations have been reported for several types of human cancer, suggesting a role for them in tumor suppression (Laurent-Puig et al. 2003; Rebouissou et al. 2004; Bluteau et al. 2002).

Agnostic Pathway Analysis

The association between the pancreas development pathway and the risk of pancreatic cancer was confirmed in an agnostic pathway analysis of PanScan data (Wei et al. 2012). In this study, a total of 197 biological pathways identified from the Kyoto

Encyclopedia of Genes and Genomes (KEGG) database were analyzed using the gene set ridge regression in association studies algorithm and the logistic kernel machine test. Two pathways were significantly associated with risk of pancreatic cancer after adjusting for multiple comparisons ($P < 0.00025$) and in replication testing: neuroactive ligand-receptor interaction, ($Ps < 0.00002$), and the olfactory transduction pathway ($P = 0.0001$). Functional enrichment analysis using the Database for Annotation, Visualization, and Integrated Discovery (DAVID) consistently found the G protein–coupled receptor signaling pathway to be the most significant pathway for pancreatic cancer in this study population. These findings need to be confirmed in separate datasets from future GWAS of pancreatic cancer. If confirmed, these novel findings will provide new perspectives on genetic susceptibility to and molecular mechanisms of pancreatic cancer.

Candidate Gene Analysis

Because obesity and diabetes are known modifiable risk factors for pancreatic cancer, there is great interest in identifying genetic factors that modify these associations. One study examined 47 genetic variants that have previously been related to type 2 diabetes, fasting glucose, or β-cell function in PanScan I data. None of the genes showed association with pancreatic cancer at the genome-wide significance level. Four genes, *FTO*, *MTNB1R*, *BCL11A* and *MADD*, were nominally associated with pancreatic cancer risk (Pierce et al. 2011).

Gene–Environment Interaction

Most human cancers are likely the consequence of joint actions of genetic and environmental factors. Identification of the interplay of gene and environment will help in understanding the biological networks underlying the complex disease risks. Yet, few studies have incorporated the known environmental or host risk factors in the analyses of GWAS data. A case–control study of 1,070 patients with pancreatic adenocarcinoma and 1,175 controls confirmed the association between *NR5A2* and risk of pancreatic cancer as observed in a GWAS (Tang et al. 2011). However, no significant interaction of *NR5A2* with BMI, diabetes, or smoking was detected. Two *FTO* gene variants were non-significantly associated with a decreased risk of pancreatic cancer in participants with a BMI < 25 kg/m^2 ($P_{interaction} = 0.0001$) but significantly associated with an increased risk of pancreatic cancer in participants with a BMI ≥ 25 kg/m^2 ($P_{interaction} = 0.0015$) (Tang et al. 2011).

Survival Analysis

Although many previous candidate gene studies have reported associations of gene variants with patient survival, no significant findings on SNPs and survival have yet been discovered from existing GWAS data. A study of 690 cases of pancreatic ductal adenocarcinoma and 1,277 healthy control subjects of German and British extraction replicated the associations of GWAS top hits with pancreatic cancer risk reported in PanScan. The *NR5A2* rs12029406_T allele and a SNP located at gene desert region of chromosome 15q14 were weakly associated with overall survival in the German population (Rizzato et al. 2011). Nevertheless, an exploratory GWAS of 550,000 SNPs conducted with 351 patients with pancreatic cancer (294 genetically European patients) identified a nonsynonymous SNP in interleukin (IL)-17F (rs763780, H161R) and an intronic SNP in strong linkage disequilibrium (rs7771466) in association with overall survival at the genome-wide significance level ($P \leq 1 \times 10^{-7}$) (Innocenti et al. 2012). The variant 161R form of IL-17F is a natural antagonist of the antiangiogenic effects of wild-type 161H IL-17F, and patients with the variant allele had significantly shorter median survival (3.1 months; 95 % CI, 2.3–4.3) than patients without this variant (6.8 months; 95 % CI, 5.8–7.3) ($P = 2.61 \times 10^{-8}$).

Summary

As observed for many complex human diseases, the identified gene variants from GWAS explain only a small proportion of the heritability of pancreatic cancer. The unexplained heritability could be due partly to gene–environment interactions or to more complex pathways involving multiple genes and exposures. Using novel statistical strategies to further mine GWAS data for gene–gene and gene–environment interactions may reveal additional gene traits that are missed in single-locus analyses (Wolpin et al. 2010b; Weinberg et al. 2011). In addition, GWAS coverage focused on SNPs with minor alleles of frequency >5 % and tagging SNPs without known functional significance may contribute to the low discovery rate. As technology advances, more coverage of rare SNPs and special selection of exome SNPs may generate more helpful information in defining the genetic susceptibility factors for pancreatic cancer. The ultimate success of using these genetic markers in risk assessment and in clinical management of the disease will also heavily depend on the understanding of the mechanistic links between the genes and the disease.

Conclusion

The field of genetic epidemiology of pancreatic cancer has made notable progress in the past 20 years. Accumulating evidence support a polygenic feature of the disease and a contributing role of common low penetrance gene variants in the development of pancreatic cancer. However, many challenges and inconsistent findings remain. Upon the establishment of consortia and completion of additional large scale GWA studies in the near future, more genetic traits are expected to be identified. The large amount of GWAS data and exposure information will be valuable in examining gene–gene and gene–environment interactions. Findings from these studies will be utilized in establishing and improving the risk prediction models for pancreatic cancer.

Functional characterization of the implicated genes should help to better define the molecular mechanisms underlie the complex etiology of this deadly disease and offer new opportunities in developing novel preventive and treatment strategies.

References

Amundadottir L, Kraft P, Stolzenberg-Solomon RZ, Fuchs CS, Petersen GM, Arslan AA et al (2009) Genome-wide association study identifies variants in the ABO locus associated with susceptibility to pancreatic cancer. Nat Genet 41:986–990

Arumugam T, Brandt W, Ramachandran V, Moore TT, Wang H, May FE et al (2011) Trefoil factor 1 stimulates both pancreatic cancer and stellate cells and increases metastasis. Pancreas 40:815–822

Asomaning K, Reid AE, Zhou W, Heist RS, Zhai R, Su L et al (2008) MDM2 promoter polymorphism and pancreatic cancer risk and prognosis. ClinCancer Res 14:4010–4015

Ayaz L, Ercan B, Dirlik M, Atik U, Tamer L (2008) The association between N-acetyltransferase 2 gene polymorphisms and pancreatic cancer. Cell BiochemFunct 26:329–333

Barbalic M, Dupuis J, Dehghan A, Bis JC, Hoogeveen RC, Schnabel RB et al (2010) Large-scale genomic studies reveal central role of ABO in sP-selectin and sICAM-1 levels. Hum Mol Genet 19:1863–1872

Bartsch H, Malaveille C, Lowenfels AB, Maisonneuve P, Hautefeuille A, Boyle P (1998) Genetic polymorphism of N-acetyltransferases, glutathione S-transferase M1 and NAD(P)H:quinone oxidoreductase in relation to malignant and benign pancreatic disease risk. The International Pancreatic Disease Study Group. Eur J Cancer Prev 7:215–223

Benod C, Vinogradova MV, Jouravel N, Kim GE, Fletterick RJ, Sablin EP (2011) Nuclear receptor liver receptor homologue 1 (LRH-1) regulates pancreatic cancer cell growth and proliferation. Proc Natl Acad Sci USA 108:16927–16931

Bhattacharyya S, Luan J, Challis B, Schmitz C, Clarkson P, Franks PW et al (2003) Association of polymorphisms in GPR10, the gene encoding the prolactin-releasing peptide receptor with blood pressure, but not obesity, in a U.K. Caucasian population. Diabetes 52:1296–1299

Bieller A, Pasche B, Frank S, Glaser B, Kunz J, Witt K et al (2001) Isolation and characterization of the human forkhead gene FOXQ1. DNA Cell Biol 20:555–561

Bluteau O, Jeannot E, Bioulac-Sage P, Marques JM, Blanc JF, Bui H et al (2002) Bi-allelic inactivation of TCF1 in hepatic adenomas. Nat Genet 32:312–315

Carette C, Vaury C, Barthelemy A, Clauin S, Grunfeld JP, Timsit J et al (2007) Exonic duplication of the hepatocyte nuclear factor-1beta gene (transcription factor 2, hepatic) as a cause of maturity onset diabetes of the young type 5. J Clin Endocrinol Metab 92:2844–2847

Chen J, Li D, Wei C, Sen S, Killary AM, Amos CI et al (2007) Aurora-A and p16 polymorphisms contribute to an earlier age at diagnosis of pancreatic cancer in Caucasians. Clin Cancer Res 13:3100–3104

Chen J, Killary AM, Sen S, Amos CI, Evans DB, Abbruzzese JL et al (2008) Polymorphisms of p21 and p27 jointly contribute to an earlier age at diagnosis of pancreatic cancer. Cancer Lett 272:32–39

Chen J, Amos CI, Merriman KW, Wei Q, Sen S, Killary AM et al (2010) Genetic variants of p21 and p27 and pancreatic cancer risk in non-Hispanic Whites: a case-control study. Pancreas 39:1–4

Chen J, Wu X, Pande M, Amos CI, Killary AM, Sen S et al (2011) Susceptibility locus for lung cancer at 15q25.1 is not associated with risk of pancreatic cancer. Pancreas 40:872–875

Couch FJ, Wang X, McWilliams RR, Bamlet WR, de AM, Petersen GM (2009) Association of breast cancer susceptibility variants with risk of pancreatic cancer. Cancer Epidemiol Biomarkers Prev 18:3044–3048

Couch FJ, Wang X, Bamlet WR, de Andrade M, Petersen GM, McWilliams RR (2010) Association of mitotic regulation pathway polymorphisms with pancreatic cancer risk and outcome. Cancer Epidemiol Biomarkers Prev 19:251–257

Dohrn L, Salles D, Siehler SY, Kaufmann J, Wiesmuller L (2012) BRCA1-mediated repression of mutagenic end-joining of DNA double-strand breaks requires complex formation with BACH1. Biochem J 441:919–926

Dong M, Ma G, Tu W, Guo KJ, Tian YL, Dong YT (2005) Clinicopathological significance of p53 and mdm2 protein expression in human pancreatic cancer. World J Gastroenterol 11: 2162–2165

Dong X, Javle M, Hess KR, Shroff R, Abbruzzese JL, Li D (2010) Insulin-like growth factor axis gene polymorphisms and clinical outcomes in pancreatic cancer. Gastroenterology 139: 464–473, 73e1–3

Dong X, Li Y, Chang P, Hess KR, Abbruzzese JL, Li D (2012) DNA mismatch repair network gene polymorphism as a susceptibility factor for pancreatic cancer. Mol Carcinog 51:491–499

Dong X, Li Y, Chang P, Tang H, Hess KR, Abbruzzese JL et al (2011b) Glucose metabolism gene variants modulate the risk of pancreatic cancer. Cancer Prev Res (Phila) 4:758–766

Duell EJ, Holly EA, Bracci PM, Wiencke JK, Kelsey KT (2002a) A population-based study of the Arg399Gln polymorphism in X-ray repair cross-complementing group 1 (XRCC1) and risk of pancreatic adenocarcinoma. Cancer Res 62:4630–4636

Duell EJ, Holly EA, Bracci PM, Liu M, Wiencke JK, Kelsey KT (2002b) A population-based, case-control study of polymorphisms in carcinogen-metabolizing genes, smoking, and pancreatic adenocarcinoma risk. J Natl Cancer Inst 94:297–306

Duell EJ, Casella DP, Burk RD, Kelsey KT, Holly EA (2006) Inflammation, genetic polymorphisms in proinflammatory genes TNF-A, RANTES, and CCR5, and risk of pancreatic adenocarcinoma. Cancer Epidemiol Biomarkers Prev 15:726–731

Duell EJ, Holly EA, Kelsey KT, Bracci PM (2010) Genetic variation in CYP17A1 and pancreatic cancer in a population-based case-control study in the San Francisco Bay Area, California. Int J Cancer 126:790–795

Farrow B, Sugiyama Y, Chen A, Uffort E, Nealon W, Mark EB (2004) Inflammatory mechanisms contributing to pancreatic cancer development. Ann Surg 239:763–769

Fayard E, Auwerx J, Schoonjans K, Fayard E, Auwerx J, Schoonjans K (2004) LRH-1: an orphan nuclear receptor involved in development, metabolism and steroidogenesis. Trends Cell Biol 14:250–260

Feuerborn A, Srivastava PK, Kuffer S, Grandy WA, Sijmonsma TP, Gretz N et al (2011) The Forkhead factor FoxQ1 influences epithelial differentiation. J Cell Physiol 226:710–719

Fong PY, Fesinmeyer MD, White E, Farin FM, Srinouanprachanh S, Afsharinejad Z et al (2010) Association of diabetes susceptibility gene calpain-10 with pancreatic cancer among smokers. J Gastrointest Cancer 41:203–208

Franks PW, Bhattacharyya S, Luan J, Montague C, Brennand J, Challis B et al (2004) Association between physical activity and blood pressure is modified by variants in the G-protein coupled receptor 10. Hypertension 43:224–228

Furuta H, Furuta M, Sanke T, Ekawa K, Hanabusa T, Nishi M et al (2002) Nonsense and missense mutations in the human hepatocyte nuclear factor-1 beta gene (TCF2) and their relation to type 2 diabetes in Japanese. J Clin Endocrinol Metab 87:3859–3863

Gargiulo S, Torrini M, Ollila S, Nasti S, Pastorino L, Cusano R et al (2009) Germline MLH1 and MSH2 mutations in Italian pancreatic cancer patients with suspected Lynch syndrome. Fam Cancer 8:547–553

Glucksmann MA, Lehto M, Tayber O, Scotti S, Berkemeier L, Pulido JC et al (1997) Novel mutations and a mutational hotspot in the MODY3 gene. Diabetes 46:1081–1086

Grochola LF, Muller TH, Bond GL, Taubert H, Udelnow A, Wurl P (2010) MDM2 SNP309 associates with accelerated pancreatic adenocarcinoma formation. Pancreas 39:76–80

Gross M, Kruisselbrink T, Anderson K, Lang N, McGovern P, Delongchamp R et al (1999) Distribution and concordance of N-acetyltransferase genotype and phenotype in an American population. Cancer Epidemiol Biomarkers Prev 8:683–692

Hamacher R, Diersch S, Scheibel M, Eckel F, Mayr M, Rad R et al (2009) Interleukin 1 beta gene promoter SNPs are associated with risk of pancreatic cancer. Cytokine 46:182–186

Harada T, Baril P, Gangeswaran R, Kelly G, Chelala C, Bhakta V et al (2007) Identification of genetic alterations in pancreatic cancer by the combined use of tissue microdissection and array-based comparative genomic hybridisation. Br J Cancer 96:373–382

Holmkvist J, Cervin C, Lyssenko V, Winckler W, Anevski D, Cilio C et al (2006) Common variants in HNF-1 alpha and risk of type 2 diabetes. Diabetologia 49:2882–2891

Huang L, Yu D, Wu C, Zhai K, Jiang G, Cao G et al (2012) Copy number variation at 6q13 functions as a long-range regulator and is associated with pancreatic cancer risk. Carcinogenesis 33:94–100

Innocenti F, Owzar K, Cox NL, Evans P, Kubo M, Zembutsu H et al (2012) A genome-wide association study of overall survival in pancreatic cancer patients treated with gemcitabine in CALGB 80303. Clin Cancer Res 18:577–584

Ionita-Laza I, Rogers AJ, Lange C, Raby BA, Lee C (2009) Genetic association analysis of copy-number variation (CNV) in human disease pathogenesis. Genomics 93:22–26

Jiao L, Bondy ML, Hassan MM, Wolff RA, Evans DB, Abbruzzese JL et al (2006) Selected polymorphisms of DNA repair genes and risk of pancreatic cancer. Cancer Detect Prev 30: 284–291

Jiao L, Doll MA, Hein DW, Bondy ML, Hassan MM, Hixson JE et al (2007a) Haplotype of N-acetyltransferase 1 and 2 and risk of pancreatic cancer. Cancer Epidemiol Biomarkers Prev 16:2379–2386

Jiao L, Bondy ML, Hassan MM, Chang DZ, Abbruzzese JL, Evans DB et al (2007b) Glutathione S-transferase gene polymorphisms and risk and survival of pancreatic cancer. Cancer 109: 840–848

Jiao L, Hassan MM, Bondy ML, Abbruzzese JL, Evans DB, Li D (2007c) The XPD, Asp312Asn and Lys751Gln polymorphisms, corresponding haplotype, and pancreatic cancer risk. Cancer Lett 245:61–68

Jiao L, Hassan MM, Bondy ML, Wolff RA, Evans DB, Abbruzzese JL et al (2008) XRCC2 and XRCC3 gene polymorphism and risk of pancreatic cancer. Am J Gastroenterol 103:360–367

Jiao L, Silverman DT, Schairer C, Thiebaut AC, Hollenbeck AR, Leitzmann MF et al (2009) Alcohol use and risk of pancreatic cancer: the NIH-AARP Diet and Health Study. Am J Epidemiol 169:1043–1051

Jyrkkanen HK, Kuosmanen S, Heinaniemi M, Laitinen H, Kansanen E, Mella-Aho E et al (2011) Novel insights into the regulation of antioxidant-response-element-mediated gene expression by electrophiles: induction of the transcriptional repressor BACH1 by Nrf2. Biochem J 440:167–174

Kanda J, Matsuo K, Suzuki T, Kawase T, Hiraki A, Watanabe M et al (2009) Impact of alcohol consumption with polymorphisms in alcohol-metabolizing enzymes on pancreatic cancer risk in Japanese. Cancer Sci 100:296–302

Krechler T, Jachymova M, Pavlikova M, Vecka M, Zeman M, Krska Z et al (2009) Polymorphism -23HPhI in the promoter of insulin gene and pancreatic cancer: a pilot study. Neoplasma 56:26–32

Kuiper RP, Ligtenberg MJ, Hoogerbrugge N, van Geurts KA (2010) Germline copy number variation and cancer risk. Curr Opin Genet Dev 20:282–289

Lang C, Chen L, Li S (2012) Cytotoxic T-lymphocyte antigen-4 +49G/A polymorphism and susceptibility to pancreatic cancer. DNA Cell Biol 31:683–687

Laurent-Puig P, Plomteux O, Bluteau O, Zinzindohoue F, Jeannot E, Dahan K et al (2003) Frequent mutations of hepatocyte nuclear factor 1 in colorectal cancer with microsatellite instability. Gastroenterology 124:1311–1314

Lee JM, Lee YK, Mamrosh JL, Busby SA, Griffin PR, Pathak MC et al (2011) A nuclear-receptor-dependent phosphatidylcholine pathway with antidiabetic effects. Nature 474:506–510

Lennon AM, Klein AP, Goggins M (2010) ABO blood group and other genetic variants associated with pancreatic cancer. Genome Med 2:39

Li D (2001) Molecular epidemiology of pancreatic cancer. Cancer J 7:259–265

Li D, Firozi PF, Zhang W, Shen J, DiGiovanni J, Lau S et al (2002) DNA adducts, genetic polymorphisms, and K-ras mutation in human pancreatic cancer. Mutat Res 513:37–48

Li D, Ahmed M, Li Y, Jiao L, Chou TH, Wolff RA et al (2005) 5,10-Methylenetetrahydrofolate reductase polymorphisms and the risk of pancreatic cancer. Cancer Epidemiol Biomarkers Prev 14:1470–1476

Li D, Jiao L, Li Y, Doll MA, Hein DW, Bondy ML et al (2006) Polymorphisms of cytochrome P4501A2 and N-acetyltransferase genes, smoking, and risk of pancreatic cancer. Carcinogenesis 27:103–111

Li D, Suzuki H, Liu B, Morris J, Liu J, Okazaki T et al (2009) DNA repair gene polymorphisms and risk of pancreatic cancer. Clin Cancer Res 15:740–746

Li D, Tanaka M, Brunicardi FC, Fisher WE, Gibbs RA, Gingras MC (2011) Association between somatostatin receptor 5 gene polymorphisms and pancreatic cancer risk and survival. Cancer 117:2863–2872

Li D, Duell EJ, Yu K, Risch HA, Olson SH, Kooperberg C et al (2012) Pathway analysis of genome-wide association study data highlights pancreatic development genes as susceptibility factors for pancreatic cancer. Carcinogenesis 33:1384–1390

Lin Y, Yagyu K, Egawa N, Ueno M, Mori M, Nakao H et al (2011) An overview of genetic polymorphisms and pancreatic cancer risk in molecular epidemiologic studies. J Epidemiol 21:2–12

Liu G, Ghadirian P, Vesprini D, Hamel N, Paradis AJ, Lal G et al (2000) Polymorphisms in GSTM1, GSTT1 and CYP1A1 and risk of pancreatic adenocarcinoma. Br J Cancer 82:1646–1649

Low S-K, Kuchiba A, Zembutsu H, Saito A, Takahashi A, Kubo M et al (2010) Genome-wide association study of pancreatic cancer in Japanese population. PLoS One (Electronic Resource) 5:e11824

Lukic S, Nikolic A, Alempijevic T, Popovic D, Sokic MA, Ugljesic M et al (2011) Angiotensin-converting enzyme gene insertion/deletion polymorphism in patients with chronic pancreatitis and pancreatic cancer. Dig Surg 28:258–262

Lynch SM, Weinstein SJ, Virtamo J, Lan Q, Liu CS, Cheng WL et al (2011) Mitochondrial DNA copy number and pancreatic cancer in the alpha-tocopherol beta-carotene cancer prevention study. Cancer Prev Res (Phila) 4:1912–1919

Lyn-Cook BD, Yan-Sanders Y, Moore S, Taylor S, Word B, Hammons GJ (2006) Increased levels of NAD(P)H: quinone oxidoreductase 1 (NQO1) in pancreatic tissues from smokers and pancreatic adenocarcinomas: a potential biomarker of early damage in the pancreas. Cell Biol Toxicol 22:73–80

Maestro MA, Cardalda C, Boj SF, Luco RF, Servitja JM, Ferrer J (2007) Distinct roles of HNF1beta, HNF1alpha, and HNF4alpha in regulating pancreas development, beta-cell function and growth. Endocr Dev 12:33–45

Mangino M, Brouilette S, Braund P, Tirmizi N, Vasa-Nicotera M, Thompson JR et al (2008) A regulatory SNP of the BICD1 gene contributes to telomere length variation in humans. Hum Mol Genet 17:2518–2523

Martin M, Hauer V, Messmer M, Orvain C, Gradwohl G (2007) Transcription factors in pancreatic development. Animal models. Endocr Dev 12:24–32

Martin JC, Herbert BS, Hocevar BA (2010) Disabled-2 downregulation promotes epithelial-to-mesenchymal transition. Br J Cancer 103:1716–1723

Matsubayashi H, Skinner HG, Iacobuzio-Donahue C, Abe T, Sato N, Riall TS et al (2005) Pancreaticobiliary cancers with deficient methylenetetrahydrofolate reductase genotypes. Clin Gastroenterol Hepatol 3:752–760

Mazaki T, Masuda H, Takayama T (2011) Polymorphisms and pancreatic cancer risk: a meta-analysis. Eur J Cancer Prev 20:169–183

McWilliams RR, Bamlet WR, Cunningham JM, Goode EL, de AM, Boardman LA et al (2008) Polymorphisms in DNA repair genes, smoking, and pancreatic adenocarcinoma risk. Cancer Res 68:4928–4935

McWilliams RR, Bamlet WR, de AM, Rider DN, Cunningham JM, Petersen GM (2009a) Nucleotide excision repair pathway polymorphisms and pancreatic cancer risk: evidence for role of MMS19L. Cancer Epidemiol Biomarkers Prev 18:1295–1302

McWilliams RR, Bamlet WR, de Andrade M, Rider DN, Couch FJ, Cunningham JM et al (2009b) Polymorphic variants in hereditary pancreatic cancer genes are not associated with pancreatic cancer risk. Cancer Epidemiol Biomarkers Prev 18:2549–2552

Melzer D, Perry JR, Hernandez D, Corsi AM, Stevens K, Rafferty I et al (2008) A genome-wide association study identifies protein quantitative trait loci (pQTLs). PLoS Genet 4:e1000072

Miyasaka K, Kawanami T, Shimokata H, Ohta S, Funakoshi A (2005) Inactive aldehyde dehydrogenase-2 increased the risk of pancreatic cancer among smokers in a Japanese male population. Pancreas 30:95–98

Miyasaka K, Hosoya H, Tanaka Y, Uegaki S, Kino K, Shimokata H et al (2010) Association of aldehyde dehydrogenase 2 gene polymorphism with pancreatic cancer but not colon cancer. Geriatr Gerontol Int 10(Suppl 1):S120–S126

Mohelnikova-Duchonova B, Vrana D, Holcatova I, Ryska M, Smerhovsky Z, Soucek P (2010) CYP2A13, ADH1B, and ADH1C gene polymorphisms and pancreatic cancer risk. Pancreas 39:144–148

Mohelnikova-Duchonova B, Marsakova L, Vrana D, Holcatova I, Ryska M, Smerhovsky Z et al (2011) Superoxide dismutase and nicotinamide adenine dinucleotide phosphate: quinone oxidoreductase polymorphisms and pancreatic cancer risk. Pancreas 40:72–78

Naccarati A, Pardini B, Polakova V, Smerhovsky Z, Vodickova L, Soucek P et al (2010) Genotype and haplotype analysis of TP53 gene and the risk of pancreatic cancer: an association study in the Czech Republic. Carcinogenesis 31:666–670

Nakao M, Hosono S, Ito H, Watanabe M, Mizuno N, Yatabe Y et al (2011a) Interaction between IGF-1 polymorphisms and overweight for the risk of pancreatic cancer in Japanese. Int J Mol Epidemiol Genet 2:354–366

Nakao M, Matsuo K, Hosono S, Ogata S, Ito H, Watanabe M et al (2011b) ABO blood group alleles and the risk of pancreatic cancer in a Japanese population. Cancer Sci 102:1076–1080

Ockenga J, Vogel A, Teich N, Keim V, Manns MP, Strassburg CP (2003) UDP glucuronosyltransferase (UGT1A7) gene polymorphisms increase the risk of chronic pancreatitis and pancreatic cancer. Gastroenterology 124:1802–1808

Ohnami S, Sato Y, Yoshimura K, Sakamoto H, Aoki K, Ueno H et al (2008) His595Tyr polymorphism in the methionine synthase reductase (MTRR) gene is associated with pancreatic cancer risk. Gastroenterology 135:477–488

Olson SH, Orlow I, Simon J, Tommasi D, Roy P, Bayuga S et al (2007) Allergies, variants in IL-4 and IL-4R alpha genes, and risk of pancreatic cancer. Cancer Detect Prev 31:345–351

Ozhan G, Lochan R, Leathart JB, Charnley R, Daly AK (2011) Cyclooxygenase-2 polymorphisms and pancreatic cancer susceptibility. Pancreas 40:1289–1294

Paterson AD, Lopes-Virella MF, Waggott D, Boright AP, Hosseini SM, Carter RE et al (2009) Genome-wide association identifies the ABO blood group as a major locus associated with serum levels of soluble E-selectin. Arterioscler Thromb Vasc Biol 29:1958–1967

Petersen GM, Boffetta P (2012) Carcinogenesis of pancreatic cancer: challenges, collaborations, progress. Mol Carcinog 51:1–2

Petersen GM, Amundadottir L, Fuchs CS, Kraft P, Stolzenberg-Solomon RZ, Jacobs KB et al (2010) A genome-wide association study identifies pancreatic cancer susceptibility loci on chromosomes 13q22.1, 1q32.1 and 5p15.33. Nat Genet 42:224–228

Piepoli A, Gentile A, Valvano MR, Barana D, Oliani C, Cotugno R et al (2006) Lack of association between UGT1A7, UGT1A9, ARP, SPINK1 and CFTR gene polymorphisms and pancreatic cancer in Italian patients. World J Gastroenterol 12:6343–6348

Pierce BL, Ahsan H (2011) Genome-wide "pleiotropy scan" identifies HNF1A region as a novel pancreatic cancer susceptibility locus. Cancer Res 71:4352–4358

Pierce BL, Austin MA, Ahsan H (2011) Association study of type 2 diabetes genetic susceptibility variants and risk of pancreatic cancer: an analysis of PanScan-I data. Cancer Causes Control 22:877–883

Prizment AE, Gross M, Rasmussen-Torvik L, Peacock JM, Anderson KE (2012) Genes related to diabetes may be associated with pancreatic cancer in a population-based case-control study in Minnesota. Pancreas 41:50–53

Qiao Y, Jiang X, Lee ST, Karuturi RK, Hooi SC, Yu Q (2011) FOXQ1 regulates epithelial-mesenchymal transition in human cancers. Cancer Res 71:3076–3086

Rebouissou S, Rosty C, Lecuru F, Boisselier S, Bui H, Le Frere-Belfa MA et al (2004) Mutation of TCF1 encoding hepatocyte nuclear factor 1alpha in gynecological cancer. Oncogene 23:7588–7592

Reid-Lombardo KM, Fridley BL, Bamlet WR, Cunningham JM, Sarr MG, Petersen GM (2011) Inflammation-related gene variants as risk factors for pancreatic cancer. Cancer Epidemiol Biomarkers Prev 20:1251–1254

Rizzato C, Campa D, Giese N, Werner J, Rachakonda PS, Kumar R et al (2011) Pancreatic cancer susceptibility loci and their role in survival. PLoS One (Electronic Resource) 6:e27921

Scola L, Giacalone A, Marasa L, Mirabile M, Vaccarino L, Forte GI et al (2009) Genetic determined downregulation of both type 1 and type 2 cytokine pathways might be protective against pancreatic cancer. Ann NY Acad Sci 1155:284–288

Sonoyama T, Sakai A, Mita Y, Yasuda Y, Kawamoto H, Yagi T et al (2011) TP53 codon 72 polymorphism is associated with pancreatic cancer risk in males, smokers and drinkers. Mol Med Report 4:489–495

Soutto M, Belkhiri A, Piazuelo MB, Schneider BG, Peng D, Jiang A et al (2011) Loss of TFF1 is associated with activation of NF-kappaB-mediated inflammation and gastric neoplasia in mice and humans. J Clin Invest 121:1753–1767

Sun T, Zhou Y, Yang M, Hu Z, Tan W, Han X et al (2008) Functional genetic variations in cytotoxic T-lymphocyte antigen 4 and susceptibility to multiple types of cancer. Cancer Res 68:7025–7034

Suzuki T, Matsuo K, Sawaki A, Mizuno N, Hiraki A, Kawase T et al (2008a) Alcohol drinking and one-carbon metabolism-related gene polymorphisms on pancreatic cancer risk. Cancer Epidemiol Biomarkers Prev 17:2742–2747

Suzuki H, Morris JS, Li Y, Doll MA, Hein DW, Liu J et al (2008b) Interaction of the cytochrome P4501A2, SULT1A1 and NAT gene polymorphisms with smoking and dietary mutagen intake in modification of the risk of pancreatic cancer. Carcinogenesis 29:1184–1191

Suzuki H, Li Y, Dong X, Hassan MM, Abbruzzese JL, Li D (2008c) Effect of insulin-like growth factor gene polymorphisms alone or in interaction with diabetes on the risk of pancreatic cancer. Cancer EpidemiolBiomarkers Prev 17:3467–3473

Swift S, Xu J, Trivedi V, Austin KM, Tressel SL, Zhang L et al (2010) A novel protease-activated receptor-1 interactor, Bicaudal D1, regulates G protein signaling and internalization. J Biol Chem 285:11402–11410

Talar-Wojnarowska R, Gasiorowska A, Olakowski M, Lampe P, Smolarz B, Romanowicz-Makowska H et al (2011) Role of cyclooxygenase-2 gene polymorphisms in pancreatic carcinogenesis. World J Gastroenterol 17:4113–4117

Tang H, Dong X, Day RS, Hassan MM, Li D (2010) Antioxidant genes, diabetes and dietary antioxidants in association with risk of pancreatic cancer. Carcinogenesis 31:607–613

Tang H, Dong X, Hassan M, Abbruzzese JL, Li D (2011) Body mass index and obesity- and diabetes-associated genotypes and risk for pancreatic cancer. Cancer Epidemiol Biomarkers Prev 20:779–792

Theodoropoulos GE, Michalopoulos NV, Panoussopoulos SG, Taka S, Gazouli M (2010a) Effects of caspase-9 and survivin gene polymorphisms in pancreatic cancer risk and tumor characteristics. Pancreas 39:976–980

Theodoropoulos GE, Panoussopoulos GS, Michalopoulos NV, Zambirinis CP, Taka S, Stamopoulos P et al (2010b) Analysis of the stromal cell-derived factor 1-3'A gene polymorphism in pancreatic cancer. Mol Med Report 3:693–698

Tom Tang Y, Emtage P, Funk WD, Hu T, Arterburn M, Park EE et al (2004) TAFA: a novel secreted family with conserved cysteine residues and restricted expression in the brain. Genomics 83:727–734

van Es MA, van Vught PW, Blauw HM, Franke L, Saris CG, Van den Bosch L et al (2008) Genetic variation in DPP6 is associated with susceptibility to amyotrophic lateral sclerosis. Nat Genet 40:29–31

Verlaan M, Drenth JP, Truninger K, Koudova M, Schulz HU, Bargetzi M et al (2005) Polymorphisms of UDP-glucuronosyltransferase 1A7 are not involved in pancreatic diseases. J Med Genet 42:e62

Voight BF, Scott LJ, Steinthorsdottir V, Morris AP, Dina C, Welch RP et al (2010) Twelve type 2 diabetes susceptibility loci identified through large-scale association analysis. Nat Genet 42:579–589

Vrana D, Pikhart H, Mohelnikova-Duchonova B, Holcatova I, Strnad R, Slamova A et al (2009) The association between glutathione S-transferase gene polymorphisms and pancreatic cancer in a central European Slavonic population. MutatRes 680:78–81

Vrana D, Novotny J, Holcatova I, Hlavata I, Soucek P (2010) CYP1B1 gene polymorphism modifies pancreatic cancer risk but not survival. Neoplasma 57:15–19

Wang Z, Tseng CP, Pong RC, Chen H, McConnell JD, Navone N et al (2002) The mechanism of growth-inhibitory effect of DOC-2/DAB2 in prostate cancer. Characterization of a novel GTPase-activating protein associated with N-terminal domain of DOC-2/DAB2. J Biol Chem 277:12622–12631

Wang L, Miao X, Tan W, Lu X, Zhao P, Zhao X et al (2005) Genetic polymorphisms in methylene-tetrahydrofolate reductase and thymidylate synthase and risk of pancreatic cancer. Clin Gastroenterol Hepatol 3:743–751

Wang L, Bamlet WR, de AM, Boardman LA, Cunningham JM, Thibodeau SN et al (2007) Mitochondrial genetic polymorphisms and pancreatic cancer risk. Cancer Epidemiol Biomarkers Prev 16:1455–1459

Wei P, Tang H, Li D (2012) Insights into pancreatic cancer etiology from pathway analysis of genome-wide association study data. PLoS One7:e46887

Weinberg CR, Shi M, Umbach DM (2011) Re.: Genetic association and gene-environment interaction: a new method for overcoming the lack of exposure information in controls. Am J Epidemiol 173:1346–1347, author reply 7–8

Wheatley-Price P, Asomaning K, Reid A, Zhai R, Su L, Zhou W et al (2008) Myeloperoxidase and superoxide dismutase polymorphisms are associated with an increased risk of developing pancreatic adenocarcinoma. Cancer 112:1037–1042

Wolpin BM, Chan AT, Hartge P, Chanock SJ, Kraft P, Hunter DJ et al (2009) ABO blood group and the risk of pancreatic cancer. J Natl Cancer Inst 101:424–431

Wolpin BM, Kraft P, Gross M, Helzlsouer K, Bueno-de-Mesquita HB, Steplowski E et al (2010a) Pancreatic cancer risk and ABO blood group alleles: results from the pancreatic cancer cohort consortium. Cancer Res 70:1015–1023

Wolpin BM, Kraft P, Xu M, Steplowski E, Olsson ML, Arslan AA et al (2010b) Variant ABO blood group alleles, secretor status, and risk of pancreatic cancer: results from the pancreatic cancer cohort consortium. Cancer Epidemiol Biomarkers Prev 19:3140–3149

Wu C, Miao X, Huang L, Che X, Jiang G, Yu D et al (2012) Genome-wide association study identifies five loci associated with susceptibility to pancreatic cancer in Chinese populations. Nat Genet 44:62–66

Xu J, Zheng SL, Isaacs SD, Wiley KE, Wiklund F, Sun J et al (2010) Inherited genetic variant predisposes to aggressive but not indolent prostate cancer. Proc Natl Acad Sci USA 107: 2136–2140

Yang M, Sun T, Wang L, Yu D, Zhang X, Miao X et al (2008) Functional variants in cell death pathway genes and risk of pancreatic cancer. Clin Cancer Res 14:3230–3236

Yang M, Sun T, Zhou Y, Wang L, Liu L, Zhang X et al (2012) The functional cytotoxic T lymphocyte-associated Protein 4 49G-to-A genetic variant and risk of pancreatic cancer. Cancer 118: 4681–4686

Zhang J, Zhang X, Dhakal IB, Gross MD, Kadlubar FF, Anderson KE (2011a) Sequence variants in antioxidant defense and DNA repair genes, dietary antioxidants, and pancreatic cancer risk. Int J Mol Epidemiol Genet 2:236–244

Zhang H, Meng F, Liu G, Zhang B, Zhu J, Wu F et al (2011b) Forkhead transcription factor foxq1 promotes epithelial-mesenchymal transition and breast cancer metastasis. Cancer Res 71:1292–1301

Zhao D, Xu D, Zhang X, Wang L, Tan W, Guo Y et al (2009) Interaction of cyclooxygenase-2 variants and smoking in pancreatic cancer: a possible role of nucleophosmin. Gastroenterology 136:1659–1668

Translational Implications of Molecular Genetics for Early Diagnosis of Pancreatic Cancer

Michael A. Hollingsworth

Abstract This chapter discusses the potential applications of molecular genetics to the early diagnosis of pancreatic cancer. The current state of the field is discussed in general terms with an emphasis on the limitations of current technologies and strategies, and the potential of molecular genetic diagnostics to impact diagnosis and management of pancreatic cancer in the future.

Molecular Genetics of Pancreatic Cancer

The altered genetic landscape of pancreatic cancer has been characterized over the past 25 years and is discussed in detail in previous review articles and in other chapters of this book. It is increasingly accepted that there is morphological progression of premalignant lesions in the pancreas (Pancreatic Intraepithelial Neoplasia or PanIN lesions, graded as I, II, and III, with the latter representing carcinoma in situ) that results from an accumulation of genetic and epigenetic events (Maitra and Hruban 2008). Earlier studies undertaken using hypothesis driven science, genomic discovery methods, and candidate gene approaches identified four genes that are mutated or modified epigenetically in a large percentage of pancreatic cancers: *KRAS2* (>90 %), *p16/CDKN2A* (>90 %), *TP53* (50–75 %), and *SMAD4* (>50 %) (Iacobuzio-Donahue 2012). A number of other genes that are altered in less than 5 % of cancers have been implicated by other studies, including two whole genome sequencing efforts (International Cancer Genome Consortium 2010; Jones et al. 2008). With the possible exception of detecting inherited genes that predispose to cancer in families that have a history of malignancy, these low frequency alleles are not currently of use for purely early diagnostic purposes (Hruban et al. 2010).

M.A. Hollingsworth (✉)
Eppley Institute for Research in Cancer,
University of Nebraska Medical Center, Omaha, NE, USA
e-mail: mahollin@unmc.edu

D.M. Simeone and A. Maitra (eds.), *Molecular Genetics of Pancreatic Cancer*,
DOI 10.1007/978-1-4614-6549-2_4,

There are currently attempts to classify the low frequency mutations into functional categories that largely fall into biochemical pathways that contribute to malignant progression (Jones et al. 2008). This is highly important to understanding the biology of pancreatic cancer and will undoubtedly contribute the development of new targeted-therapies for the disease. As such diagnostic tests for these may contribute to future clinical decisions regarding therapy; however, direct detection of these mutations are unlikely to be useful for early detection of cancer in the near future.

Though much attention has been paid to whole genome exome (that part of the genome that encodes expressed proteins) sequencing of tumors, this strategy is limited since only 1 % of the human genome encodes expressed proteins. More recent efforts have begun to investigate the role of the remaining 99 % of the genome in cancer progression, including expression and role of multiple types of noncoding RNAs. The analysis of mutations in noncoding RNAs, DNA structural elements, and other features of nucleic acids in tumors is only beginning, and will hopefully provide additional insight and the potential for new molecular diagnostic tools in the future.

Difficulties of Early Detection of Pancreatic Cancer

Early detection for cancer has improved the survival of patients with many types of cancer and is critical for future improvements in effectively treating the disease. The rationale for this is that early detection of cancer allows for cures, usually by surgical resection. Pancreatic cancer, however, presents special challenges. Currently, pancreatic cancer is highly lethal, even when detected in early stages. The 5-year survival rate for surgically resected patients with stage 1 disease (tumors less than 2.5 cm, confined to the pancreas) is less than 30 % (Witkowski et al. 2013). One implication of this fact is that even when there is resection of small tumors with negative margins by pathological examination, some tumor cells have escaped to colonize other organ sites. This undoubtedly results in part from aggressive biological properties of pancreatic cancer. It is proposed by some that pancreatic cancer invades and metastasizes relatively early in disease progression, perhaps even before there is full transformation of the cells (Rhim et al. 2012). There is also evidence that clonal evolution of cancer occurs over a longer period of time. Mathematical modeling of the rate of mutation acquisition (as revealed by exome sequence analysis of spatially distinct and presumably progressive lesions from seven pancreatic cancer cases) suggested that there is as an 11.7-year period of time from an initiating mutation in a pancreatic cell to the acquisition of additional mutations that confer fully transformed growth properties on a cell, and that there is another 6.8 years until the first metastatic subclone is derived. In this model death occurred at about 2.7 years after of the appearance of the putative metastatic subclone (Iacobuzio-Donahue 2012). If this model is correct, it suggests that there is a window of time in which early detection may impact disease outcome.

Nonetheless, at this time, early detection of pancreatic cancer followed by surgical resection will not be curative for most patients. This has led some to suggest that early detection of pancreatic cancer will not be useful for improving survival. Many clinicians believe that surgical resection should be accompanied by neoadjuvant or adjuvant therapy; however, the currently available therapies are also generally not curative. Part of the reason for this deficiency is the conundrum presented by the fact that we cannot identify early cancers at a rate that is sufficient to undertake clinical trials of large numbers of these patients so that we can identify curative therapies. It is anticipated that early detection will provide more opportunities for clinical trials in the future. Moreover, the advent and anticipated improvement of cancer therapies that target molecular defects that arise from genetic mutations, epigenetic alterations, and other factors will provide treatment options in the future. Thus, diagnostic tests of the future should attempt to identify the presence of malignancy and characterize the molecular defects that are responsible for driving the biological properties of each malignancy.

Assays for Early Detection of Pancreatic Cancer

Several factors must be taken into account regarding the development of diagnostic tests for pancreatic cancer. These include the development of accurate and molecularly sensitive tests for appropriate clinical samples that show performance characteristics (diagnostic sensitivity and specificity) that will be helpful in making clinical decisions. The assays must also be economically feasible. A starting point for most diagnostic tests that are currently under development include assay of bloods—serum, plasma, or cellular content. A second source of diagnostic samples includes stool, or samples obtained by endoscopic sampling (pancreatic juice, fine needle aspirate, or biopsy). A third type of diagnostic test would be an imaging test, presumably that included a targeted imaging agent that improved discrimination of malignant cells from benign conditions. One problem for early detection is that of the molecular or cellular sensitivity of the test. Early small lesions are unlikely to produce a sufficient amount of material or cells to be detected when diluted into the large volume of blood that is in circulation. This is further complicated in pancreatic cancer by the fact that up to 90 % of each pancreatic tumor is comprised of a desmoplastic reaction, which does not include tumor cells that are the targets of molecular diagnostic tests.

Regarding assays for blood or other body fluids, there are efforts to develop tests to detect nucleic acids (DNA and RNA) and protein in blood, stool or other clinical samples, and there are assays to capture circulating tumor cells. There are practical positives and negatives for translational application of all of these tests. There is evidence that nucleic acids derived from tumors can be detected in blood, although the molecular forms of these are not well established. Possibilities include free nucleic acids or forms bound to proteins, exosomes, cellular debris, or material carried by immune cells such as macrophages or dendritic cells. One problem encountered in analyzing nucleic acids in blood or other body fluids is the issue of identifying mutated sequences against a background of normal sequence. Newly developed economical technologies such as ICE-COLD PCR that selectively amplify mutated sequences (Milbury et al. 2011) may increase the sensitivity of these assays and enable practical diagnostic tests in the future.

Circulating tumor cells represent an important source of potential diagnostic material. Unfortunately, by definition, the presence of circulating tumor cells implies that the tumor is metastatic, and so it is anticipated that analysis of these cells will not aid early detection, but instead will provide a potential source of a "peripheral biopsy" that will allow for molecular characterization of the genotype and phenotype of the parental tumor (Yu et al. 2012). Whether circulating tumor cells are representative of the parental tumor and all metastatic deposits remains to be determined.

An important component of diagnosis of early stage pancreatic cancer is the analysis of fine needle aspirate (FNA) samples of pancreatic lesions obtained by endoscopic ultrasound. Unfortunately, cytopathological analysis of these samples is often difficult and results in indeterminate findings (Payne et al. 2009). The addition of molecular genetic analysis to these samples should enhance the accuracy of diagnosis, and in the future may aid in directing therapeutic approaches; however, this area of diagnostic endeavor requires further development.

The imaging of pancreatic lesions has improved with the development of pancreas specific CT protocols; however, the detection of small lesions in the pancreas remains a problem that could be improved by the addition of imaging techniques that identify alterations associated with malignancy (Fisher et al. 2008). Efforts to develop agents that target molecules expressed as a consequence of malignant transformation are underway, but are nascent at this point in time (Bausch et al. 2011). Ultimately, early diagnosis of pancreatic cancer will need to include improved imaging techniques.

Molecular Discrimination of Disease by Molecular Genetics

If an acceptable assay is developed that will accurately and sensitively detect mutated, methylated or expressed nucleic acid sequences, it is likely there will be problems with the performance characteristics of these tests with respect to sensitivity and specificity for detecting pancreatic cancer. Consider the genes known to be commonly affected in pancreatic cancer. There are mutations in *KRAS2* in virtually all pancreatic cancers, but similar mutations have been observed in patients with pancreatitis and in normal individuals (Lohr et al. 2000, 2005). In fact, 50 % of the relatively common PanIN 1 lesions are predicted to contain *KRAS2* mutations (Feldmann et al. 2007). This suggests that detection of *KRAS2* mutations alone does not predict the presence of cancer and would lead to numerous false positives. Thus, it would be desirable to add another test to detection of mutations in *KRAS2*. *p16/CDKN2A* is inactivated in almost all pancreatic cancers, and is apparent in PanIn 2 lesions. As such *p16/CDKN2A* would be a good candidate as a marker, except that in 40 % of cases there is homozygous deletion of the allele. Another 40 % have a mutant allele that is accompanied by deletion or methylation of the second allele and 10–15 % show methylation of the promoter. A similar scenario is evident for *SMAD4*, which is lost by homozygous deletion in 30 % of pancreatic cancers and is mutated and inactivated by loss or methylation in 25 % of cases (Iacobuzio-Donahue et al. 2012). Detection of loss of alleles is feasible if not optimal for tissue biopsies, but assaying for loss of alleles is not practical as a test for circulating DNA derived from cancer. Inactivation of *TP53* may be more promising as a test, as mutant alleles can be found in up to 75 % of pancreatic cancers and this is usually accompanied by loss of the second allele directly or through methylation. Mutations in *TP53* are found in PanIN 3 lesions, which is an appropriate diagnostic target. Thus, one early detection strategy that should be evaluated carefully would be to detect the presence of mutations in both *KRAS2* and *TP53* in blood products (plasma or cells). One limitation of this approach that has prevented a comprehensive analysis to date is the fact that there are numerous sites of mutation in TP53, which complicates the development of economical detection assays for many clinical samples.

Besides detecting mutated genes directly, it should be possible to develop diagnostic tests to detect consequences of mutated genotypes. This is being evaluated by detecting mutated proteins directly, or by detecting alterations in expression of cellular products (or biochemical pathways) whose levels or features (e.g., posttranslational processing) are altered as a consequence of the mutated genome. There are a number of efforts to develop biomarkers based on this approach; however, none have achieved performance characteristics that warrant deployment as an early diagnostic test to date.

Another possibility for early detection that is currently being explored is analysis of the specificity of autoantibodies that are produced in cancer patients (Raedle et al. 1996). The rationale is that many individuals develop autoantibodies to mutated proteins, overexpressed proteins, or altered forms of proteins that arise during malignant progression of cancer. An advantage of this strategy, at least in theory, is that it may allow for detection of alterations that occur in very early lesions. For example, a small focus of malignant cells may not make a sufficient level of any compound to be detected in blood or body fluids. However, the development of antibodies to proteins in those cells would be amplified and detectable as stable compounds in serum.

An increasing clinical dilemma with respect to diagnosis of pancreatic cancer is the increased incidental findings of pancreatic cystic lesions that result from increased use of imaging techniques (CT and MRI). It is often difficult to discriminate malignant precursors (intraductal papillary mucinous neoplasms, mucinous cystadenomas) from other benign cystic lesions. Molecular diagnostic strategies that could evaluate FNA samples or other tissue samples from these lesions or blood based assays of that would discriminate those lesions with aggressive biological properties would impact disease management. Initial exome sequencing studies of several cystic lesions has begun to reveal candidate mutations that should be further explored (Wu et al. 2011).

Summary and Future Directions

Molecular genetic studies have provided a great deal of insight into the biochemical underpinnings of pancreatic cancer; however, this characterization is far from over, as only a fraction of the genome has been analyzed and the results obtained so far have not yet translated into improved diagnostics or therapeutics. The early detection of pancreatic cancer remains possible if the premalignant lesions are present for several years; however, the small size and relatively inaccessible location of these lesions present a significant barrier to many diagnostic modalities. It would be desirable to develop economical screening assays that detect the presence of mutations and could be deployed annually in at risk or aging populations. An example of this would be tests that detect mutated genes, their products, or autoantibodies to mutated gene products or products that are altered or uniquely expressed as consequences of the acquisition of mutations. It is unlikely that a defined set of mutations will be solely diagnostic for pancreatic cancer, but these tests may indicate the presence of premalignant or malignant cells in certain organ types. It should be possible to develop secondary screening protocols (such as cancer-specific imaging) to detect the locations of the neoplasms. Prohibitive costs associated with endoscopic and imaging studies will preclude their use as early diagnostic tools, though these modalities should be used and enhanced as secondary screening protocols. Molecular diagnostic tests should be developed to aid in detecting malignancies in the small quantities of clinical samples obtained by FNA or biopsy. Ultimately, molecular characterization of tumors should be used to direct appropriate targeted therapies at the time of diagnosis, or to direct prevention strategies aimed at blocking the progression of premalignant lesions.

References

Bausch D, Thomas S, Mino-Kenudson M, Fernandez-del CC, Bauer TW, Williams M, Warshaw AL, Thayer SP, Kelly KA (2011) Plectin-1 as a novel biomarker for pancreatic cancer. Clin Cancer Res 17(2):302–309

Feldmann G, Beaty R, Hruban RH, Maitra A (2007) Molecular genetics of pancreatic intraepithelial neoplasia. J Hepatobiliary Pancreat Surg 14(3):224–232

Fisher WE, Hodges SE, Yagnik V, Moron FE, WU MF, Hilsenbeck SG, Raijman IL, Brunicardi FC (2008) Accuracy of CT in predicting malignant potential of cystic pancreatic neoplasms. HPB (Oxford) 10(6):483–490

Hruban RH, Canto MI, Goggins M, Schulick R, Klein AP (2010) Update on familial pancreatic cancer. Adv Surg 44:293–311

Iacobuzio-Donahue CA (2012) Genetic evolution of pancreatic cancer: lessons learnt from the pancreatic cancer genome sequencing project. Gut 61(7):1085–1094

Iacobuzio-Donahue CA, Velculescu VE, Wolfgang CL, Hruban RH (2012) Genetic basis of pancreas cancer development and progression: insights from whole-exome and whole-genome sequencing. Clin Cancer Res 18(16):4257–4265

International Cancer Genome Consortium, Hudson TJ, Anderson W, Artez A, Barker AD, Bell C, Bernabe RR, Bhan MK, Calvo F, Eerola I, Gerhard DS, Guttmacher A, Guyer M, Hemsley FM,

Jennings JL, Kerr D, Klatt P, Kolar P, Kusada J, Lane DP, Laplace F, Youyong L, Nettekoven G, Ozenberger B, Peterson J, Rao TS, Remacle J, Schafer AJ, Shibata T, Stratton MR, Vockley JG, Watanabe K, Yang H, Yuen MM, Knoppers BM, Bobrow M, Cambon-Thomsen A, Dressler LG, Dyke SO, Joly Y, Kato K, Kennedy KL, Nicolas P, Parker MJ, Rial-Sebbag E, Romeo-Casabona CM, Shaw KM, Wallace S, Wiesner GL, Zeps N, Lichter P, Biankin AV, Chabannon C, Chin L, Clement B, de Alava E, Degos F, Ferguson ML, Geary P, Hayes DN, Hudson TJ, Johns AL, Kasprzyk A, Nakagawa H, Penny R, Piris MA, Sarin R, Scarpa A, Shibata T, Van De Vijver M, Futreal PA, Aburatani H, Bayes M, Botwell DD, Campbell PJ, Estivill X, Gerhard DS, Grimmond SM, Gut I, Hirst M, Lopez-Otin C, Majumder P, Marra M, McPherson JD, Nakagawa H, Ning Z, Puente XS, Ruan Y, Shibata T, Stratton MR, Stunnenberg HG, Swerdlow H, Velculescu VE, Wilson RK, Xue HH, Yang L, Spellman PT, Bader GD, Boutros PC, Campbell PJ, Flicek P, Getz G, Guigo R, Guo G, Haussler D, Heath S, Hubbard TJ, Jiang T, Jones SM, Li Q, Lopez-Bigas N, Luo R, Muthuswamy L, Ouellette BF, Pearson JV, Puente XS, Quesada V, Raphael BJ, Sander C, Shibata T, Speed TP, Stein LD, Stuart JM, Teague JW, Totoki Y, Tsunoda T, Valencia A, Wheeler DA, Wu H, Zhao S, Zhou G, Stein LD, Guigo R, Hubbard TJ, Joly Y, Jones SM, Kasprzyk A, Lathrop M, Lopez-Bigas N, Ouellette BF, Spellman PT, Teague JW, Thomas G, Valencia A, Yoshida T, Kennedy KL, Axton M, Dyke SO, Futreal PA, Gerhard DS, Gunter C, Guyer M, Hudson TJ, McPherson JD, Miller LJ, Ozenberger B, Shaw KM, Kasprzyk A, Stein LD, Zhang J, Haider SA, Wang J, Yung CK, Cros A, Liang Y, Gnaneshan S, Guberman J, Hsu J, Bobrow M, Chalmers DR, Hasel KW, Joly Y, Kaan TS, Kennedy KL, Knoppers BM, Lowrance WW, Masui T, Nicolas P, Rial-Sebbag E, Rodriguez LL, Vergely C, Yoshida T, Grimmond SM, Biankin AV, Bowtell DD, Cloonan N, Defazio A, Eshleman JR, Etemadmoghadam D, Gardiner BB, Kench JG, Scarpa A, Sutherland RL, Tempero MA, Waddell NJ, Wilson PJ, McPherson JD, Gallinger S, Tsao MS, Shaw PA, Petersen GM, Mukhopadhyay D, Chin L, Depinho RA, Thayer S, Muthuswamy L, Shazand K, Beck T, Sam M, Timms L, Ballin V, Lu Y, Ji J, Zhang X, Chen F, Hu X, Zhou G, Yang Q, Tian G, Zhang L, Xing X, Li X, Zhu Z, Yu Y, Yu J, Yang H, Lathrop M, Tost J, Brennan P, Holcatova I, Zaridze D, Brazma A, Egevard L, Prokhortchouk E, Banks RE, Uhlen M, Cambon-Thomsen A, Viksna J, Ponten F, Skryabin K, Stratton MR, Futreal PA, Birney E, Borg A, Borresen-Dale AL, Caldas C, Foekens JA, Martin S, Reis-Filho JS, Richardson AL, Sotiriou C, Stunnenberg HG, Thoms G, van de Vijver M, van't Veer L, Calvo F, Birnbaum D, Blanche H, Boucher P, Boyault S, Chabannon C, Gut I, Masson-Jacquemier JD, Lathrop M, Pauporte I, Pivot X, Vincent-Salomon A, Tabone E, Theillet C, Thomas G, Tost J, Treilleux I, Calvo F, Bioulac-Sage P, Clement B, Decaens T, Degos F, Franco D, Gut I, Gut M, Heath S, Lathrop M, Samuel D, Thomas G, Zucman-Rossi J, Lichter P, Eils R, Brors B, Korbel JO, Korshunov A, Landgraf P, Lehrach H, Pfister S, Radlwimmer B, Reifenberger G, Taylor MD, von Kalle C, Majumder PP, Sarin R, Rao TS, Bhan MK, Scarpa A, Pederzoli P, Lawlor RA, Delledonne M, Bardelli A, Biankin AV, Grimmond SM, Gress T, Klimstra D, Zamboni G, Shibata T, Nakamura Y, Nakagawa H, Kusada J, Tsunoda T, Miyano S, Aburatani H, Kato K, Fujimoto A, Yoshida T, Campo E, Lopez-Otin C, Estivill X, Guigo R, de Sanjose S, Piris MA, Montserrat E, Gonzalez-Diaz M, Puente XS, Jares P, Valencia A, Himmelbauer H, Quesada V, Bea S, Stratton MR, Futreal PA, Campbell PJ, Vincent-Salomon A, Richardson AL, Reis-Filho JS, van de Vijver M, Thomas G, Masson-Jacquemier JD, Aparicio S, Borg A, Borresen-Dale AL, Caldas C, Foekens JA, Stunnenberg HG, van't Veer L, Easton DF, Spellman PT, Martin S, Barker AD, Chin L, Collins FS, Compton CC, Ferguson ML, Gerhard DS, Getz G, Gunter C, Guttmacher A, Guyer M, Hayes DN, Lander ES, Ozenberger B, Penny R, Peterson J, Sander C, Shaw KM, Speed TP, Spellman PT, Vockley JG, Wheeler DA, Wilson RK, Hudson TJ, Chin L, Knoppers BM, Lander ES, Lichter P, Stein LD, Stratton MR, Anderson W, Barker AD, Bell C, Bobrow M, Burke W, Collins FS, Compton CC, Depinho RA, Easton DF, Futreal PA, Gerhard DS, Green AR, Guyer M, Hamilton SR, Hubbard TJ, Kallioniemi OP, Kennedy KL, Ley TJ, Liu ET, Lu Y, Majumder P, Marra M, Ozenberger B, Peterson J, Schafer AJ, Spellman PT, Stunnenberg HG, Wainwright BJ, Wilson RK, Yang H (2010) International network of cancer genome projects. Nature 464(7291): 993–998

Jones S, Zhang X, Parsons DW, Lin JC, Leary RJ, Angenendt P, Mankoo P, Carter H, Kamiyama H, Jimeno A, Hong SM, Fu B, Lin MT, Calhoun ES, Kamiyama M, Walter K, Nikolskaya T, Nikolsky Y, Hartigan J, Smith DR, Hidalgo M, Leach SD, Klein AP, Jaffee EM, Goggins M, Maitra A, Iacobuzio-Donahue C, Eshleman JR, Kern SE, Hruban RH, Karchin R, Papadopoulos N, Parmigiani G, Vogelstein B, Velculescu VE, Kinzler KW (2008) Core signaling pathways in human pancreatic cancers revealed by global genomic analyses. Science 321(5897):1801–1806

Lohr M, Maisonneuve P, Lowenfels AB (2000) K-Ras mutations and benign pancreatic disease. Int J Pancreatol 27(2):93–103

Lohr M, Kloppel G, Maisonneuve P, Lowenfels AB, Luttges J (2005) Frequency of K-ras mutations in pancreatic intraductal neoplasias associated with pancreatic ductal adenocarcinoma and chronic pancreatitis: a meta-analysis. Neoplasia 7(1):17–23

Maitra A, Hruban RH (2008) Pancreatic cancer. Annu Rev Pathol 3:157–188

Milbury CA, Li J, Makrigiorgos GM (2011) Ice-COLD-PCR enables rapid amplification and robust enrichment for low-abundance unknown DNA mutations. Nucleic Acids Res 39(1):e2

Payne M, Staerkel G, Gong Y (2009) Indeterminate diagnosis in fine-needle aspiration of the pancreas: reasons and clinical implications. Diagn Cytopathol 37(1):21–29

Raedle J, Oremek G, Welker M, Roth WK, Caspary WF, Zeuzem S (1996) P53 autoantibodies in patients with pancreatitis and pancreatic carcinoma. Pancreas 13(3):241–246

Rhim AD, Mirek ET, Aiello NM, Maitra A, Bailey JM, McAllister F, Reichert M, Beatty GL, Rustgi AK, Vonderheide RH, Leach SD, Stanger BZ (2012) EMT and dissemination precede pancreatic tumor formation. Cell 148(1–2):349–361

Witkowski ER, Smith JK, Tseng JF (2013) Outcomes following resection of pancreatic cancer. J Surg Oncol 107(1):97–103

Wu J, Jiao Y, Dal Molin M, Maitra A, de Wilde RF, Wood LD, Eshleman JR, Goggins MG, Wolfgang CL, Canto MI, Schulick RD, Edil BH, Choti MA, Adsay V, Klimstra DS, Offerhaus GJ, Klein AP, Kopelovich L, Carter H, Karchin R, Allen PJ, Schmidt CM, Naito Y, Diaz LA Jr, Kinzler KW, Papadopoulos N, Hruban RH, Vogelstein B (2011) Whole-exome sequencing of neoplastic cysts of the pancreas reveals recurrent mutations in components of ubiquitin-dependent pathways. Proc Natl Acad Sci USA 108(52):21188–21193

Yu M, Ting DT, Stott SL, Wittner BS, Ozsolak F, Paul S, Ciciliano JC, Smas ME, Winokur D, Gilman AJ, Ulman MJ, Xega K, Contino G, Alagesan B, Brannigan BW, Milos PM, Ryan DP, Sequist LV, Bardeesy N, Ramaswamy S, Toner M, Maheswaran S, Haber DA (2012) RNA sequencing of pancreatic circulating tumour cells implicates WNT signalling in metastasis. Nature 487(7408):510–513

The Biology of K-Ras Signaling Pathways in Pancreatic Cancer

Helen Court, Mark R. Philips, and Dafna Bar-Sagi

Abstract Activating mutations in the K-Ras oncogene occur in approximately 90 % of cases of pancreatic ductal adenocarcinoma, and tumors containing mutant K-Ras often acquire a dependency on the expression of the oncogene. Therapies that block the oncogenic functions of K-Ras could have clinical efficacy for a disease that is currently refractory to all forms of treatment. This chapter describes the evidence, from both *in vitro* studies and studies using genetic mouse models, of the importance of oncogenic K-Ras and its downstream signaling pathways in driving pancreatic tumor formation and cancer cell growth.

Introduction

The decades-old observation that pancreatic ductal adenocarcinoma (PDA) is almost always associated with an activating mutation in the *KRAS* gene has focused attention on this oncogene as a key therapeutic target for this lethal disease. Many tumor cells containing *KRAS* mutations are considered to be K-Ras "addicted," meaning that they depend on the oncogene in order to survive. Therapies that block K-Ras

H. Court • M.R. Philips
Department of Medicine, NYU School of Medicine, 550 First Avenue, New York, NY, USA

Department of Cell Biology, NYU School of Medicine, 550 First Avenue, New York, NY, USA

Department of Biochemistry and Molecular Pharmacology, NYU School of Medicine, 550 First Avenue, New York, NY, USA

The Cancer Institute, NYU School of Medicine, 550 First Avenue, New York, NY, USA

D. Bar-Sagi (✉)
Department of Biochemistry and Molecular Pharmacology, NYU School of Medicine, 550 First Avenue, New York, NY, USA
e-mail: dafna.bar-sagi@nyumc.org

D.M. Simeone and A. Maitra (eds.), *Molecular Genetics of Pancreatic Cancer*, DOI 10.1007/978-1-4614-6549-2_5, signaling could therefore potentially benefit almost all patients with PDA. Thirty years of study of the cell biology of K-Ras has yielded a wealth of information but an effective treatment for PDA is still elusive. This chapter describes how oncogenic K-Ras signaling is involved in almost every aspect of the initiation and progression of PDA from precursor lesions to metastatic disease. Oncogenic K-Ras induces numerous alterations that drive normal pancreatic cells to become invasive cancer cells and this provides many opportunities for therapeutic intervention. Understanding as much as possible about K-Ras signaling should allow us to determine where such intervention would prove the most effective in treating this disease.

The K-Ras Oncogene

Ras oncogenes were first identified in the 1970s as the sequences responsible for the transforming properties of the Harvey (Ha-MSV) and Kirsten (Ki-MSV) rodent tumor viruses. It was discovered that these recombinant retroviruses contained DNA sequences derived from the rat genome that encoded the *Hras* (Ha-MSV) and *Kras* (Ki-MSV) genes. Later *HRAS* became the first oncogene isolated from human cancer cells by its ability to transform NIH3T3 mouse fibroblasts (Shih and Weinberg 1982; Goldfarb et al. 1982; Pulciani et al. 1982). The transforming sequences from human tumor cells were found to be homologs of the viral *v-h-ras* and *v-k-ras* genes (Parada et al. 1982; Der et al. 1982; Santos et al. 1982).

Molecular cloning and sequencing showed that the oncogenes derived from tumor cells and the normal cellular *HRAS* and *KRAS* genes differed by only a single point mutation, most commonly in codons 12, 13, or 61 (Reddy et al. 1982; Tabin et al. 1982; Taparowsky et al. 1982; Santos et al. 1984). This information was used to screen a wide variety of different human tumors for the presence of oncogenic *RAS* mutations (Bos 1989). It is estimated that up to 30 % of human tumors contain *RAS* mutations making it the most frequently mutated oncogene in human cancer (Barbacid 1987).

There are three main isoforms of Ras proteins that are highly homologous. In addition to H-Ras and K-Ras, the third isoform N-Ras was identified after the other two isoforms as the transforming gene present in a neuroblastoma cell line (Shimizu et al. 1983a, b; Hall et al. 1983). K-Ras in fact exists as two alternatively spliced isoforms, K-Ras4A and K-Ras4B, which differ only in the sequence encoded by the fourth exon. K-Ras4B is considered to be the more relevant isoform to human cancer due to its more ubiquitous expression in tissues in both mice and humans (Pells et al. 1997; Plowman et al. 2006a), and due to the fact that a *KRAS* knockout mouse has an embryonic lethal phenotype (Johnson et al. 1997; Koera et al. 1997) whereas a targeted knockout of exon 4A has no phenotype (Plowman et al. 2003). However, because the oncogenic mutations in *KRAS* occur in the shared first and second exons, these mutations results in the production of oncogenic versions of both splice variants. Data regarding the importance of the K-Ras4A isoform *in vivo* is somewhat contradictory and more research is required to determine what role this isoform may play in oncogenesis (Patek et al. 2008a, b; To et al. 2008; Plowman et al. 2006b; Abubaker et al. 2009).

The Ras proteins are prototypical small GTPases (Scolnick et al. 1979; Shih et al. 1980; Tamanoi et al. 1984; Temeles et al. 1985) that act as molecular switches cycling between an active GTP-bound and an inactive GDP-bound state (Field et al. 1987; Satoh et al. 1987). When GTP-bound, Ras can interact with downstream effectors involved in numerous cellular pathways that control cell growth, differentiation, and survival. The GTP/GDP cycle is controlled by guanine nucleotide exchange factors (GEFs) that activate Ras by promoting the release of GDP allowing the more abundant GTP to bind (Wolfman and Macara 1990), and GTPase activating proteins (GAPs) that dramatically accelerate the intrinsic rate of GTP hydrolysis, thereby inactivating Ras and curtailing signaling (Trahey and McCormick 1987). Oncogenic mutations in the Ras protein render it locked constitutively in the active GTP-bound state by reducing the intrinsic GTP hydrolysis rate and rendering the protein insensitive to the action of GAPs (McGrath et al. 1984; Sweet et al. 1984; Gibbs et al. 1984; Manne et al. 1985; Trahey and McCormick 1987).

A defining feature of Ras proteins is that they are peripheral membrane proteins that associate with cellular membranes by virtue of a series of posttranslational modifications (Wright and Philips 2006). The extreme C terminus of Ras ends with a "CaaX motif" in which C is a cysteine, "a" is generally an aliphatic residue and X is one of a number of amino acids (Fu and Casey 1999). This CaaX motif renders Ras a substrate for modification by farnesyltransferase, which catalyzes the addition of a 15-carbon farnesyl lipid to the cysteine of the CaaX motif (Schafer et al. 1989, 1990). Subsequently the aaX amino acids following the farnesylcysteine are cleaved off by a protease, Ras converting enzyme 1 (Rce1) (Boyartchuk et al. 1997; Freije et al. 1999; Otto et al. 1999). The α-carboxyl-group on the farnesylcysteine is then methylated by isoprenylcysteine carboxyl methyltransferase (Icmt) (Clarke et al. 1988; Gutierrez et al. 1989; Hrycyna et al. 1991; Pillinger et al. 1994). This methyl esterification neutralizes the negative charge of the carboxyl group and is therefore thought to increase the affinity of the farnesylcysteine for the plasma membrane by reducing the repulsion of the carboxyl group by the negatively charged head groups of the inner leaflet of the phospholipid bilayer (Hancock et al. 1991). Correct membrane association has been shown to be essential for both the biological and oncogenic functions of Ras proteins (Hancock et al. 1989; Gutierrez et al. 1989; Willumsen et al. 1984). Therefore, disrupting the addition of the modifications that enable Ras to associate with membranes has been seen as an attractive way to inhibit the function of oncogenic Ras in cancer (Downward 2003).

Oncogenic K-Ras Effector Pathways and Pancreatic Cancer

The exchange of the nucleotide bound to Ras from GDP to GTP results in a conformational change in the Ras protein that affects the affinity of binding to effector molecules (Ito et al. 1997; Geyer et al. 1996). Conformational changes in Ras occur in two areas of the protein within the highly conserved GTPase domain termed switch I and switch II (Milburn et al. 1990). A Ras effector is defined as a protein that preferentially binds to the GTP-bound form of Ras. Effectors interact with Ras via a Ras-binding domain (RBD). While no sequence homology exists between RBDs from different effectors, they all share an ubiquitin superfold topology (ββαββαβ) (Nassar et al. 1995; Geyer et al. 1997; Walker et al. 1999). Oncogenic mutations in Ras, such as the substitution of valine or aspartic acid for glycine at codon 12 (G12V or G12D), render the protein constitutively GTP-bound because residues with a side chain in this position sterically interfere with the geometry of the transition state of GTP hydrolysis in the presence of GAPs (Scheffzek et al. 1997; Krengel et al. 1990; Tong et al. 1991). Mutation of the glutamine at position 61 is also oncogenic because this residue forms a hydrogen bond with the arginine at position 789 in GAP p120 (Scheffzek et al. 1997) and positions a catalytic water molecule for nucleophilic attack on the γ-phosphate of GTP (Buhrman et al. 2010; Scheidig et al. 1999), which is essential for GTP hydrolysis. These mutations therefore enable Ras to constitutively interact and activate downstream effectors. Thus, the oncogenic nature of Ras results from its ability to promote unchecked signaling down a variety of pathways that induce cell growth, proliferation, and survival.

The importance of Ras signaling in pancreatic cancer is highlighted by the fact that mutations in K-Ras are found extremely frequently in patient tumors. Early analysis of tumors revealed a prevalence of oncogenic mutations of K-Ras in pancreatic ductal adenocarcinoma (PDA) in excess of 90 % (Almoguera et al. 1988). However, recent evidence from analysis of the catalogue of somatic mutations in cancer (COSMIC) database (Forbes et al. 2011) suggests that the percentage of mutations in pancreatic cancer is 60 % (Prior et al. 2012). K-Ras mutations have been found to be present in early PanIN lesions and in surrounding areas of acinar-ductal metaplasia (ADM) (Shi et al. 2009; Kanda et al. 2012) consistent with the hypothesis that this mutation is an initiating event in PanIN formation. PDA is believed to originate from somatic mutations in *KRAS* during adulthood rather than during embryonic development. Indeed, although germ line mutations in *HRAS* and *KRAS* as well as other components of the downstream MAPK cascade have been found to be responsible for Noonan, LEOPARD, cardio-facio-cutaneous and Costello syndromes, that share similar features including facial abnormalities, heart defects, impaired growth and development, and, in some cases, cancer predisposition (Schubbert et al. 2007a, b), none of these syndromes appear to predispose to the development of PDA.

While the requirement for K-Ras signaling in pancreatic cancer is clear, what is not fully understood is what effector pathways downstream of Ras are necessary and sufficient to transmit its oncogenic signals. There are at least ten distinct functional classes of putative Ras effectors (Fig. 1) (Repasky et al. 2004). Raf-1 kinase was the first Ras effector to be discovered and remains the best characterized (Moodie et al. 1993; Warne et al. 1993; Zhang et al. 1993; Vojtek et al. 1993). The canonical pathway of Raf-1 activation occurs downstream of receptor tyrosine kinases (RTKs), such as the epidermal growth factor receptor (EGFR). When growth factors (such as EGF) bind to their cognate RTK, this induces dimerization

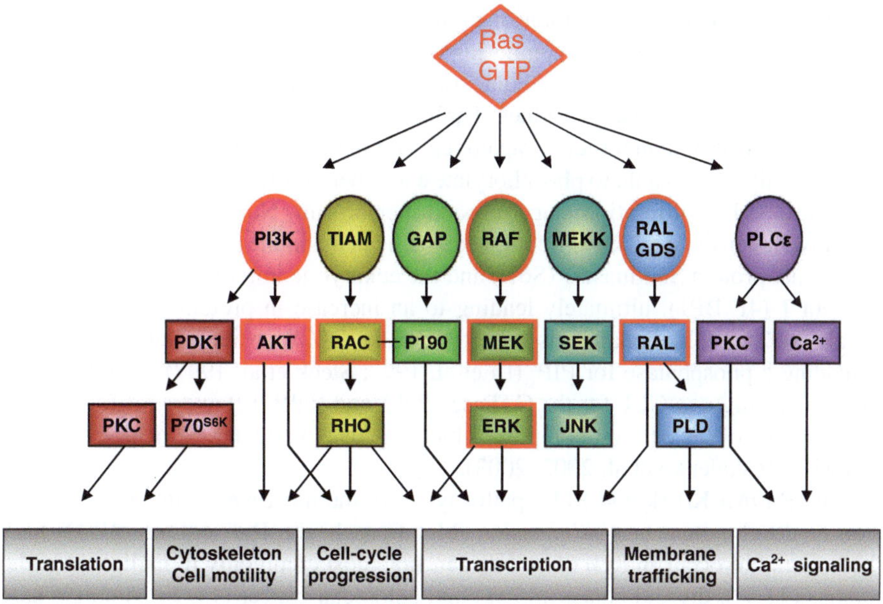

Fig. 1 Effector pathways downstream of oncogenic Ras stimulate many cellular processes. The signaling pathways shown have known or speculated roles in oncogenesis. Outlined in *red* are pathways involved in pancreatic cancer

and cross-phosphorylation of tyrosine residues in the cytosolic domain of the RTK (Schreiber et al. 1983; Ushiro and Cohen 1980; Yarden and Schlessinger 1987a, b; Zhang et al. 2006). The SH2 domain of the adapter protein Grb2 then binds to the phosphotyrosine residues in the RTK, and Grb2 in turn recruits the Ras GEF SOS to the plasma membrane via an SH3 domain in the Grb2 protein (Buday 1999). This recruitment enables SOS to interact with and activate Ras on the plasma membrane (Boriack-Sjodin et al. 1998). Ras-GTP is then able to bind and activate the effector Raf-1 by a mechanism that is not yet completely understood (Marais et al. 1995). Downstream of Raf-1 is the mitogen activated protein kinase (MAPK) cascade that includes MEK (MAPK/Erk kinase), Erk-1 and Erk-2. The Erk proteins are serine/threonine kinases with a variety of different substrates. Once phosphorylated, the Erk proteins form dimers that translocate into the nucleus where their substrates include proteins in the Ets family of transcription factors.

In addition to Raf-1, two other well-characterized effectors of Ras are phosphatidylinositol 3-kinase (PI3K) (Rodriguez-Viciana et al. 1994) and a group of exchange factors for the small GTPase Ral which includes RalGDS. PI3Ks are lipid kinases that phosphorylate the 3′ hydroxyl group of the inositol ring of phosphatidylinositol phosphates. Class 1A PI3Ks are activated downstream of RTKs and function primarily to generate the lipid second messenger phosphatidylinositol-3,4,5-trisphosphate (PIP_3) by phosphorylating phosphatidylinositol-4,5-bisphosphate (PIP_2) in the plasma membrane. The presence of PIP_3 at the plasma membrane results in the recruitment and activation of proteins containing plekstrin homology (PH) domains including serine/threonine kinases of the Akt family (Akt1, Akt2, and Akt3) and Pdk1 kinase (3-phosphoinositide-dependent kinase). Akt is activated by phosphorylation of two key residues, Thr308 by Pdk1 and Ser473 by the rapamycin-insensitive mammalian target of rapamycin complex 2 (mTORC2) (Sarbassov et al. 2005). Active Akt is able to phosphorylate a number of different downstream targets to control cell proliferation, survival, and metabolism. Notably, Akt activates the rapamycin-sensitive mTORC1 complex which results in the phosphorylation of p70 ribosomal protein S6 kinase 1 (S6K) and the eukaryotic initiation factor 4E binding protein 1 (4E-BP1), ultimately leading to an increase in protein synthesis (Inoki et al. 2002). PI3K signaling is antagonized by *PTEN*, a tumor suppressor gene encoding a phosphatase for PIP_3 (Li et al. 1997; Steck et al. 1997). RalGDS functions by acting as a GEF for the GTPases RalA and RalB. Effectors for Ral include components of the exocyst complex, which regulates vesicular trafficking and exocytosis (Moskalenko et al. 2002, 2003).

The Raf/MEK/Erk and PI3K pathways have the most well established roles in cancer development and progression. Mutations in the Raf isoform BRAF have been found to occur in 8 % of human cancers, most commonly in malignant melanomas (41 %), thyroid cancer (45 %) and colorectal cancer (14 %). A single base missense mutation that results in the replacement of valine for glutamic acid at codon 600 (V600E, previously described as V599E (Kumar et al. 2003)) in the activation segment of the kinase domain is responsible for at least 80 % of the BRAF mutations found in human cancer (Davies et al. 2002). The kinase activity of this mutant is greatly elevated; it is able to potently transform NIH3T3 cells and constitutively stimulates Erk activity *in vivo* independent of RAS. Gain-of-function mutations in the catalytic subunit of PI3K p110 (PI3KCA) also occur frequently in cancer. These mutations increase enzymatic function, enhance downstream signaling elements and promote oncogenic transformation (Kang et al. 2005; Samuels et al. 2005). However, mutations in effectors downstream of Ras are infrequent in pancreatic cancer, presumably because the pathways are sufficiently activated through oncogenic Ras signaling. *BRAF* mutations are rare in pancreatic cancer (Jones et al. 2008). They have been reported to occur in tumors that also had a K-Ras mutation with a frequency of around 10 % (Ishimura et al. 2003). Mutations in *PI3KCA* have been found to occur in 9 % of patients with PDA (Janku et al. 2011). Amplifications and overexpression of *AKT2* were found in 10-20 % of pancreatic cancer cell lines and tumors (Cheng et al. 1996; Ruggeri et al. 1998). *EGFR* mutations are also rare, occurring in less than 3 % of patients, but have also been found to coexist with K-Ras mutations (Oliveira-Cunha et al. 2012). Point mutations in the tumor suppressor *PTEN* are infrequently found in pancreatic cancer but functional inactivation of the gene occurs commonly by promoter methylation or inhibition of protein or mRNA synthesis (Ebert et al. 2002; Altomare et al. 2002; Asano et al. 2004).

EGFR genomic amplifications and overexpression are a common event in pancreatic cancer (Tzeng et al. 2007; Bloomston et al. 2006; Tobita et al. 2003; Fjallskog et al. 2003), as is expression of some of its ligands (Kobrin et al. 1994; Zhu et al. 2000). This observation is a little surprising as activating mutations in K-Ras, being downstream, would be expected to a certain degree to circumvent the requirement for EGFR signaling. This appears to be the case in some other tumor types such as non small cell lung cancers where mutation in *KRAS* and *EGFR* are mutually exclusive (Shigematsu et al. 2005). It has been suggested that signaling through EGFR may still be necessary in the presence of oncogenic K-Ras to activate the other isoforms of Ras and also possibly any remaining wild type alleles of K-Ras (Ardito et al. 2012). In contrast, there are also studies that demonstrate a selective loss of the wild type allele of K-Ras in human tumors. Mutant allele specific imbalance, which can occur by either copy number gains or uniparental disomy, was found in 58 % of tumors including pancreatic cancers (Soh et al. 2009). Similar findings have also been found in mouse models (Qiu et al. 2011) and there is a growing body of evidence that suggests the wild type allele of K-Ras may function as a tumor suppressor (Zhang et al. 2001; Li et al. 2003, 2007; Hegi et al. 1994; Bremner and Balmain 1990).

Despite there being no known activating mutations found in the Ral pathway in cancer, it has been suggested that in human cells the Ral pathway may be the most important pathway downstream of Ras for cellular transformation (Hamad et al. 2002; Rangarajan et al. 2004). RalGDS appears to be required for the survival of Ras transformed cells in a mouse model (Gonzalez-Garcia et al. 2005). The two main substrates of RalGDS appear however to have different roles in oncogenesis. Ectopically expressed RalA is transforming and is required for K-Ras^{G12V} transformation, whereas RalB impedes transformation (Lim et al. 2005). However, RalB was found to be required for invasion and metastasis of two pancreatic cancer cell lines *in vivo* (Lim et al. 2006). In addition to this, RalA was found to be activated in a panel of pancreatic cancer cell lines (Lim et al. 2005) and both RalA and RalB were more frequently activated in pancreatic tumor samples than either Erk or Akt (Lim et al. 2006).

Mouse Models of Oncogenic K-Ras Driven Pancreatic Cancer

The importance of oncogenic K-Ras mutations in pancreatic cancer initiation and maintenance has now been verified with several mouse models. In 2003, David Tuveson utilized a mouse harboring a conditional oncogenic allele of K-Ras^{G12D} under the control of the endogenous K-Ras promoter (Jackson et al. 2001). Expression of the oncogene was blocked by a STOP element flanked by LoxP sites upstream of the gene. Crossing of the *Lox-STOP-Lox-KRAS*^{G12D} mouse (*LSL-KRAS*^{G12D}) to mice containing *Cre* recombinase under the control of pancreas specific promoters (*PDX-1-Cre* and *p48-Cre*) allowed for recombination of the STOP element and expression of the oncogene in a pancreas-specific manner (Hingorani et al. 2003). This was the first example of the expression of oncogenic K-Ras from its endogenous locus in a mouse model of pancreatic cancer. These animals showed a phenotype that recapitulated the progression of human pancreatic ductal adenocarcinoma from early stage PanIN lesions to invasive metastatic disease. This result was important because it helped to confirm the PanIN progression model that had been put forward from examination of human specimens (Brat et al. 1998; Hruban et al. 1999; Maitra et al. 2003). In addition, the result was groundbreaking because previous attempts to develop mouse models that targeted K-Ras^{G12D} to the pancreas with a variety of different pancreas-specific promoters (e.g., cytokeratin-19, Elastase, Mist1) had failed to produce pancreatic lesions that resembled those seen in human PDA (Brembeck et al. 2003; Grippo et al. 2003; Tuveson et al. 2006).

The *LSL-KRAS^{G12D};PDX-1-Cre/p48-Cre* mouse models have been subsequently combined with a variety of different floxed, loss of function and dominant negative alleles of tumor supressors (Hingorani et al. 2005; Izeradjene et al. 2007; Bardeesy et al. 2006; Aguirre et al. 2003; Vincent et al. 2009). The rapidly accelerated disease progression in these models helps confirm the hypothesis that tumor suppressor genes such as p53, p16^{INK4A}, and Smad4 help keep oncogenic K-Ras-driven neoplasia in check.

These studies provided compelling evidence that K-Ras^{G12D} is required for PanIN formation; however, the requirement for PanIN progression and PDA maintenance had not been tested. To address this question, a mouse was created that contained an oncogenic allele of K-Ras that could be turned on or off by the administration or removal of doxycycline in the drinking water of adult mice (*p48-Cre;R26-rtTa-IRES-EGFP;TetO-Kras^{G12D}*, referred to as iKras) (Collins et al. 2012). Removal of doxycycline from these animals after 23 weeks of K-Ras^{G12D} expression resulted in an almost complete reversion of PanINs after 2 weeks and a regeneration of the acinar cell compartment. Similar results were also observed when K-Ras^{G12D} was expressed for 3 weeks with concomitant cerulein treatment to induce pancreatitis (see next section). PanIN reversion was associated with a down-regulation in phospho-Erk1/2 levels. Surprisingly, however, switching off oncogenic K-Ras expression did not cause an increase in apoptotic cells as shown by staining for cleaved caspase-3. Instead, loss of PanIN and acinar regeneration appeared to occur by a process of ductal-acinar metaplasia (DAM), as cells co-expressing the acinar cell marker amylase and the ductal maker cytokeratin-19 were frequently observed. However, if K-Ras^{G12D} expression was induced for 5 weeks with concomitant cerulein treatment, while removal of doxycycline resulted in PanIN regression there was an incomplete regeneration of the acinar cell compartment leaving a small fibrotic pancreas with fewer acini than expected. In these pancreata there was a dramatic increase in apoptotic cells upon doxycycline removal suggesting that either the regenerative capability of the pancreas decreases with the age of the mice or that more advanced stage PanIN lesions are not able to undergo DAM. Importantly, iKras mice crossed with p53 null mice produced disease that progressed to PDA and doxycycline removal resulted in complete regression of all tumors (Collins et al. 2012; Ying et al. 2012).

In some of these models, such as the *LSL-KRAS^{G12D};p48-Cre* model (Hingorani et al. 2003), K-Ras^{G12D} expression occurs in every cell of the pancreas raising the question of why some cells undergo neoplastic transformation while other cells remain normal. This observation led to speculation as to what is the precise cell of origin of the PanINs observed. Although PanINs have an obvious ductal morphology it is possible that they arise from another cell type by a process of transdifferentiation.

One study sought to address this question by targeting K-Ras^{G12D} expression to different cell types in the adult pancreas using *Cre* drivers with different expression patterns (Gidekel Friedlander et al. 2009). Expression in *Pdx1*$^+$ cells, which includes adult endocrine β cells, some ductal cells, acinar cells, and possibly adult progenitor/stem cells induced transformation resulting in PanIN formation. However, *proCPA1*$^+$ cells were not efficiently transformed by K-Ras^{G12D}. *ProCPA1* encodes for the pancreas specific pro-carboxypeptidase A expressed mostly in acinar cells and possibly some centroacinar cells. The same result was observed for *insulin*$^+$ cells. These results suggest that a *Pdx1*$^+$ cell is the most likely cell of origin for PDA. However, *insulin*$^+$ cells of the endocrine lineage of the adult pancreas were able transdifferentiate and give rise to PDA under certain conditions, highlighting the plasticity of the pancreas and complicating the question of the cell of origin in human PDA.

The difference in the efficiency of transformation of different cell types in the pancreas could occur because the threshold of Ras signaling required to transform is higher in some cells relative to others. One study showed that expression of a K-Ras^{G12D} transgene in adult acinar cells at higher levels than from the endogenous promoter was sufficient to induce PanINs that progressed to PDA whereas endogenous levels of expression was not. This study found higher levels of active Ras in pancreatic tumor samples than in untransformed areas of pancreas expressing K-Ras^{G12D} from the endogenous promoter suggesting that upregulation of Ras activity is necessary to bypass a transformation barrier in the pancreas (Ji et al. 2009). However, interpretation of these results is hindered by the in vitro assay used to determine the amount of active Ras that may not fully reflect the level of Ras signaling in intact cells. Two recent studies highlighting the importance of EGFR in the development of K-Ras driven pancreatic cancer lend some credence to this hypothesis (Navas et al. 2012; Ardito et al. 2012). EGFR was found to be required for pancreatitis-dependent acinar cell-derived tumorigenesis and ADM following cerulein treatment both *in vivo* and *in vitro*. One of these studies implicated Erk activation downstream of EGFR signaling in this process, implying that the signaling downstream of K-Ras^{G12D} alone was insufficient to transform cells whereas in combination with signaling through EGFR, a critical threshold could be reached to promote neoplasia (Ardito et al. 2012). However, a second study instead implicated signaling through Akt and Stat3 downstream of EGFR (Navas et al. 2012). Both studies agreed that mutations in p53 bypassed the requirement for EGFR signaling in tumor development, which may explain why the EGFR inhibitor erlotinib has shown poor efficacy when combined with gemcitabine in clinical trials (Moore et al. 2007). A third study showed that concomitant expression of TGFα, a ligand for EGFR, and K-Ras^{G12D} accelerates the progression of PanIN lesions in a *p48-Cre;LSL-KRAS*^{G12D} mouse model (Siveke et al. 2007), suggesting that signaling through EGFR in combination with oncogenic K-Ras signaling may indeed help to bypass a transformation barrier in the pancreas.

Mouse models have also been used to address the question of which pathways downstream of Ras are the most important for malignancy. Upregulation of nuclear phospho-Erk (pErk) staining downstream of K-Ras^{G12D} expression is an early feature of mouse PanIN lesions, whereas normal pancreatic tissue is negative for pErk staining (Ijichi et al. 2006; Guerra et al. 2007). However, paradoxically some cell lines and tumor samples have low levels of pErk due to a negative feedback mechanism involving MAPK phosphatase 2 (Yip-Schneider et al. 1999, 2001). Activation of Akt has been found in up to 59 % of tumor samples (Altomare et al. 2002; Schlieman et al. 2003; Yamamoto et al. 2004). High levels of both pErk and phospho-Akt (pAkt) have been associated with reduced survival in patients following surgical resection (Chadha et al. 2006). Recently it has been shown that expression of $BRAF^{V600E}$, but not $PI3KCA^{H1047R}$, in the adult mouse pancreas can induce PanIN formation (Collisson et al. 2012), and when combined with gain-of-function $p53^{R270H}$ the PanINs progress to PDA. However, a pancreatic specific deletion of *PTEN* during embryogenesis in mice did result in the formation of some PanINs and papillary ductal adenocarcinomas in a subset of animals (Stanger et al. 2005), and was able to synergize with K-Ras^{G12D} to accelerate the development of PDA (Hill et al. 2010). Rac1 is another small GTPase that is activated downstream of oncogenic Ras, either via PI3K signaling or via the Ras effector Tiam, and is a key component in the reorganization of the actin cytoskeleton induced by Ras oncogenes (Bar-Sagi and Feramisco 1986; Ridley et al. 1992; Qiu et al. 1995; Nimnual et al. 1998; Rodriguez-Viciana et al. 1997; Lambert et al. 2002). Active Rac1 functions to induce actin polymerization, and its overexpression has been detected in human patient samples of pancreatic cancer (Crnogorac-Jurcevic et al. 2001). Rac1 has long been found to be to be required for Ras transformation and recently conditional loss of *Rac1* in the pancreas was found to impair PanIN formation, early metaplastic changes and neoplasia-associated actin rearrangements in the *LSL-KRAS*^{G12D}*;p48-Cre* mouse model (Heid et al. 2011). It was suggested that Rac1 may be required for F-actin rearrangements that take place during the ADM that precedes PanIN formation in this mouse model (Bi et al. 2005), and the PanINs that form in the absence of Rac1 may develop from an alternative cell type that does not require ADM (Heid et al. 2011).

Oncogenic K-Ras and Pancreatitis

Chronic pancreatitis is a significant risk factor for PDA in humans (Lowenfels et al. 1999), which suggests that inflammation plays a role in the progression of the disease. Mouse models have been used to show that inflammation can act synergistically with oncogenic K-Ras^{G12D} in driving carcinogenesis. Cerulein is an analog of cholecystokinin which, when administered to rodents in supraphysiologic doses, stimulates the premature intracellular activation of pancreatic digestive enzymes, which causes tissue damage resulting in pancreatitis (Lampel and Kern 1977; Watanabe et al. 1984; Ohshio et al. 1989; Silverman et al. 1989; Niederau et al. 1985). Cerulein-induced acute pancreatitis is a well-studied animal model that has been used to examine the effect of acute pancreatitis on PanIN progression in the *LSL-KRAS*^{G12D}*;PDX-1-Cre* mouse model (Carriere et al. 2009). Two brief episodes of acute pancreatitis were sufficient to accelerate pancreatic cancer development.

Thus, a brief inflammatory insult to the pancreas, when occurring in the context of oncogenic K-Ras^{G12D}, can enhance pancreatic malignant transformation.

Interestingly it has been shown that turning on K-Ras^{G12D} expression in adult pancreatic cells of mice or rats fails to induce the development of PanINs or PDA without concomitant or previous treatment with cerulein to induce pancreatitis (Guerra et al. 2007, 2011; Tanaka et al. 2010; Habbe et al. 2008; De La et al. 2008), whereas K-Ras^{G12D} expression during embryogenesis or early adulthood alone is sufficient to induce PanINs that are able to progress to PDA (Guerra et al. 2007). Thus, these studies in mouse models suggest that adult cells of the exocrine pancreas may be refractory to transformation by oncogenic K-Ras and that pancreatitis produces a permissive environment that enhances transformation.

Although the molecular mechanism underlying the cooperation between oncogenic K-Ras and pancreatitis remains to be established, one hypothesis is that pancreatic injury may induce a trans-differentiation or de-differentiation of cells to a less mature differentiated state similar to an embryonic progenitor cell that is more permissive to transformation. Cerulein treatment strongly induces ADM in the regenerating pancreas and could represent such a trans-differentiation event (Willemer et al. 1987). Pancreatitis and pancreatic regeneration have been found to induce expression of genes normally associated with undifferentiated pancreatic progenitor cells such as Sox9, Pdx1, E-cadherin, β-catenin, Notch components and Hedgehog components (Jensen et al. 2005; Fendrich et al. 2008; Sharma et al. 1999; Yoshida et al. 2008; Siveke et al. 2008). However, in a wild type pancreas, this response and the ADM observed is transient and the acinar cells rapidly regenerate. Somehow oncogenic K-Ras signaling seems to alter the fate of the regenerating cells so that they form PanINs instead of acini. Consistent with this, many of these pathways associated with the progenitor cell population remain active in PanINs and PDA including Sox9 (Prevot et al. 2012) and Notch (Miyamoto et al. 2003; Hingorani et al. 2003). Despite overwhelming evidence that oncogenic K-Ras signaling and inflammation synergize to promote pancreatic cancer development, there is some controversy regarding the contribution that cellular senescence plays in this process. It has been suggested that K-Ras^{G12D} expression in early PanINs either promotes oncogene-induced senescence that can be relieved by limited episodes of pancreatitis (Guerra et al. 2011), or inhibits senescence induced in normal ductal cells by pancreatitis (Lee and Bar-Sagi 2010). It remains to be seen what is the reason for these differences but it is possible that the age of animals used or the stage of PanINs observed could account for such discrepancies.

Oncogenic K-Ras and Developmental Reprogramming

It is not uncommon for tumors to display a reactivation of embryonic signaling pathways that are essential for development, such as the Notch, Hedgehog, and Wnt pathways. Indeed, pancreatic cancer exhibits several examples of this. Upregulated expression of Notch receptors and ligands has been observed in human pancreatic cancer samples as has expression of the Notch target gene Hes1, which is usually restricted to centroacinar cells in the normal pancreas (Miyamoto et al. 2003). Aberrant cytoplasmic and nuclear expression of β-Catenin has been observed in human PanIN and PDA (Al-Aynati et al. 2004; Lowy et al. 2003), and canonical Wnt signaling has been found to be active in pancreatic cancer cell lines (Pasca di Magliano et al. 2007). Additionally, sonic hedgehog is abnormally expressed in pancreatic adenocarcinoma and PanINs (Thayer et al. 2003). The functional relationships between oncogenic K-Ras and these pathways have therefore been a subject of great interest. Activation of the Notch pathway by expression of the Notch1 intracellular domain (NICD) in adult acinar cells has also been found to synergize with oncogenic K-Ras expression in the pancreas to accelerate PanIN progression (De La et al. 2008). In contrast, another study suggested that Notch1 functions as a tumor suppressor in the mouse pancreas (Hanlon et al. 2010). One explanation for these differing results could be due to the difference in timing of the Notch activation and loss in these models being either in adulthood or during embryonic development. These pathways are extremely complex and changes in the specific roles or activity level of individual components or alterations in the balance of activity of components could have unpredictable effects. Notch signaling inhibits progenitor cell differentiation in the embryonic pancreas (Hald et al. 2003), so reactivation of Notch signaling may function to induce a more embryonic-like state in the pancreas that can synergize with K-Ras to enhance transformation. However, it is as yet unclear the precise role Notch signaling plays in pancreatic cancer development and progression, be it oncogenic or tumor suppressive. Another developmentally important pathway that is reactivated in pancreatic cancer is the Wnt pathway. Despite this, stabilized β-catenin was found to impair $\text{K-Ras}^{\text{G12D}}$ induced PanIN development following cerulein-induced pancreatitis in mice. In contrast β-catenin signaling was found to be important for acinar cell regeneration following cerulein-induced pancreatitis: a *p48-Cre;*β*-catenin*$^{flx/flx}$ mouse was found to have a significant decrease in the acinar cell area 3 and 5 days following cerulein treatment (Morris et al. 2010). This suggests that oncogenic K-Ras signaling may function to suppress a β-catenin-driven acinar cell regeneration program in favor of neoplastic transformation and PanIN formation and emphasizes how important the timing of pathway activation may be. Hedgehog ligands secreted from pancreatic cancer cells seem to have an important role in paracrine signaling to the adjacent stroma (Tian et al. 2009). Autocrine signaling which occurs via secreted sonic hedgehog binding to the 12 trans-membrane domain receptor Patched (Ptch), resulting in the activation of the Smoothened (Smo) seven trans-membrane domain protein, does not appear to be required for PDA development in mice. Despite this, expression of the downstream target Gli1 is required for survival of mouse and human pancreatic cancer cell lines (Nolan-Stevaux et al. 2009). In contrast to the stroma, Gli expression in mouse PDA cells may depend on K-Ras signaling in a Smo independent manner, as depleting 80 % of K-Ras expression with *Kras*-targeted siRNAs resulted in a significant down-regulation of the Gli1 and Ptch1 mRNAs in PDA lines.

Oncogenic K-Ras and the Tumor Microenvironment

The microenvironment surrounding tumor cells consists of other cell types, soluble factors, signaling molecules, extracellular matrix, and mechanical cues (Swartz et al. 2012). It is becoming increasingly apparent how specific interactions with the microenvironment affect all aspects of tumor biology. In pancreatic cancer there is increasing evidence that the inflammatory response to tissue damage following pancreatitis synergizes with oncogenic K-Ras and promotes cancer development (Fig. 2). An abundant desmoplastic stroma is one of the characteristic histological features of PDA (Chu et al. 2007; Neesse et al. 2011; Korc 2007; Mahadevan and

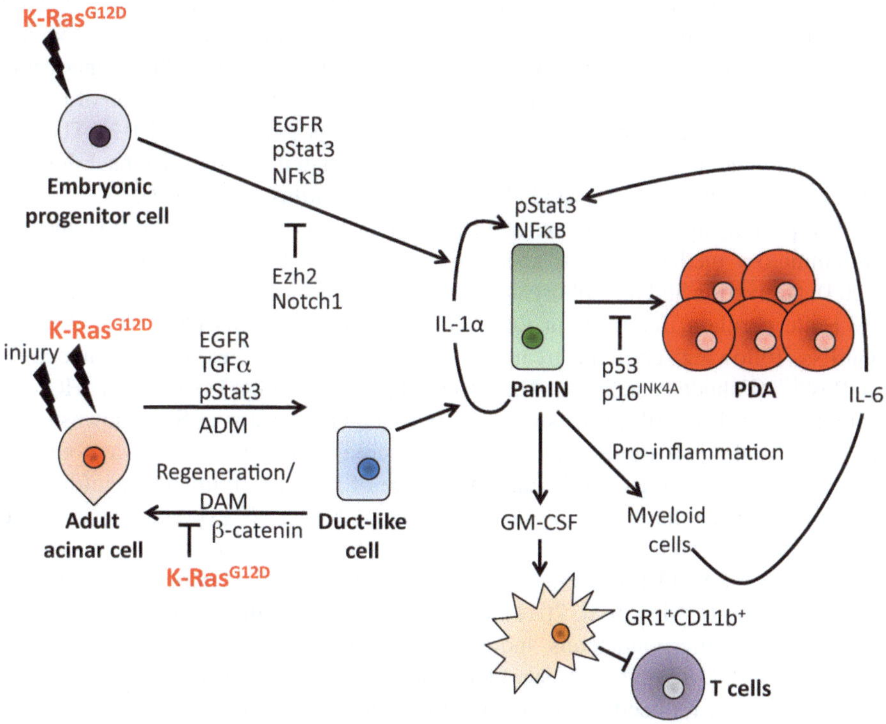

Fig. 2 Oncogenic K-Ras and injury in the form of pancreatitis synergize to induce development of PanINs that progress to PDA. If K-Ras^{G12D} is expressed during embryogenesis in an as yet unidentified progenitor cell, PanINs form that progress to PDA with a long latency but do not require pancreatic injury. This process may or may not proceed through ADM. However, K-Ras^{G12D} expression in adult acinar cells requires pancreatitis to develop into PDA. Injury induces ADM in the pancreas and K-Ras^{G12D} signaling diverts the metaplastic cells away from regenerative expansion of the acinar cell population in favor of PanIN formation. PanINs promote expression and activation of inflammatory mediators including GM-CSF, NFκB, Stat3, IL-6, IL-1α, and Cox2, which further synergize with K-Ras^{G12D} signaling and promote an immunosuppressive environment, which allows progression to PDA. Expression of tumor suppressors such as p53 and p16^{INK4A} is frequently lost during this progression

Von Hoff 2007). The desmoplastic stroma consists of extracellular matrix (ECM), activated fibroblasts, inflammatory cells and tumor vasculature. Importantly, K-Ras^{G12D} expression in the pancreas in mouse models also induces a desmoplastic response that is found in association with PanINs and areas of PDA (Hingorani et al. 2003, 2005). Cyclooxygenase-2 (Cox-2) promotes inflammation, and the expression of Cox-2 has been found to be upregulated in human PanINs and PDA (Maitra et al. 2002; Albazaz et al. 2005). Additionally, an anti-inflammatory selective Cox-2 inhibitor has been found to delay PanIN progression in the *PDX-1-Cre;LSL-KRAS^{G12D}* mouse model (Funahashi et al. 2007). Recently, the pro-inflammatory NF-κB pathway has been shown to be required for PDA development in the *PDX-1-Cre;LSL-KRAS^{G12D}* mouse model (Maniati et al. 2011; Ling et al. 2012) as conditional deletion of *IKK2* in the pancreas was found to inhibit both PanIN progression and K-Ras^{G12D} induced inflammatory responses. NF-κB is constitutively activated in human pancreatic adenocarcinoma and human pancreatic cancer cell lines but not in normal pancreatic tissues (Wang et al. 1999; Fujioka et al. 2003). Oncogenic K-Ras^{G12D} expression in the pancreas has been shown to induce expression of IL-1α, which in turn results in constitutive activation of NF-κB (Ling et al. 2012). There is also some evidence to suggest that an NF-κB-mediated positive feedback loop is able to further enhance oncogenic Ras signaling (Daniluk et al. 2012).

The protein signal transducer and activator of transcription 3 (Stat3) is another inflammatory mediator that is aberrantly activated in human PDA (Scholz et al. 2003). Activation and phosphorylation of Stat3 was found to be transiently induced by acute cerulein treatment in the mouse pancreas and this pStat3 persisted in PanINs following cerulein treatment in pancreata that expressed oncogenic K-Ras^{G12D} (Fukuda et al. 2011). The observed pattern of pStat3 staining by IHC was found to correlate with expression of IL-6, a known activator of Stat3 downstream of Ras signaling (Ancrile et al. 2007). An increase in IL-6 mRNA was found in pancreata expressing K-Ras^{G12D} and the source of IL-6 was found to be infiltrating macrophages (Lesina et al. 2011). Treatment of K-Ras^{G12D} expressing pancreatic acinar cells with an IL-6R/IL-6 complex but not IL-6 alone was able to induce phophorylation of Stat3 however, implying IL-6 transsignaling rather than classical IL-6 signaling (Lesina et al. 2011). Pancreatic Stat3 deletion in a *Stat3$^{flx/flx}$* mouse ameliorated both spontaneous and pancreatitis-induced PanIN formation in the *PDX-1-Cre;LSL-KRAS^{G12D}* mouse model and the PanIN formed in the absence of Stat3 displayed reduced inflammatory infiltrates (Corcoran et al. 2011; Fukuda et al. 2011). Similar results were seen in an IL6$^{-/-}$ mouse strain (Lesina et al. 2011). Consistent with this, Stat3 deficient acini were found to secrete less cytokines and inflammatory mediators that are known Stat3 target genes in response to cerulein in vitro. Knockdown of *Stat3* in mouse pancreatic cancer cells dramatically reduced PDAC formation compared with control shRNA following orthotopic injection into syngenic recipient mice (Corcoran et al. 2011). Stat3 signaling has also been implicated in controlling expression of matrix metalloproteinase 7 (MMP7), which has been found to be associated with metastatic disease in both humans and mouse models (Fukuda et al. 2011). This evidence all suggests that inflammation plays an important role in the progression from PanIN to PDA.

Recently oncogenic K-Ras signaling in the pancreas has been found to modulate the immune response in order to evade immune surveillance (Clark et al. 2007). The extensive stromal reaction surrounding PanINs and areas of PDA may provide an immunosuppressive environment that protects the transformed cells from T cells. Oncogenic K-Ras expressing PDECs, PanINs and PDA have been found to express GM-CSF (Pylayeva-Gupta et al. 2012; Bayne et al. 2012), which has been implicated in the regulation of proliferation and maturation of putative immunosuppressive $Gr1^{+}CD11b^{+}$ myeloid cells (Barreda et al. 2004) that have been implicated in tumor-induced immune tolerance (Dolcetti et al. 2010; Bronte et al. 1999; Gabrilovich and Nagaraj 2009; Marigo et al. 2010). $K\text{-}Ras^{G12D}$ expressing PDECs and cancer cells were found to induce the differentiation of progenitor $Gr1^{-}CD11b^{-}$ cells to $Gr1^{+}CD11b^{+}$ cells that were able to inhibit the proliferation of $CD3^{+}$ splenic T cells, and knockdown of GM-CSF in PDECs was found to both inhibit growth when engrafted into a wild type pancreas and increase the accumulation of $CD8^{+}$ cytotoxic T cells into the pancreas (Pylayeva-Gupta et al. 2012; Bayne et al. 2012).

To confirm that inflammation in the pancreas promotes PDA, conditional knockout animals that have impaired regeneration of the pancreas following cerulein-induced injury have been found to display accelerated PanIN progression. It has been shown that Ezh2, a polycomb group protein and a member of the polycomb repressor complex 2, is transiently upregulated during pancreatic regeneration, where it functions to suppress expression of $p16^{INK4A}$ and thereby promote cellular proliferation and regeneration. In the absence of pancreatic Ezh2, regeneration is impaired and the pancreas has a reduced ability to resolve cerulein-induced inflammation. The ability of Ezh2 to inhibit expression of $p16^{INK4A}$ makes it a good candidate for a tumor suppressor gene. However, loss of Ezh2 in the pancreas accelerated PanIN progression in the *p48Cre;LSL-KRAS^{G12D}* model (Mallen-St Clair et al. 2012). Thus, genetic alterations that enhance the inflamed state of the pancreas following damage are able to accelerate oncogenesis.

Oncogenic K-Ras and Pancreatic Cancer Cell Metabolism

One area of tumor biology that is receiving a lot of recent interest is alterations in metabolic pathways seen in cancer cells compared to normal cells. The Warburg effect was an observation made in the 1920s that under aerobic conditions, tumor tissues metabolize approximately tenfold more glucose to lactate in a given time than normal tissues (Warburg et al. 1924; Minami 1923). That is, the Pasteur effect, which is the inhibition of fermentation by oxygen, tends not to apply in tumor cells. Aerobic glycolysis is not an efficient method of producing ATP so there has been much confusion and debate regarding the advantages upregulating this pathway might have to cancer cells. It has been suggested that the Warburg effect occurs because proliferating cancer cells require not only ATP but also an abundant quantity of NADPH and macromolecular precursors needed to generate new cells such as acetyl-CoA for fatty acids, glycolytic intermediates for nonessential amino acids, and ribose for nucleotides (Vander Heiden et al. 2009). Oncogenic Ras has been shown to promote glycolysis (Yun et al. 2009; Racker et al. 1985) and pancreatic cancer cells have been found by proteomic analysis to have increased expression of glycolytic enzymes (Zhou et al. 2011, 2012) compared to normal ductal cells. Recently the iKras p53 null mouse has been used to study the effects of oncogenic K-Ras on cancer cell metabolism in the pancreas (Ying et al. 2012). Withdrawal of K-Ras^{G12D} expression was found to significantly affect intermediates in glucose metabolism including glucose-6-phosphate (G6P), fructose-6-phosphate (F6P), and fructose-1,6-bisphosphate (FBP), as determined by targeted liquid chromatography-tandem mass spectrometry (LC-MS/MS) metabolomic studies. This was accompanied by a decrease in glucose uptake and lactate production and down-regulation of expression of genes for glucose transporters and rate-limiting glycolytic enzymes. As expected, steady-state metabolite profiling and other methods showed that these changes in glycolytic flux were associated with a decrease in several intermediates of biosynthetic pathways such as hexosamine biosynthesis, protein glycosylation and ribose biogenesis through the nonoxidative arm of the pentose phosphate pathway. The effects observed of removal of K-Ras^{G12D} were recapitulated by treatment with the MEK inhibitor AZD8330, highlighting the importance of MAPK signaling downstream of K-Ras in this phenomenon.

Autophagy is a process that mediates the lysosomal degradation of cytoplasmic components such as damaged organelles and unused proteins. It is a vital contributor to cellular metabolism as it provides nutrients from internal sources when external sources are limited. Autophagy is considered to be a programmed pro-survival mechanism and therefore has a pro-tumor effect. However, there is some evidence to suggest that under certain conditions an "autophagic cell death" pathway may come into play to limit tumor growth (Levine and Yuan 2005; Hippert et al. 2006). It is known that pancreatic cancers have elevated levels of autophagy under basal conditions, despite the presence of abundant nutrients, and this has been correlated with poor outcome (Fujii et al. 2008; Yang et al. 2011). Also, genetic and chemical inhibition of autophagy was able to suppress the growth of pancreatic cancer cells *in vitro* and induce tumor regression in both pancreatic cancer xenografts and genetic mouse models (Yang et al. 2011). Data suggest that oncogenic Ras expression alters the requirement for autophagy within a cell and this may be attributable to an increase in the need for autophagic substrates for mitochondrial metabolism to preserve mitochondrial function (Guo et al. 2011). Another study suggested that the requirement for autophagy for the optimal growth and survival of K-Ras transformed cells was to impair mitochondrial respiration by mitophagy thereby facilitating the induction of the Warburg effect (Kim et al. 2011). This hypothesis is supported by studies which show a reduction in glucose metabolism in autophagy deficient MEFs (Lock et al. 2011) and that knockdown of K-Ras in a pancreatic cancer AsPC-1 cell line resulted in increased expression of mitochondrial genes (Ohnami et al. 1999).

K-Ras Signaling *In Vitro*

While the majority of insight into the role of K-Ras signaling in pancreatic cancer development and progression has been garnered from *in vivo* studies using mouse models, there is a significant contribution from *in vitro* experiments utilizing established pancreatic cancer cell lines and RNAi technology. The concept of oncogene addiction suggests that cancer cells become dependent on signaling from one particular oncogene in order to survive. Knocking down K-Ras has been found to induce apoptosis in pancreatic cancer cell lines in agreement with this model (Fleming et al. 2005). The extent of addiction to K-Ras signaling has been thoroughly tested in a panel of pancreatic cancer cell lines containing K-Ras mutations. Surprisingly the effect of knocking down K-Ras in these cell lines was found to vary significantly with some of the cell lines tested having very little dependency on K-Ras. Many of the K-Ras-dependent cells contained *KRAS* genomic amplifications, exhibited a classic epithelial morphology, and expressed E-Cadherin, whereas most K-Ras-independent cells appeared less uniformly epithelial and expressed little or no E-cadherin, suggesting that they may have undergone an epithelial to mesenchymal transition (EMT). From this study it was possible to identify a gene expression signature that can be used to accurately predict the K-Ras dependency of tumors in different tissue types (Singh et al. 2009). Such signature could prove useful in the future to predict what patients would benefit from therapies that target the Ras signaling pathway.

In one recent study a high-throughput loss-of-function RNAi screen was carried out to find genes with synthetic lethal interactions with oncogenic K-Ras, where knockdown of the gene would affect the viability of cell lines with oncogenic K-Ras mutations but not those without (Scholl et al. 2009). The screen was carried out with a panel of cell lines both with and without K-Ras mutations including the pancreatic cell lines Panc-1 that contains a K-Ras^{G12D} mutation and BxPC3 that is wild type for K-Ras. The screen identified *STK33*, a putative member of the calcium/calmodulin-dependent protein kinase subfamily of serine/threonine protein kinases. Knockdown of *STK33* in Panc-1 cells impaired colony formation in semisolid medium and decreased their ability to form tumors in immunocompromised mice but had no effect on BxPc3 cells. Despite the apparent importance of *STK33* in these cancer cell lines, no amplifications of the gene or significant increases in gene expression were observed in cell lines with oncogenic K-Ras mutations. Knockdown of *STK33* was also found to decrease the phosphorylation of S6K1 serine/threonine protein kinase and its downstream substrate RPS6 in an oncogenic K-Ras dependent manner. There is evidence to suggest that this pathway may be involved in controlling apoptosis via the proapoptotic BH3-only protein BAD which is known to be phosphorylated and inactivated by S6K1 resulting in an inhibition of mitochondrial apoptosis (Scholl et al. 2009; Azoitei et al. 2012). Subsequent studies targeting STK33 both by siRNA and inhibitors in K-Ras mutant cancer cells were unable however to confirm the observed synthetic lethality (Babij et al. 2011; Luo et al. 2012). These discrepancies highlight the drawback to using siRNAs, where the risks for off-target effects and false positive results are high and the need for these studies to be carefully controlled.

Conclusions

Oncogenic K-Ras and several of its downstream effector pathways have been shown to have essential roles in all aspects of pancreatic cancer initiation, progression, invasion, and metastasis. The evidence suggests that any pharmacological agents able to completely block K-Ras signaling in pancreatic cancer should result in significant tumor shrinkage and cell death and therefore have a significant clinical impact on a disease that is so refractory to all currently available treatments. Despite substantial effort, all attempts to therapeutically target the mutated Ras protein directly with small molecules that could promote the hydrolysis of GTP have been unsuccessful. Therefore, the focus of drug discovery has concentrated on either downstream components of the Ras signaling pathway or the upstream pathway involved in the posttranslational modification of the Ras protein. Effective inhibitors specific for many of the key components of the Ras/Raf/MEK/Erk and Ras/PI3K/PTEN/mTOR pathways have been developed. Some, such as the orally available MEK1 inhibitor Selumetinib, have been tested in phase I and phase II clinical trials (Chappell et al. 2011). However, there are many more pathways downstream of Ras than just these two, and it is as yet unclear the specific importance of these individual pathways in tumorigenesis. We do not know how many of these pathways will need to be inhibited to completely block oncogenic K-Ras signaling, and it seems likely that mutiple inhibitors would produce intolerable significant side effects. The failure of inhibitors to farnesyl transferase (FTIs), the enzyme that catalyzes the addition of a 15-carbon prenyl group to Ras, to show any efficacy in clinical trials serves as a cautionary tale to rational drug design. These FTIs, despite being very effective inhibitors of farnesyl transferase, failed because K-Ras was able to be alternatively prenylated by geranylgeranyltransferase (GGT), an enzyme that was not affected by FTIs (Whyte et al. 1997). Preclinical testing of FTIs was carried out using cells and tumors transformed with H-Ras, an isoform that is not a substrate for GGT (Appels et al. 2005; Brunner et al. 2003). The other enzymes in the posttranslational modification pathway of Ras, Rce1, and Icmt are now of interest as potential drug targets and have shown some promise in preclinical studies (Wahlstrom et al. 2008). Due to the potential difficulties of targeting K-Ras itself, another approach has been to look for other signaling pathways specifically required for cell survival only in the presence of oncogenic K-Ras. Screens for such synthetic lethal interactions have identified a number of potential drug targets (Scholl et al. 2009; Barbie et al. 2009), so there is hope that in the future these studies can generate effective therapies for K-Ras driven cancers.

References

Abubaker J, Bavi P, Al-Haqawi W, Sultana M, Al-Harbi S, Al-Sanea N, Abduljabbar A, Ashari LH, Alhomoud S, Al-Dayel F, Uddin S, Al-Kuraya KS (2009) Prognostic significance of alterations in KRAS isoforms KRAS-4A/4B and KRAS mutations in colorectal carcinoma. J Pathol 219(4):435–445. doi:10.1002/path.2625

Aguirre AJ, Bardeesy N, Sinha M, Lopez L, Tuveson DA, Horner J, Redston MS, DePinho RA (2003) Activated Kras and Ink4a/Arf deficiency cooperate to produce metastatic pancreatic ductal adenocarcinoma. Genes Dev 17(24):3112–3126, doi:10.1101/gad.1158703. 1158703 [pii]

Al-Aynati MM, Radulovich N, Riddell RH, Tsao MS (2004) Epithelial-cadherin and beta-catenin expression changes in pancreatic intraepithelial neoplasia. Clin Cancer Res 10(4):1235–1240

Albazaz R, Verbeke CS, Rahman SH, McMahon MJ (2005) Cyclooxygenase-2 expression associated with severity of PanIN lesions: a possible link between chronic pancreatitis and pancreatic cancer. Pancreatology 5(4–5):361–369, doi:PAN20050054_5361 [pii]. 10.1159/000086536

Almoguera C, Shibata D, Forrester K, Martin J, Arnheim N, Perucho M (1988) Most human carcinomas of the exocrine pancreas contain mutant c-K-ras genes. Cell 53(4):549–554. doi:0092-8674(88)90571-5 [pii]

Altomare DA, Tanno S, De Rienzo A, Klein-Szanto AJ, Skele KL, Hoffman JP, Testa JR (2002) Frequent activation of AKT2 kinase in human pancreatic carcinomas. J Cell Biochem 87(4):470–476

Ancrile B, Lim KH, Counter CM (2007) Oncogenic Ras-induced secretion of IL6 is required for tumorigenesis. Genes Dev 21(14):1714–1719, doi: 21/14/1714 [pii]. 10.1101/gad.1549407

Appels NM, Beijnen JH, Schellens JH (2005) Development of farnesyl transferase inhibitors: a review. Oncologist 10(8):565–578, doi: 10/8/565 [pii]. 10.1634/theoncologist.10-8-565

Ardito CM, Gruner BM, Takeuchi KK, Lubeseder-Martellato C, Teichmann N, Mazur PK, Delgiorno KE, Carpenter ES, Halbrook CJ, Hall JC, Pal D, Briel T, Herner A, Trajkovic-Arsic M, Sipos B, Liou GY, Storz P, Murray NR, Threadgill DW, Sibilia M, Washington MK, Wilson CL, Schmid RM, Raines EW, Crawford HC, Siveke JT (2012) EGF receptor is required for KRAS-induced pancreatic tumorigenesis. Cancer Cell 22(3):304–317, doi:S1535-6108(12)00337-6 [pii]. 10.1016/j.ccr.2012.07.024

Asano T, Yao Y, Zhu J, Li D, Abbruzzese JL, Reddy SA (2004) The PI 3-kinase/Akt signaling pathway is activated due to aberrant Pten expression and targets transcription factors NF-kappaB and c-Myc in pancreatic cancer cells. Oncogene 23(53):8571–8580, doi:1207902 [pii]. 10.1038/sj.onc.1207902

Azoitei N, Hoffmann CM, Ellegast JM, Ball CR, Obermayer K, Gossele U, Koch B, Faber K, Genze F, Schrader M, Kestler HA, Dohner H, Chiosis G, Glimm H, Frohling S, Scholl C (2012) Targeting of KRAS mutant tumors by HSP90 inhibitors involves degradation of STK33. J Exp Med 209(4):697–711, doi:jem.20111910 [pii]. 10.1084/jem.20111910

Babij C, Zhang Y, Kurzeja RJ, Munzli A, Shehabeldin A, Fernando M, Quon K, Kassner PD, Ruefli-Brasse AA, Watson VJ, Fajardo F, Jackson A, Zondlo J, Sun Y, Ellison AR, Plewa CA, San MT, Robinson J, McCarter J, Schwandner R, Judd T, Carnahan J, Dussault I (2011) STK33 kinase activity is nonessential in KRAS-dependent cancer cells. Cancer Res 71(17):5818–5826, doi:0008-5472.CAN-11-0778 [pii]. 10.1158/0008-5472.CAN-11-0778

Barbacid M (1987) Ras genes. Annu Rev Biochem 56:779–827. doi:10.1146/annurev.bi.56.070187.004023

Barbie DA, Tamayo P, Boehm JS, Kim SY, Moody SE, Dunn IF, Schinzel AC, Sandy P, Meylan E, Scholl C, Frohling S, Chan EM, Sos ML, Michel K, Mermel C, Silver SJ, Weir BA, Reiling JH, Sheng Q, Gupta PB, Wadlow RC, Le H, Hoersch S, Wittner BS, Ramaswamy S, Livingston DM, Sabatini DM, Meyerson M, Thomas RK, Lander ES, Mesirov JP, Root DE, Gilliland DG, Jacks T, Hahn WC (2009) Systematic RNA interference reveals that oncogenic KRAS-driven cancers require TBK1. Nature 462(7269):108–112, doi:nature08460 [pii]. 10.1038/nature08460

Bardeesy N, Aguirre AJ, Chu GC, Cheng KH, Lopez LV, Hezel AF, Feng B, Brennan C, Weissleder R, Mahmood U, Hanahan D, Redston MS, Chin L, Depinho RA (2006) Both p16(Ink4a) and the p19(Arf)-p53 pathway constrain progression of pancreatic adenocarcinoma in the mouse. Proc Natl Acad Sci USA 103(15):5947–5952, doi:0601273103 [pii]. 10.1073/pnas.0601273103

Barreda DR, Hanington PC, Belosevic M (2004) Regulation of myeloid development and function by colony stimulating factors. Dev Comp Immunol 28(5):509–554, doi:10.1016/j.dci.2003.09.010. S0145305X03001848 [pii]

Bar-Sagi D, Feramisco JR (1986) Induction of membrane ruffling and fluid-phase pinocytosis in quiescent fibroblasts by ras proteins. Science 233(4768):1061–1068

Bayne LJ, Beatty GL, Jhala N, Clark CE, Rhim AD, Stanger BZ, Vonderheide RH (2012) Tumor-derived granulocyte-macrophage colony-stimulating factor regulates myeloid inflammation and T cell immunity in pancreatic cancer. Cancer Cell 21(6):822–835, doi:S1535-6108(12)00167-5 [pii]. 10.1016/j.ccr.2012.04.025

Bi Y, Page SL, Williams JA (2005) Rho and Rac promote acinar morphological changes, actin reorganization, and amylase secretion. Am J Physiol Gastrointest Liver Physiol 289(3):G561–G570, doi:00508.2004 [pii]. 10.1152/ajpgi.00508.2004

Bloomston M, Bhardwaj A, Ellison EC, Frankel WL (2006) Epidermal growth factor receptor expression in pancreatic carcinoma using tissue microarray technique. Dig Surg 23(1–2):74–79, doi:93497 [pii]. 10.1159/000093497

Boriack-Sjodin PA, Margarit SM, Bar-Sagi D, Kuriyan J (1998) The structural basis of the activation of Ras by Sos. Nature 394(6691):337–343. doi:10.1038/28548

Bos JL (1989) ras oncogenes in human cancer: a review. Cancer Res 49(17):4682–4689

Boyartchuk VL, Ashby MN, Rine J (1997) Modulation of Ras and a-factor function by carboxyl-terminal proteolysis. Science 275(5307):1796–1800

Brat DJ, Lillemoe KD, Yeo CJ, Warfield PB, Hruban RH (1998) Progression of pancreatic intraductal neoplasias to infiltrating adenocarcinoma of the pancreas. Am J Surg Pathol 22(2):163–169

Brembeck FH, Schreiber FS, Deramaudt TB, Craig L, Rhoades B, Swain G, Grippo P, Stoffers DA, Silberg DG, Rustgi AK (2003) The mutant K-ras oncogene causes pancreatic periductal lymphocytic infiltration and gastric mucous neck cell hyperplasia in transgenic mice. Cancer Res 63(9):2005–2009

Bremner R, Balmain A (1990) Genetic changes in skin tumor progression: correlation between presence of a mutant ras gene and loss of heterozygosity on mouse chromosome 7. Cell 61(3):407–417. doi:0092-8674(90)90523-H [pii]

Bronte V, Chappell DB, Apolloni E, Cabrelle A, Wang M, Hwu P, Restifo NP (1999) Unopposed production of granulocyte-macrophage colony-stimulating factor by tumors inhibits CD8+ T cell responses by dysregulating antigen-presenting cell maturation. J Immunol 162(10):5728–5737

Brunner TB, Hahn SM, Gupta AK, Muschel RJ, McKenna WG, Bernhard EJ (2003) Farnesyltransferase inhibitors: an overview of the results of preclinical and clinical investigations. Cancer Res 63(18):5656–5668

Buday L (1999) Membrane-targeting of signalling molecules by SH2/SH3 domain-containing adaptor proteins. Biochim Biophys Acta 1422(2):187–204. doi:S0304-4157(99)00005-2 [pii]

Buhrman G, Holzapfel G, Fetics S, Mattos C (2010) Allosteric modulation of Ras positions Q61 for a direct role in catalysis. Proc Natl Acad Sci USA 107(11):4931–4936, doi:0912226107 [pii]. 10.1073/pnas.0912226107

Carriere C, Young AL, Gunn JR, Longnecker DS, Korc M (2009) Acute pancreatitis markedly accelerates pancreatic cancer progression in mice expressing oncogenic Kras. Biochem Biophys Res Commun 382(3):561–565, doi:S0006-291X(09)00533-6 [pii]. 10.1016/j.bbrc.2009.03.068

Chadha KS, Khoury T, Yu J, Black JD, Gibbs JF, Kuvshinoff BW, Tan D, Brattain MG, Javle MM (2006) Activated Akt and Erk expression and survival after surgery in pancreatic carcinoma. Ann Surg Oncol 13(7):933–939. doi:10.1245/ASO.2006.07.011

Chappell WH, Steelman LS, Long JM, Kempf RC, Abrams SL, Franklin RA, Basecke J, Stivala F, Donia M, Fagone P, Malaponte G, Mazzarino MC, Nicoletti F, Libra M, Maksimovic-Ivanic D, Mijatovic S, Montalto G, Cervello M, Laidler P, Milella M, Tafuri A, Bonati A, Evangelisti C, Cocco L, Martelli AM, McCubrey JA (2011) Ras/Raf/MEK/ERK and PI3K/PTEN/Akt/mTOR inhibitors: rationale and importance to inhibiting these pathways in human health. Oncotarget 2(3):135–164. doi:240 [pii]

Cheng JQ, Ruggeri B, Klein WM, Sonoda G, Altomare DA, Watson DK, Testa JR (1996) Amplification of AKT2 in human pancreatic cells and inhibition of AKT2 expression and tumorigenicity by antisense RNA. Proc Natl Acad Sci USA 93(8):3636–3641

Chu GC, Kimmelman AC, Hezel AF, DePinho RA (2007) Stromal biology of pancreatic cancer. J Cell Biochem 101(4):887–907. doi:10.1002/jcb.21209

Clark CE, Hingorani SR, Mick R, Combs C, Tuveson DA, Vonderheide RH (2007) Dynamics of the immune reaction to pancreatic cancer from inception to invasion. Cancer Res 67(19):9518–9527, doi:67/19/9518 [pii]. 10.1158/0008-5472.CAN-07-0175

Clarke S, Vogel JP, Deschenes RJ, Stock J (1988) Posttranslational modification of the Ha-ras oncogene protein: evidence for a third class of protein carboxyl methyltransferases. Proc Natl Acad Sci USA 85(13):4643–4647

Collins MA, Bednar F, Zhang Y, Brisset JC, Galban S, Galban CJ, Rakshit S, Flannagan KS, Adsay NV, Pasca di Magliano M (2012) Oncogenic Kras is required for both the initiation and maintenance of pancreatic cancer in mice. J Clin Invest 122(2):639–653, doi:59227 [pii]. 10.1172/JCI59227

Collisson EA, Trejo CL, Silva JM, Gu S, Korkola JE, Heiser LM, Charles RP, Rabinovich BA, Hann B, Dankort D, Spellman PT, Phillips WA, Gray JW, McMahon M (2012) A central role for RAF MEK ERK signaling in the genesis of pancreatic ductal adenocarcinoma. Cancer Discov 2(8):685–693, doi:2159-8290.CD-11-0347 [pii]. 10.1158/2159-8290.CD-11-0347

Corcoran RB, Contino G, Deshpande V, Tzatsos A, Conrad C, Benes CH, Levy DE, Settleman J, Engelman JA, Bardeesy N (2011) STAT3 plays a critical role in KRAS-induced pancreatic tumorigenesis. Cancer Res 71(14):5020–5029, doi:0008-5472.CAN-11-0908 [pii]. 10.1158/0008-5472.CAN-11-0908

Crnogorac-Jurcevic T, Efthimiou E, Capelli P, Blaveri E, Baron A, Terris B, Jones M, Tyson K, Bassi C, Scarpa A, Lemoine NR (2001) Gene expression profiles of pancreatic cancer and stromal desmoplasia. Oncogene 20(50):7437–7446. doi:10.1038/sj.onc.1204935

Daniluk J, Liu Y, Deng D, Chu J, Huang H, Gaiser S, Cruz-Monserrate Z, Wang H, Ji B, Logsdon CD (2012) An NF-kappaB pathway-mediated positive feedback loop amplifies Ras activity to pathological levels in mice. J Clin Invest 122(4):1519–1528, doi:59743 [pii]. 10.1172/JCI59743

Davies H, Bignell GR, Cox C, Stephens P, Edkins S, Clegg S, Teague J, Woffendin H, Garnett MJ, Bottomley W, Davis N, Dicks E, Ewing R, Floyd Y, Gray K, Hall S, Hawes R, Hughes J, Kosmidou V, Menzies A, Mould C, Parker A, Stevens C, Watt S, Hooper S, Wilson R, Jayatilake H, Gusterson BA, Cooper C, Shipley J, Hargrave D, Pritchard-Jones K, Maitland N, Chenevix-Trench G, Riggins GJ, Bigner DD, Palmieri G, Cossu A, Flanagan A, Nicholson A, Ho JW, Leung SY, Yuen ST, Weber BL, Seigler HF, Darrow TL, Paterson H, Marais R, Marshall CJ, Wooster R, Stratton MR, Futreal PA (2002) Mutations of the BRAF gene in human cancer. Nature 417(6892):949–954, doi:10.1038/nature00766. nature00766 [pii]

De La OJ, Emerson LL, Goodman JL, Froebe SC, Illum BE, Curtis AB, Murtaugh LC (2008) Notch and Kras reprogram pancreatic acinar cells to ductal intraepithelial neoplasia. Proc Natl Acad Sci USA 105(48):18907–18912, doi:0810111105 [pii]. 10.1073/pnas.0810111105

Der CJ, Krontiris TG, Cooper GM (1982) Transforming genes of human bladder and lung carcinoma cell lines are homologous to the ras genes of Harvey and Kirsten sarcoma viruses. Proc Natl Acad Sci U S A 79(11):3637–3640

Dolcetti L, Peranzoni E, Ugel S, Marigo I, Fernandez Gomez A, Mesa C, Geilich M, Winkels G, Traggiai E, Casati A, Grassi F, Bronte V (2010) Hierarchy of immunosuppressive strength among myeloid-derived suppressor cell subsets is determined by GM-CSF. Eur J Immunol 40(1):22–35. doi:10.1002/eji.200939903

Downward J (2003) Targeting RAS signalling pathways in cancer therapy. Nat Rev Cancer 3(1):11–22. doi:10.1038/nrc969. nrc969 [pii]

Ebert MP, Fei G, Schandl L, Mawrin C, Dietzmann K, Herrera P, Friess H, Gress TM, Malfertheiner P (2002) Reduced PTEN expression in the pancreas overexpressing transforming growth factor-beta 1. Br J Cancer 86(2):257–262. doi:10.1038/sj.bjc.6600031

Fendrich V, Esni F, Garay MV, Feldmann G, Habbe N, Jensen JN, Dor Y, Stoffers D, Jensen J, Leach SD, Maitra A (2008) Hedgehog signaling is required for effective regeneration of exocrine pancreas. Gastroenterology 135(2):621–631, doi:S0016-5085(08)00640-9 [pii]. 10.1053/j.gastro.2008.04.011

Field J, Broek D, Kataoka T, Wigler M (1987) Guanine nucleotide activation of, and competition between, RAS proteins from Saccharomyces cerevisiae. Mol Cell Biol 7(6):2128–2133

Fjallskog ML, Lejonklou MH, Oberg KE, Eriksson BK, Janson ET (2003) Expression of molecular targets for tyrosine kinase receptor antagonists in malignant endocrine pancreatic tumors. Clin Cancer Res 9(4):1469–1473

Fleming JB, Shen GL, Holloway SE, Davis M, Brekken RA (2005) Molecular consequences of silencing mutant K-ras in pancreatic cancer cells: justification for K-ras-directed therapy. Mol Cancer Res 3(7):413–423, doi:3/7/413 [pii]. 10.1158/1541-7786.MCR-04-0206

Forbes SA, Bindal N, Bamford S, Cole C, Kok CY, Beare D, Jia M, Shepherd R, Leung K, Menzies A, Teague JW, Campbell PJ, Stratton MR, Futreal PA (2011) COSMIC: mining complete cancer genomes in the Catalogue of Somatic Mutations in Cancer. Nucleic Acids Res 39(Database issue):D945–D950, doi:gkq929 [pii]. 0.1093/nar/gkq929

Freije JM, Blay P, Pendas AM, Cadinanos J, Crespo P, Lopez-Otin C (1999) Identification and chromosomal location of two human genes encoding enzymes potentially involved in proteolytic maturation of farnesylated proteins. Genomics 58(3):270–280, doi:10.1006/geno.1999.5834. S0888754399958342 [pii]

Fu HW, Casey PJ (1999) Enzymology and biology of CaaX protein prenylation. Recent Prog Horm Res 54:315–342, discussion 342–313

Fujii S, Mitsunaga S, Yamazaki M, Hasebe T, Ishii G, Kojima M, Kinoshita T, Ueno T, Esumi H, Ochiai A (2008) Autophagy is activated in pancreatic cancer cells and correlates with poor patient outcome. Cancer Sci 99(9):1813–1819, doi:CAS893 [pii]. 10.1111/j.1349-7006.2008.00893.x

Fujioka S, Sclabas GM, Schmidt C, Niu J, Frederick WA, Dong QG, Abbruzzese JL, Evans DB, Baker C, Chiao PJ (2003) Inhibition of constitutive NF-kappa B activity by I kappa B alpha M suppresses tumorigenesis. Oncogene 22(9):1365–1370, doi:10.1038/sj.onc.1206323. 1206323 [pii]

Fukuda A, Wang SC, Morris JP, Folias AE, Liou A, Kim GE, Akira S, Boucher KM, Firpo MA, Mulvihill SJ, Hebrok M (2011) Stat3 and MMP7 contribute to pancreatic ductal adenocarcinoma initiation and progression. Cancer Cell 19(4):441–455, doi:S1535-6108(11)00091-2 [pii]. 10.1016/j.ccr.2011.03.002

Funahashi H, Satake M, Dawson D, Huynh NA, Reber HA, Hines OJ, Eibl G (2007) Delayed progression of pancreatic intraepithelial neoplasia in a conditional Kras(G12D) mouse model by a selective cyclooxygenase-2 inhibitor. Cancer Res 67(15):7068–7071, doi:0008-5472.CAN-07-0970 [pii]. 10.1158/0008-5472.CAN-07-0970

Gabrilovich DI, Nagaraj S (2009) Myeloid-derived suppressor cells as regulators of the immune system. Nat Rev Immunol 9(3):162–174, doi:nri2506 [pii]. 10.1038/nri2506

Geyer M, Schweins T, Herrmann C, Prisner T, Wittinghofer A, Kalbitzer HR (1996) Conformational transitions in p21ras and in its complexes with the effector protein Raf-RBD and the GTPase activating protein GAP. Biochemistry 35(32):10308–10320, doi:10.1021/bi952858k. bi952858k [pii]

Geyer M, Herrmann C, Wohlgemuth S, Wittinghofer A, Kalbitzer HR (1997) Structure of the Ras-binding domain of RalGEF and implications for Ras binding and signalling. Nat Struct Biol 4(9):694–699

Gibbs JB, Sigal IS, Poe M, Scolnick EM (1984) Intrinsic GTPase activity distinguishes normal and oncogenic ras p21 molecules. Proc Natl Acad Sci USA 81(18):5704–5708

Gidekel Friedlander SY, Chu GC, Snyder EL, Girnius N, Dibelius G, Crowley D, Vasile E, DePinho RA, Jacks T (2009) Context-dependent transformation of adult pancreatic cells by oncogenic K-Ras. Cancer Cell 16(5):379–389, doi:S1535-6108(09)00338-9 [pii]. 10.1016/j.ccr.2009.09.027

Goldfarb M, Shimizu K, Perucho M, Wigler M (1982) Isolation and preliminary characterization of a human transforming gene from T24 bladder carcinoma cells. Nature 296(5856): 404–409

Gonzalez-Garcia A, Pritchard CA, Paterson HF, Mavria G, Stamp G, Marshall CJ (2005) RalGDS is required for tumor formation in a model of skin carcinogenesis. Cancer Cell 7(3):219–226, doi:S1535-6108(05)00059-0 [pii]. 10.1016/j.ccr.2005.01.029

Grippo PJ, Nowlin PS, Demeure MJ, Longnecker DS, Sandgren EP (2003) Preinvasive pancreatic neoplasia of ductal phenotype induced by acinar cell targeting of mutant Kras in transgenic mice. Cancer Res 63(9):2016–2019

Guerra C, Schuhmacher AJ, Canamero M, Grippo PJ, Verdaguer L, Perez-Gallego L, Dubus P, Sandgren EP, Barbacid M (2007) Chronic pancreatitis is essential for induction of pancreatic ductal adenocarcinoma by K-Ras oncogenes in adult mice. Cancer Cell 11(3):291–302, doi:S1535-6108(07)00027-X [pii]. 10.1016/j.ccr.2007.01.012

Guerra C, Collado M, Navas C, Schuhmacher AJ, Hernandez-Porras I, Canamero M, Rodriguez-Justo M, Serrano M, Barbacid M (2011) Pancreatitis-induced inflammation contributes to pancreatic cancer by inhibiting oncogene-induced senescence. Cancer Cell 19(6):728–739, doi:S1535-6108(11)00189-9 [pii]. 10.1016/j.ccr.2011.05.011

Guo JY, Chen HY, Mathew R, Fan J, Strohecker AM, Karsli-Uzunbas G, Kamphorst JJ, Chen G, Lemons JM, Karantza V, Coller HA, Dipaola RS, Gelinas C, Rabinowitz JD, White E (2011) Activated Ras requires autophagy to maintain oxidative metabolism and tumorigenesis. Genes Dev 25(5):460–470, doi:gad.2016311 [pii]. 10.1101/gad.2016311

Gutierrez L, Magee AI, Marshall CJ, Hancock JF (1989) Post-translational processing of p21ras is two-step and involves carboxyl-methylation and carboxy-terminal proteolysis. EMBO J 8(4):1093–1098

Habbe N, Shi G, Meguid RA, Fendrich V, Esni F, Chen H, Feldmann G, Stoffers DA, Konieczny SF, Leach SD, Maitra A (2008) Spontaneous induction of murine pancreatic intraepithelial neoplasia (mPanIN) by acinar cell targeting of oncogenic Kras in adult mice. Proc Natl Acad Sci U S A 105(48):18913–18918, doi:0810097105 [pii]. 10.1073/pnas.0810097105

Hald J, Hjorth JP, German MS, Madsen OD, Serup P, Jensen J (2003) Activated Notch1 prevents differentiation of pancreatic acinar cells and attenuate endocrine development. Dev Biol 260(2):426–437. doi:S0012160603003269 [pii]

Hall A, Marshall CJ, Spurr NK, Weiss RA (1983) Identification of transforming gene in two human sarcoma cell lines as a new member of the ras gene family located on chromosome 1. Nature 303(5916):396–400

Hamad NM, Elconin JH, Karnoub AE, Bai W, Rich JN, Abraham RT, Der CJ, Counter CM (2002) Distinct requirements for Ras oncogenesis in human versus mouse cells. Genes Dev 16(16):2045–2057. doi:10.1101/gad.993902

Hancock JF, Magee AI, Childs JE, Marshall CJ (1989) All ras proteins are polyisoprenylated but only some are palmitoylated. Cell 57(7):1167–1177. doi:0092-8674(89)90054-8 [pii]

Hancock JF, Cadwallader K, Marshall CJ (1991) Methylation and proteolysis are essential for efficient membrane binding of prenylated p21K-ras(B). EMBO J 10(3):641–646

Hanlon L, Avila JL, Demarest RM, Troutman S, Allen M, Ratti F, Rustgi AK, Stanger BZ, Radtke F, Adsay V, Long F, Capobianco AJ, Kissil JL (2010) Notch1 functions as a tumor suppressor in a model of K-ras-induced pancreatic ductal adenocarcinoma. Cancer Res 70(11):4280–4286, doi:0008-5472.CAN-09-4645 [pii]. 10.1158/0008-5472.CAN-09-4645

Hegi ME, Devereux TR, Dietrich WF, Cochran CJ, Lander ES, Foley JF, Maronpot RR, Anderson MW, Wiseman RW (1994) Allelotype analysis of mouse lung carcinomas reveals frequent allelic losses on chromosome 4 and an association between allelic imbalances on chromosome 6 and K-ras activation. Cancer Res 54(23):6257–6264

Heid I, Lubeseder-Martellato C, Sipos B, Mazur PK, Lesina M, Schmid RM, Siveke JT (2011) Early requirement of Rac1 in a mouse model of pancreatic cancer. Gastroenterology 141(2):719–730, 730 e711-717. doi:S0016-5085(11)00598-1 [pii]. 10.1053/j.gastro.2011.04.043

Hill R, Calvopina JH, Kim C, Wang Y, Dawson DW, Donahue TR, Dry S, Wu H (2010) PTEN loss accelerates KrasG12D-induced pancreatic cancer development. Cancer Res 70(18):7114–7124, doi:0008-5472.CAN-10-1649 [pii]. 10.1158/0008-5472.CAN-10-1649

Hingorani SR, Petricoin EF, Maitra A, Rajapakse V, King C, Jacobetz MA, Ross S, Conrads TP, Veenstra TD, Hitt BA, Kawaguchi Y, Johann D, Liotta LA, Crawford HC, Putt ME, Jacks T, Wright CVE, Hruban RH, Lowy AM, Tuveson DA (2003) Preinvasive and invasive ductal pancreatic cancer and its early detection in the mouse. Cancer Cell 4(6):437–450

Hingorani SR, Wang L, Multani AS, Combs C, Deramaudt TB, Hruban RH, Rustgi AK, Chang S, Tuveson DA (2005) Trp53R172H and KrasG12D cooperate to promote chromosomal instability and widely metastatic pancreatic ductal adenocarcinoma in mice. Cancer Cell 7(5):469–483. doi:10.1016/j.ccr.2005.04.023

Hippert MM, O'Toole PS, Thorburn A (2006) Autophagy in cancer: good, bad, or both? Cancer Res 66(19):9349–9351, doi:66/19/9349 [pii]. 10.1158/0008-5472.CAN-06-1597

Hruban RH, Wilentz RE, Goggins M, Offerhaus GJ, Yeo CJ, Kern SE (1999) Pathology of incipient pancreatic cancer. Ann Oncol 10(Suppl 4):9–11

Hrycyna CA, Sapperstein SK, Clarke S, Michaelis S (1991) The Saccharomyces cerevisiae STE14 gene encodes a methyltransferase that mediates C-terminal methylation of a-factor and RAS proteins. EMBO J 10(7):1699–1709

Ijichi H, Chytil A, Gorska AE, Aakre ME, Fujitani Y, Fujitani S, Wright CV, Moses HL (2006) Aggressive pancreatic ductal adenocarcinoma in mice caused by pancreas-specific blockade of transforming growth factor-beta signaling in cooperation with active Kras expression. Genes Dev 20(22):3147–3160, doi:20/22/3147 [pii]. 10.1101/gad.1475506

Inoki K, Li Y, Zhu T, Wu J, Guan KL (2002) TSC2 is phosphorylated and inhibited by Akt and suppresses mTOR signalling. Nat Cell Biol 4(9):648–657, doi:10.1038/ncb839. ncb839 [pii]

Ishimura N, Yamasawa K, Karim Rumi MA, Kadowaki Y, Ishihara S, Amano Y, Nio Y, Higami T, Kinoshita Y (2003) BRAF and K-ras gene mutations in human pancreatic cancers. Cancer Lett 199(2):169–173. doi:S0304383503003847 [pii]

Ito Y, Yamasaki K, Iwahara J, Terada T, Kamiya A, Shirouzu M, Muto Y, Kawai G, Yokoyama S, Laue ED, Walchli M, Shibata T, Nishimura S, Miyazawa T (1997) Regional polysterism in the GTP-bound form of the human c-Ha-Ras protein. Biochemistry 36(30):9109–9119, doi:10.1021/bi970296u. bi970296u [pii]

Izeradjene K, Combs C, Best M, Gopinathan A, Wagner A, Grady WM, Deng C-X, Hruban RH, Adsay NV, Tuveson DA, Hingorani SR (2007) Kras(G12D) and Smad4/Dpc4 haploinsufficiency cooperate to induce mucinous cystic neoplasms and invasive adenocarcinoma of the pancreas. Cancer Cell 11(3):229–243. doi:10.1016/j.ccr.2007.01.017

Jackson EL, Willis N, Mercer K, Bronson RT, Crowley D, Montoya R, Jacks T, Tuveson DA (2001) Analysis of lung tumor initiation and progression using conditional expression of oncogenic K-ras. Genes Dev 15(24):3243–3248. doi:10.1101/gad.943001

Janku F, Lee JJ, Tsimberidou AM, Hong DS, Naing A, Falchook GS, Fu S, Luthra R, Garrido-Laguna I, Kurzrock R (2011) PIK3CA mutations frequently coexist with RAS and BRAF mutations in patients with advanced cancers. PLoS One 6(7):e22769, doi:10.1371/journal.pone.0022769. PONE-D-11-08638 [pii]

Jensen JN, Cameron E, Garay MV, Starkey TW, Gianani R, Jensen J (2005) Recapitulation of elements of embryonic development in adult mouse pancreatic regeneration. Gastroenterology 128(3):728–741. doi:S0016508504021997 [pii]

Ji B, Tsou L, Wang H, Gaiser S, Chang DZ, Daniluk J, Bi Y, Grote T, Longnecker DS, Logsdon CD (2009) Ras activity levels control the development of pancreatic diseases. Gastroenterology 137(3):1072–1082, 1082 e1071-1076. doi:S0016-5085(09)00900-7 [pii]. 10.1053/j.gastro.2009.05.052

Johnson L, Greenbaum D, Cichowski K, Mercer K, Murphy E, Schmitt E, Bronson RT, Umanoff H, Edelmann W, Kucherlapati R, Jacks T (1997) K-ras is an essential gene in the mouse with partial functional overlap with N-ras. Genes Dev 11(19):2468–2481

Jones S, Zhang X, Parsons DW, Lin JC, Leary RJ, Angenendt P, Mankoo P, Carter H, Kamiyama H, Jimeno A, Hong SM, Fu B, Lin MT, Calhoun ES, Kamiyama M, Walter K, Nikolskaya T,

Nikolsky Y, Hartigan J, Smith DR, Hidalgo M, Leach SD, Klein AP, Jaffee EM, Goggins M, Maitra A, Iacobuzio-Donahue C, Eshleman JR, Kern SE, Hruban RH, Karchin R, Papadopoulos N, Parmigiani G, Vogelstein B, Velculescu VE, Kinzler KW (2008) Core signaling pathways in human pancreatic cancers revealed by global genomic analyses. Science 321(5897):1801–1806, doi:1164368 [pii]. 10.1126/science.1164368

Kanda M, Matthaei H, Wu J, Hong SM, Yu J, Borges M, Hruban RH, Maitra A, Kinzler K, Vogelstein B, Goggins M (2012) Presence of somatic mutations in most early-stage pancreatic intraepithelial neoplasia. Gastroenterology 142(4):730–733 e739, doi:S0016-5085(12)00007-8 [pii]. 10.1053/j.gastro.2011.12.042

Kang S, Bader AG, Vogt PK (2005) Phosphatidylinositol 3-kinase mutations identified in human cancer are oncogenic. Proc Natl Acad Sci U S A 102(3):802–807, doi:0408864102 [pii]. 10.1073/pnas.0408864102

Kim JH, Kim HY, Lee YK, Yoon YS, Xu WG, Yoon JK, Choi SE, Ko YG, Kim MJ, Lee SJ, Wang HJ, Yoon G (2011) Involvement of mitophagy in oncogenic K-Ras-induced transformation: overcoming a cellular energy deficit from glucose deficiency. Autophagy 7(10):1187–1198, doi:16643 [pii]. 10.4161/auto.7.10.16643

Kobrin MS, Funatomi H, Friess H, Buchler MW, Stathis P, Korc M (1994) Induction and expression of heparin-binding EGF-like growth factor in human pancreatic cancer. Biochem Biophys Res Commun 202(3):1705–1709, doi:S0006-291X(84)72131-0 [pii]. 10.1006/bbrc.1994.2131

Koera K, Nakamura K, Nakao K, Miyoshi J, Toyoshima K, Hatta T, Otani H, Aiba A, Katsuki M (1997) K-ras is essential for the development of the mouse embryo. Oncogene 15(10):1151–1159. doi:10.1038/sj.onc.1201284

Korc M (2007) Pancreatic cancer-associated stroma production. Am J Surg 194(4 Suppl):S84–S86, doi:S0002-9610(07)00348-0 [pii]. 10.1016/j.amjsurg.2007.05.004

Krengel U, Schlichting I, Scherer A, Schumann R, Frech M, John J, Kabsch W, Pai EF, Wittinghofer A (1990) Three-dimensional structures of H-ras p21 mutants: molecular basis for their inability to function as signal switch molecules. Cell 62(3):539–548. doi:0092-8674(90)90018-A [pii]

Kumar R, Angelini S, Czene K, Sauroja I, Hahka-Kemppinen M, Pyrhonen S, Hemminki K (2003) BRAF mutations in metastatic melanoma: a possible association with clinical outcome. Clin Cancer Res 9(9):3362–3368

Lambert JM, Lambert QT, Reuther GW, Malliri A, Siderovski DP, Sondek J, Collard JG, Der CJ (2002) Tiam1 mediates Ras activation of Rac by a PI(3)K-independent mechanism. Nat Cell Biol 4(8):621–625, doi:10.1038/ncb833. ncb833 [pii]

Lampel M, Kern HF (1977) Acute interstitial pancreatitis in the rat induced by excessive doses of a pancreatic secretagogue. Virchows Arch A Pathol Anat Histol 373(2):97–117

Lee KE, Bar-Sagi D (2010) Oncogenic KRas suppresses inflammation-associated senescence of pancreatic ductal cells. Cancer Cell 18(5):448–458, doi:S1535-6108(10)00422-8 [pii]. 10.1016/j.ccr.2010.10.020

Lesina M, Kurkowski MU, Ludes K, Rose-John S, Treiber M, Kloppel G, Yoshimura A, Reindl W, Sipos B, Akira S, Schmid RM, Algul H (2011) Stat3/Socs3 activation by IL-6 transsignaling promotes progression of pancreatic intraepithelial neoplasia and development of pancreatic cancer. Cancer Cell 19(4):456–469, doi:S1535-6108(11)00119-X [pii]. 10.1016/j.ccr.2011.03.009

Levine B, Yuan J (2005) Autophagy in cell death: an innocent convict? J Clin Invest 115(10):2679–2688. doi:10.1172/JCI26390

Li J, Yen C, Liaw D, Podsypanina K, Bose S, Wang SI, Puc J, Miliaresis C, Rodgers L, McCombie R, Bigner SH, Giovanella BC, Ittmann M, Tycko B, Hibshoosh H, Wigler MH, Parsons R (1997) PTEN, a putative protein tyrosine phosphatase gene mutated in human brain, breast, and prostate cancer. Science 275(5308):1943–1947

Li J, Zhang Z, Dai Z, Plass C, Morrison C, Wang Y, Wiest JS, Anderson MW, You M (2003) LOH of chromosome 12p correlates with Kras2 mutation in non-small cell lung cancer. Oncogene 22(8):1243–1246, doi:10.1038/sj.onc.1206192. 1206192 [pii]

Li H, Cao HF, Wan J, Li Y, Zhu ML, Zhao P (2007) Growth inhibitory effect of wild-type Kras2 gene on a colonic adenocarcinoma cell line. World J Gastroenterol 13(6):934–938

Lim KH, Baines AT, Fiordalisi JJ, Shipitsin M, Feig LA, Cox AD, Der CJ, Counter CM (2005) Activation of RalA is critical for Ras-induced tumorigenesis of human cells. Cancer Cell 7(6):533–545, doi:S1535-6108(05)00157-1 [pii]. 10.1016/j.ccr.2005.04.030

Lim KH, O'Hayer K, Adam SJ, Kendall SD, Campbell PM, Der CJ, Counter CM (2006) Divergent roles for RalA and RalB in malignant growth of human pancreatic carcinoma cells. Curr Biol 16(24):2385–2394, doi:S0960-9822(06)02359-1 [pii]. 10.1016/j.cub.2006.10.023

Ling J, Kang Y, Zhao R, Xia Q, Lee DF, Chang Z, Li J, Peng B, Fleming JB, Wang H, Liu J, Lemischka IR, Hung MC, Chiao PJ (2012) KrasG12D-induced IKK2/beta/NF-kappaB activation by IL-1alpha and p62 feedforward loops is required for development of pancreatic ductal adenocarcinoma. Cancer Cell 21(1):105–120, doi:S1535-6108(11)00475-2 [pii]. 10.1016/j.ccr.2011.12.006

Lock R, Roy S, Kenific CM, Su JS, Salas E, Ronen SM, Debnath J (2011) Autophagy facilitates glycolysis during Ras-mediated oncogenic transformation. Mol Biol Cell 22(2):165–178, doi:mbc.E10-06-0500 [pii]. 10.1091/mbc.E10-06-0500

Lowenfels AB, Maisonneuve P, Lankisch PG (1999) Chronic pancreatitis and other risk factors for pancreatic cancer. Gastroenterol Clin North Am 28(3):673–685

Lowy AM, Fenoglio-Preiser C, Kim OJ, Kordich J, Gomez A, Knight J, James L, Groden J (2003) Dysregulation of beta-catenin expression correlates with tumor differentiation in pancreatic duct adenocarcinoma. Ann Surg Oncol 10(3):284–290

Luo T, Masson K, Jaffe JD, Silkworth W, Ross NT, Scherer CA, Scholl C, Frohling S, Carr SA, Stern AM, Schreiber SL, Golub TR (2012) STK33 kinase inhibitor BRD-8899 has no effect on KRAS-dependent cancer cell viability. Proc Natl Acad Sci U S A 109(8):2860–2865, doi:1120589109 [pii]. 10.1073/pnas.1120589109

Mahadevan D, Von Hoff DD (2007) Tumor-stroma interactions in pancreatic ductal adenocarcinoma. Mol Cancer Ther 6(4):1186–1197, doi:1535-7163.MCT-06-0686 [pii]. 10.1158/1535-7163.MCT-06-0686

Maitra A, Ashfaq R, Gunn CR, Rahman A, Yeo CJ, Sohn TA, Cameron JL, Hruban RH, Wilentz RE (2002) Cyclooxygenase 2 expression in pancreatic adenocarcinoma and pancreatic intraepithelial neoplasia: an immunohistochemical analysis with automated cellular imaging. Am J Clin Pathol 118(2):194–201. doi:10.1309/TPG4-CK1C-9V8V-8AWC

Maitra A, Adsay NV, Argani P, Iacobuzio-Donahue C, De Marzo A, Cameron JL, Yeo CJ, Hruban RH (2003) Multicomponent analysis of the pancreatic adenocarcinoma progression model using a pancreatic intraepithelial neoplasia tissue microarray. Mod Pathol 16(9):902–912. doi:10.1097/01.MP.0000086072.56290.FB

Mallen-St Clair J, Soydaner-Azeloglu R, Lee KE, Taylor L, Livanos A, Pylayeva-Gupta Y, Miller G, Margueron R, Reinberg D, Bar-Sagi D (2012) EZH2 couples pancreatic regeneration to neoplastic progression. Genes Dev 26(5):439–444, doi:26/5/439 [pii]. 10.1101/gad.181800.111

Maniati E, Bossard M, Cook N, Candido JB, Emami-Shahri N, Nedospasov SA, Balkwill FR, Tuveson DA, Hagemann T (2011) Crosstalk between the canonical NF-kappaB and Notch signaling pathways inhibits Ppargamma expression and promotes pancreatic cancer progression in mice. J Clin Invest 121(12):4685–4699, doi:45797 [pii]. 10.1172/JCI45797

Manne V, Bekesi E, Kung HF (1985) Ha-ras proteins exhibit GTPase activity: point mutations that activate Ha-ras gene products result in decreased GTPase activity. Proc Natl Acad Sci USA 82(2):376–380

Marais R, Light Y, Paterson HF, Marshall CJ (1995) Ras recruits Raf-1 to the plasma membrane for activation by tyrosine phosphorylation. EMBO J 14(13):3136–3145

Marigo I, Bosio E, Solito S, Mesa C, Fernandez A, Dolcetti L, Ugel S, Sonda N, Bicciato S, Falisi E, Calabrese F, Basso G, Zanovello P, Cozzi E, Mandruzzato S, Bronte V (2010) Tumor-induced tolerance and immune suppression depend on the C/EBPbeta transcription factor. Immunity 32(6):790–802, doi:S1074-7613(10)00202-5 [pii]. 10.1016/j.immuni.2010.05.010

McGrath JP, Capon DJ, Goeddel DV, Levinson AD (1984) Comparative biochemical properties of normal and activated human ras p21 protein. Nature 310(5979):644–649

Milburn MV, Tong L, deVos AM, Brunger A, Yamaizumi Z, Nishimura S, Kim SH (1990) Molecular switch for signal transduction: structural differences between active and inactive forms of protooncogenic ras proteins. Science 247(4945):939–945

Minami S (1923) Experiments on the surviving carcinoma tissues. (Respiration and glycolysis). Biochem Z 142:334–350

Miyamoto Y, Maitra A, Ghosh B, Zechner U, Argani P, Iacobuzio-Donahue CA, Sriuranpong V, Iso T, Meszoely IM, Wolfe MS, Hruban RH, Ball DW, Schmid RM, Leach SD (2003) Notch mediates TGF alpha-induced changes in epithelial differentiation during pancreatic tumorigenesis. Cancer Cell 3(6):565–576. doi:S1535610803001405 [pii]

Moodie SA, Willumsen BM, Weber MJ, Wolfman A (1993) Complexes of Ras.GTP with Raf-1 and mitogen-activated protein kinase kinase. Science 260(5114):1658–1661

Moore MJ, Goldstein D, Hamm J, Figer A, Hecht JR, Gallinger S, Au HJ, Murawa P, Walde D, Wolff RA, Campos D, Lim R, Ding K, Clark G, Voskoglou-Nomikos T, Ptasynski M, Parulekar W (2007) Erlotinib plus gemcitabine compared with gemcitabine alone in patients with advanced pancreatic cancer: a phase III trial of the National Cancer Institute of Canada Clinical Trials Group. J Clin Oncol 25(15):1960–1966, doi:JCO.2006.07.9525 [pii]. 10.1200/JCO.2006.07.9525

Morris JP, Cano DA, Sekine S, Wang SC, Hebrok M (2010) Beta-catenin blocks Kras-dependent reprogramming of acini into pancreatic cancer precursor lesions in mice. J Clin Invest 120(2):508–520, doi:40045 [pii]. 10.1172/JCI40045

Moskalenko S, Henry DO, Rosse C, Mirey G, Camonis JH, White MA (2002) The exocyst is a Ral effector complex. Nat Cell Biol 4(1):66–72, doi:10.1038/ncb728. ncb728 [pii]

Moskalenko S, Tong C, Rosse C, Mirey G, Formstecher E, Daviet L, Camonis J, White MA (2003) Ral GTPases regulate exocyst assembly through dual subunit interactions. J Biol Chem 278(51):51743–51748, doi:10.1074/jbc.M308702200. M308702200 [pii]

Nassar N, Horn G, Herrmann C, Scherer A, McCormick F, Wittinghofer A (1995) The 2.2 A crystal structure of the Ras-binding domain of the serine/threonine kinase c-Raf1 in complex with Rap1A and a GTP analogue. Nature 375(6532):554–560. doi:10.1038/375554a0

Navas C, Hernandez-Porras I, Schuhmacher AJ, Sibilia M, Guerra C, Barbacid M (2012) EGF receptor signaling is essential for k-ras oncogene-driven pancreatic ductal adenocarcinoma. Cancer Cell 22(3):318–330, doi:S1535-6108(12)00338-8 [pii]. 10.1016/j.ccr.2012.08.001

Neesse A, Michl P, Frese KK, Feig C, Cook N, Jacobetz MA, Lolkema MP, Buchholz M, Olive KP, Gress TM, Tuveson DA (2011) Stromal biology and therapy in pancreatic cancer. Gut 60(6):861–868, doi:gut.2010.226092 [pii]. 10.1136/gut.2010.226092

Niederau C, Ferrell LD, Grendell JH (1985) Caerulein-induced acute necrotizing pancreatitis in mice: protective effects of proglumide, benzotript, and secretin. Gastroenterology 88(5 Pt 1): 1192–1204. doi:S0016508585001457 [pii]

Nimnual AS, Yatsula BA, Bar-Sagi D (1998) Coupling of Ras and Rac guanosine triphosphatases through the Ras exchanger Sos. Science 279(5350):560–563

Nolan-Stevaux O, Lau J, Truitt ML, Chu GC, Hebrok M, Fernandez-Zapico ME, Hanahan D (2009) GLI1 is regulated through Smoothened-independent mechanisms in neoplastic pancreatic ducts and mediates PDAC cell survival and transformation. Genes Dev 23(1):24–36, doi:23/1/24 [pii]. 10.1101/gad.1753809

Ohnami S, Matsumoto N, Nakano M, Aoki K, Nagasaki K, Sugimura T, Terada M, Yoshida T (1999) Identification of genes showing differential expression in antisense K-ras-transduced pancreatic cancer cells with suppressed tumorigenicity. Cancer Res 59(21):5565–5571

Ohshio G, Saluja A, Leli U, Sengupta A, Steer ML (1989) Failure of a potent cholecystokinin antagonist to protect against diet-induced pancreatitis in mice. Pancreas 4(6):739–743

Oliveira-Cunha M, Hadfield KD, Siriwardena AK, Newman W (2012) EGFR and KRAS mutational analysis and their correlation to survival in pancreatic and periampullary cancer. Pancreas 41(3):428–434, doi:10.1097/MPA.0b013e3182327a03. 00006676-201204000-00011 [pii]

Otto JC, Kim E, Young SG, Casey PJ (1999) Cloning and characterization of a mammalian prenyl protein-specific protease. J Biol Chem 274(13):8379–8382

Parada LF, Tabin CJ, Shih C, Weinberg RA (1982) Human EJ bladder carcinoma oncogene is homologue of Harvey sarcoma virus ras gene. Nature 297(5866):474–478

Pasca di Magliano M, Biankin AV, Heiser PW, Cano DA, Gutierrez PJ, Deramaudt T, Segara D, Dawson AC, Kench JG, Henshall SM, Sutherland RL, Dlugosz A, Rustgi AK, Hebrok M (2007) Common activation of canonical Wnt signaling in pancreatic adenocarcinoma. PLoS One 2(11):e1155. doi:10.1371/journal.pone.0001155

Patek CE, Arends MJ, Wallace WA, Luo F, Hagan S, Brownstein DG, Rose L, Devenney PS, Walker M, Plowman SJ, Berry RL, Kolch W, Sansom OJ, Harrison DJ, Hooper ML (2008a) Mutationally activated K-ras 4A and 4B both mediate lung carcinogenesis. Exp Cell Res 314(5):1105–1114, doi:S0014-4827(07)00533-2 [pii]. 10.1016/j.yexcr.2007.11.004

Patek CE, Arends MJ, Rose L, Luo F, Walker M, Devenney PS, Berry RL, Lawrence NJ, Ridgway RA, Sansom OJ, Hooper ML (2008b) The pro-apoptotic K-Ras 4A proto-oncoprotein does not affect tumorigenesis in the ApcMin/+ mouse small intestine. BMC Gastroenterol 8:24, doi:1471-230X-8-24 [pii]. 10.1186/1471-230X-8-24

Pells S, Divjak M, Romanowski P, Impey H, Hawkins NJ, Clarke AR, Hooper ML, Williamson DJ (1997) Developmentally-regulated expression of murine K-ras isoforms. Oncogene 15(15):1781–1786. doi:10.1038/sj.onc.1201354

Pillinger MH, Volker C, Stock JB, Weissmann G, Philips MR (1994) Characterization of a plasma membrane-associated prenylcysteine-directed alpha carboxyl methyltransferase in human neutrophils. J Biol Chem 269(2):1486–1492

Plowman SJ, Williamson DJ, O'Sullivan MJ, Doig J, Ritchie AM, Harrison DJ, Melton DW, Arends MJ, Hooper ML, Patek CE (2003) While K-ras is essential for mouse development, expression of the K-ras 4A splice variant is dispensable. Mol Cell Biol 23(24):9245–9250

Plowman SJ, Berry RL, Bader SA, Luo F, Arends MJ, Harrison DJ, Hooper ML, Patek CE (2006a) K-ras 4A and 4B are co-expressed widely in human tissues, and their ratio is altered in sporadic colorectal cancer. J Exp Clin Cancer Res 25(2):259–267

Plowman SJ, Arends MJ, Brownstein DG, Luo F, Devenney PS, Rose L, Ritchie AM, Berry RL, Harrison DJ, Hooper ML, Patek CE (2006b) The K-Ras 4A isoform promotes apoptosis but does not affect either lifespan or spontaneous tumor incidence in aging mice. Exp Cell Res 312(1):16–26, doi:S0014-4827(05)00460-X [pii]. 10.1016/j.yexcr.2005.10.004

Prevot PP, Simion A, Grimont A, Colletti M, Khalaileh A, Van den Steen G, Sempoux C, Xu X, Roelants V, Hald J, Bertrand L, Heimberg H, Konieczny SF, Dor Y, Lemaigre FP, Jacquemin P (2012) Role of the ductal transcription factors HNF6 and Sox9 in pancreatic acinar-to-ductal metaplasia. Gut, doi:gutjnl-2011-300266 [pii]. 10.1136/gutjnl-2011-300266

Prior IA, Lewis PD, Mattos C (2012) A comprehensive survey of Ras mutations in cancer. Cancer Res 72(10):2457–2467, doi:72/10/2457 [pii]. 10.1158/0008-5472.CAN-11-2612

Pulciani S, Santos E, Lauver AV, Long LK, Robbins KC, Barbacid M (1982) Oncogenes in human tumor cell lines: molecular cloning of a transforming gene from human bladder carcinoma cells. Proc Natl Acad Sci U S A 79(9):2845–2849

Pylayeva-Gupta Y, Lee KE, Hajdu CH, Miller G, Bar-Sagi D (2012) Oncogenic Kras-induced GM-CSF production promotes the development of pancreatic neoplasia. Cancer Cell 21(6):836–847, doi:S1535-6108(12)00166-3 [pii]. 10.1016/j.ccr.2012.04.024

Qiu RG, Chen J, Kirn D, McCormick F, Symons M (1995) An essential role for Rac in Ras transformation. Nature 374(6521):457–459. doi:10.1038/374457a0

Qiu W, Sahin F, Iacobuzio-Donahue CA, Garcia-Carracedo D, Wang WM, Kuo CY, Chen D, Arking DE, Lowy AM, Hruban RH, Remotti HE, Su GH (2011) Disruption of p16 and activation of Kras in pancreas increase ductal adenocarcinoma formation and metastasis in vivo. Oncotarget 2(11):862–873. doi:357 [pii]

Racker E, Resnick RJ, Feldman R (1985) Glycolysis and methylaminoisobutyrate uptake in rat-1 cells transfected with ras or myc oncogenes. Proc Natl Acad Sci USA 82(11):3535–3538

Rangarajan A, Hong SJ, Gifford A, Weinberg RA (2004) Species- and cell type-specific requirements for cellular transformation. Cancer Cell 6(2):171–183, doi:10.1016/j.ccr.2004.07.009. S1535610804002053 [pii]

Reddy EP, Reynolds RK, Santos E, Barbacid M (1982) A point mutation is responsible for the acquisition of transforming properties by the T24 human bladder carcinoma oncogene. Nature 300(5888):149–152

Repasky GA, Chenette EJ, Der CJ (2004) Renewing the conspiracy theory debate: does Raf function alone to mediate Ras oncogenesis? Trends Cell Biol 14(11):639–647, doi:S0962-8924(04)00265-X [pii]. 10.1016/j.tcb.2004.09.014

Ridley AJ, Paterson HF, Johnston CL, Diekmann D, Hall A (1992) The small GTP-binding protein rac regulates growth factor-induced membrane ruffling. Cell 70(3):401–410. doi:0092-8674(92)90164-8 [pii]

Rodriguez-Viciana P, Warne PH, Dhand R, Vanhaesebroeck B, Gout I, Fry MJ, Waterfield MD, Downward J (1994) Phosphatidylinositol-3-OH kinase as a direct target of Ras. Nature 370(6490):527–532. doi:10.1038/370527a0

Rodriguez-Viciana P, Warne PH, Khwaja A, Marte BM, Pappin D, Das P, Waterfield MD, Ridley A, Downward J (1997) Role of phosphoinositide 3-OH kinase in cell transformation and control of the actin cytoskeleton by Ras. Cell 89(3):457–467. doi:S0092-8674(00)80226-3 [pii]

Ruggeri BA, Huang L, Wood M, Cheng JQ, Testa JR (1998) Amplification and overexpression of the AKT2 oncogene in a subset of human pancreatic ductal adenocarcinomas. Mol Carcinog 21(2):81–86. doi:10.1002/(SICI)1098-2744(199802)21:2<81::AID-MC1>3.0.CO;2-R [pii]

Samuels Y, Diaz LA Jr, Schmidt-Kittler O, Cummins JM, Delong L, Cheong I, Rago C, Huso DL, Lengauer C, Kinzler KW, Vogelstein B, Velculescu VE (2005) Mutant PIK3CA promotes cell growth and invasion of human cancer cells. Cancer Cell 7(6):561–573, doi:S1535-6108(05)00160-1 [pii]. 10.1016/j.ccr.2005.05.014

Santos E, Tronick SR, Aaronson SA, Pulciani S, Barbacid M (1982) T24 human bladder carcinoma oncogene is an activated form of the normal human homologue of BALB- and Harvey-MSV transforming genes. Nature 298(5872):343–347

Santos E, Martin-Zanca D, Reddy EP, Pierotti MA, Della Porta G, Barbacid M (1984) Malignant activation of a K-ras oncogene in lung carcinoma but not in normal tissue of the same patient. Science 223(4637):661–664

Sarbassov DD, Guertin DA, Ali SM, Sabatini DM (2005) Phosphorylation and regulation of Akt/PKB by the rictor-mTOR complex. Science 307(5712):1098–1101, doi:307/5712/1098 [pii]. 10.1126/science.1106148

Satoh T, Nakamura S, Kaziro Y (1987) Induction of neurite formation in PC12 cells by microinjection of proto-oncogenic Ha-ras protein preincubated with guanosine-5′-O-(3-thiotriphosphate). Mol Cell Biol 7(12):4553–4556

Schafer WR, Kim R, Sterne R, Thorner J, Kim SH, Rine J (1989) Genetic and pharmacological suppression of oncogenic mutations in ras genes of yeast and humans. Science 245(4916):379–385

Schafer WR, Trueblood CE, Yang CC, Mayer MP, Rosenberg S, Poulter CD, Kim SH, Rine J (1990) Enzymatic coupling of cholesterol intermediates to a mating pheromone precursor and to the ras protein. Science 249(4973):1133–1139

Scheffzek K, Ahmadian MR, Kabsch W, Wiesmuller L, Lautwein A, Schmitz F, Wittinghofer A (1997) The Ras-RasGAP complex: structural basis for GTPase activation and its loss in oncogenic Ras mutants. Science 277(5324):333–338

Scheidig AJ, Burmester C, Goody RS (1999) The pre-hydrolysis state of p21(ras) in complex with GTP: new insights into the role of water molecules in the GTP hydrolysis reaction of ras-like proteins. Structure 7(11):1311–1324. doi:st7b01 [pii]

Schlieman MG, Fahy BN, Ramsamooj R, Beckett L, Bold RJ (2003) Incidence, mechanism and prognostic value of activated AKT in pancreas cancer. Br J Cancer 89(11):2110–2115, doi:10.1038/sj.bjc.6601396. 6601396 [pii]

Scholl C, Frohling S, Dunn IF, Schinzel AC, Barbie DA, Kim SY, Silver SJ, Tamayo P, Wadlow RC, Ramaswamy S, Dohner K, Bullinger L, Sandy P, Boehm JS, Root DE, Jacks T, Hahn WC, Gilliland DG (2009) Synthetic lethal interaction between oncogenic KRAS dependency and STK33 suppression in human cancer cells. Cell 137(5):821–834, doi:S0092-8674(09)00316-X [pii]. 10.1016/j.cell.2009.03.017

Scholz A, Heinze S, Detjen KM, Peters M, Welzel M, Hauff P, Schirner M, Wiedenmann B, Rosewicz S (2003) Activated signal transducer and activator of transcription 3 (STAT3) supports the malignant phenotype of human pancreatic cancer. Gastroenterology 125(3):891–905. doi:S0016508503010643 [pii]

Schreiber AB, Libermann TA, Lax I, Yarden Y, Schlessinger J (1983) Biological role of epidermal growth factor-receptor clustering. Investigation with monoclonal anti-receptor antibodies. J Biol Chem 258(2):846–853

Schubbert S, Shannon K, Bollag G (2007a) Hyperactive Ras in developmental disorders and cancer. Nat Rev Cancer 7(4):295–308, doi:nrc2109 [pii]. 10.1038/nrc2109

Schubbert S, Bollag G, Shannon K (2007b) Deregulated Ras signaling in developmental disorders: new tricks for an old dog. Curr Opin Genet Dev 17(1):15–22, doi:S0959-437X(06)00239-5 [pii]. 10.1016/j.gde.2006.12.004

Scolnick EM, Papageorge AG, Shih TY (1979) Guanine nucleotide-binding activity as an assay for src protein of rat-derived murine sarcoma viruses. Proc Natl Acad Sci U S A 76(10):5355–5359

Sharma A, Zangen DH, Reitz P, Taneja M, Lissauer ME, Miller CP, Weir GC, Habener JF, Bonner-Weir S (1999) The homeodomain protein IDX-1 increases after an early burst of proliferation during pancreatic regeneration. Diabetes 48(3):507–513

Shi C, Hong SM, Lim P, Kamiyama H, Khan M, Anders RA, Goggins M, Hruban RH, Eshleman JR (2009) KRAS2 mutations in human pancreatic acinar-ductal metaplastic lesions are limited to those with PanIN: implications for the human pancreatic cancer cell of origin. Mol Cancer Res 7(2):230–236, doi:1541-7786.MCR-08-0206 [pii]. 10.1158/1541-7786.MCR-08-0206

Shigematsu H, Lin L, Takahashi T, Nomura M, Suzuki M, Wistuba II, Fong KM, Lee H, Toyooka S, Shimizu N, Fujisawa T, Feng Z, Roth JA, Herz J, Minna JD, Gazdar AF (2005) Clinical and biological features associated with epidermal growth factor receptor gene mutations in lung cancers. J Natl Cancer Inst 97(5):339–346, doi:97/5/339 [pii]. 10.1093/jnci/dji055

Shih C, Weinberg RA (1982) Isolation of a transforming sequence from a human bladder carcinoma cell line. Cell 29(1):161–169. doi:0092-8674(82)90100-3 [pii]

Shih TY, Papageorge AG, Stokes PE, Weeks MO, Scolnick EM (1980) Guanine nucleotide-binding and autophosphorylating activities associated with the p21src protein of Harvey murine sarcoma virus. Nature 287(5784):686–691

Shimizu K, Goldfarb M, Perucho M, Wigler M (1983a) Isolation and preliminary characterization of the transforming gene of a human neuroblastoma cell line. Proc Natl Acad Sci U S A 80(2):383–387

Shimizu K, Goldfarb M, Suard Y, Perucho M, Li Y, Kamata T, Feramisco J, Stavnezer E, Fogh J, Wigler MH (1983b) Three human transforming genes are related to the viral ras oncogenes. Proc Natl Acad Sci U S A 80(8):2112–2116

Silverman M, Ilardi C, Bank S, Kranz V, Lendvai S (1989) Effects of the cholecystokinin receptor antagonist L-364,718 on experimental pancreatitis in mice. Gastroenterology 96(1):186–192. doi:S0016508589000004 [pii]

Singh A, Greninger P, Rhodes D, Koopman L, Violette S, Bardeesy N, Settleman J (2009) A gene expression signature associated with "K-Ras addiction" reveals regulators of EMT and tumor cell survival. Cancer Cell 15(6):489–500, doi:S1535-6108(09)00111-1 [pii]. 10.1016/j.ccr.2009.03.022

Siveke JT, Einwachter H, Sipos B, Lubeseder-Martellato C, Kloppel G, Schmid RM (2007) Concomitant pancreatic activation of Kras(G12D) and Tgfa results in cystic papillary neoplasms reminiscent of human IPMN. Cancer Cell 12(3):266–279, doi:S1535-6108(07)00231-0 [pii]. 10.1016/j.ccr.2007.08.002

Siveke JT, Lubeseder-Martellato C, Lee M, Mazur PK, Nakhai H, Radtke F, Schmid RM (2008) Notch signaling is required for exocrine regeneration after acute pancreatitis. Gastroenterology 134(2):544–555, doi:S0016-5085(07)01993-2 [pii]. 10.1053/j.gastro.2007.11.003

Soh J, Okumura N, Lockwood WW, Yamamoto H, Shigematsu H, Zhang W, Chari R, Shames DS, Tang X, MacAulay C, Varella-Garcia M, Vooder T, Wistuba II, Lam S, Brekken R, Toyooka S, Minna JD, Lam WL, Gazdar AF (2009) Oncogene mutations, copy number gains and mutant allele specific imbalance (MASI) frequently occur together in tumor cells. PLoS One 4(10):e7464. doi:10.1371/journal.pone.0007464

Stanger BZ, Stiles B, Lauwers GY, Bardeesy N, Mendoza M, Wang Y, Greenwood A, Cheng KH, McLaughlin M, Brown D, Depinho RA, Wu H, Melton DA, Dor Y (2005) Pten constrains centroacinar cell expansion and malignant transformation in the pancreas. Cancer Cell 8(3):185–195, doi:S1535-6108(05)00236-9 [pii]. 10.1016/j.ccr.2005.07.015

Steck PA, Pershouse MA, Jasser SA, Yung WK, Lin H, Ligon AH, Langford LA, Baumgard ML, Hattier T, Davis T, Frye C, Hu R, Swedlund B, Teng DH, Tavtigian SV (1997) Identification of a candidate tumour suppressor gene, MMAC1, at chromosome 10q23.3 that is mutated in multiple advanced cancers. Nat Genet 15(4):356–362. doi:10.1038/ng0497-356

Swartz MA, Iida N, Roberts EW, Sangaletti S, Wong MH, Yull FE, Coussens LM, DeClerck YA (2012) Tumor microenvironment complexity: emerging roles in cancer therapy. Cancer Res 72(10):2473–2480, doi:0008-5472.CAN-12-0122 [pii]. 10.1158/0008-5472.CAN-12-0122

Sweet RW, Yokoyama S, Kamata T, Feramisco JR, Rosenberg M, Gross M (1984) The product of ras is a GTPase and the T24 oncogenic mutant is deficient in this activity. Nature 311(5983): 273–275

Tabin CJ, Bradley SM, Bargmann CI, Weinberg RA, Papageorge AG, Scolnick EM, Dhar R, Lowy DR, Chang EH (1982) Mechanism of activation of a human oncogene. Nature 300(5888):143–149

Tamanoi F, Walsh M, Kataoka T, Wigler M (1984) A product of yeast RAS2 gene is a guanine nucleotide binding protein. Proc Natl Acad Sci U S A 81(22):6924–6928

Tanaka H, Fukamachi K, Futakuchi M, Alexander DB, Long N, Tamamushi S, Minami K, Seino S, Ohara H, Joh T, Tsuda H (2010) Mature acinar cells are refractory to carcinoma development by targeted activation of Ras oncogene in adult rats. Cancer Sci 101(2):341–346, doi:CAS1410 [pii]. 10.1111/j.1349-7006.2009.01410.x

Taparowsky E, Suard Y, Fasano O, Shimizu K, Goldfarb M, Wigler M (1982) Activation of the T24 bladder carcinoma transforming gene is linked to a single amino acid change. Nature 300(5894):762–765

Temeles GL, Gibbs JB, D'Alonzo JS, Sigal IS, Scolnick EM (1985) Yeast and mammalian ras proteins have conserved biochemical properties. Nature 313(6004):700–703

Thayer SP, di Magliano MP, Heiser PW, Nielsen CM, Roberts DJ, Lauwers GY, Qi YP, Gysin S, Fernandez-del Castillo C, Yajnik V, Antoniu B, McMahon M, Warshaw AL, Hebrok M (2003) Hedgehog is an early and late mediator of pancreatic cancer tumorigenesis. Nature 425(6960):851–856, doi:10.1038/nature02009. nature02009 [pii]

Tian H, Callahan CA, DuPree KJ, Darbonne WC, Ahn CP, Scales SJ, de Sauvage FJ (2009) Hedgehog signaling is restricted to the stromal compartment during pancreatic carcinogenesis. Proc Natl Acad Sci U S A 106(11):4254–4259, doi:0813203106 [pii]. 10.1073/pnas.0813203106

To MD, Wong CE, Karnezis AN, Del Rosario R, Di Lauro R, Balmain A (2008) Kras regulatory elements and exon 4A determine mutation specificity in lung cancer. Nat Genet 40(10):1240–1244, doi:ng.211 [pii]. 10.1038/ng.211

Tobita K, Kijima H, Dowaki S, Kashiwagi H, Ohtani Y, Oida Y, Yamazaki H, Nakamura M, Ueyama Y, Tanaka M, Inokuchi S, Makuuchi H (2003) Epidermal growth factor receptor expression in human pancreatic cancer: significance for liver metastasis. Int J Mol Med 11(3):305–309

Tong LA, de Vos AM, Milburn MV, Kim SH (1991) Crystal structures at 2.2 A resolution of the catalytic domains of normal ras protein and an oncogenic mutant complexed with GDP. J Mol Biol 217(3):503–516

Trahey M, McCormick F (1987) A cytoplasmic protein stimulates normal N-ras p21 GTPase, but does not affect oncogenic mutants. Science 238(4826):542–545

Tuveson DA, Zhu L, Gopinathan A, Willis NA, Kachatrian L, Grochow R, Pin CL, Mitin NY, Taparowsky EJ, Gimotty PA, Hruban RH, Jacks T, Konieczny SF (2006) Mist1-KrasG12D knock-in mice develop mixed differentiation metastatic exocrine pancreatic carcinoma and hepatocellular carcinoma. Cancer Res 66(1):242–247, doi:66/1/242 [pii]. 10.1158/0008-5472. CAN-05-2305

Tzeng CW, Frolov A, Frolova N, Jhala NC, Howard JH, Vickers SM, Buchsbaum DJ, Heslin MJ, Arnoletti JP (2007) EGFR genomic gain and aberrant pathway signaling in pancreatic cancer patients. J Surg Res 143(1):20–26, doi:S0022-4804(07)00315-0 [pii]. 10.1016/j.jss.2007.01.051

Ushiro H, Cohen S (1980) Identification of phosphotyrosine as a product of epidermal growth factor-activated protein kinase in A-431 cell membranes. J Biol Chem 255(18):8363–8365

Vander Heiden MG, Cantley LC, Thompson CB (2009) Understanding the Warburg effect: the metabolic requirements of cell proliferation. Science 324(5930):1029–1033, doi:324/5930/1029 [pii]. 10.1126/science.1160809

Vincent DF, Yan KP, Treilleux I, Gay F, Arfi V, Kaniewski B, Marie JC, Lepinasse F, Martel S, Goddard-Leon S, Iovanna JL, Dubus P, Garcia S, Puisieux A, Rimokh R, Bardeesy N, Scoazec JY, Losson R, Bartholin L (2009) Inactivation of TIF1gamma cooperates with Kras to induce cystic tumors of the pancreas. PLoS Genet 5(7):e1000575. doi:10.1371/journal.pgen.1000575

Vojtek AB, Hollenberg SM, Cooper JA (1993) Mammalian Ras interacts directly with the serine/threonine kinase Raf. Cell 74(1):205–214. doi:0092-8674(93)90307-C [pii]

Wahlstrom AM, Cutts BA, Liu M, Lindskog A, Karlsson C, Sjogren AK, Andersson KM, Young SG, Bergo MO (2008) Inactivating Icmt ameliorates K-RAS-induced myeloproliferative disease. Blood 112(4):1357–1365, doi:blood-2007-06-094060 [pii]. 10.1182/blood-2007-06-094060

Walker EH, Perisic O, Ried C, Stephens L, Williams RL (1999) Structural insights into phosphoinositide 3-kinase catalysis and signalling. Nature 402(6759):313–320. doi:10.1038/46319

Wang W, Abbruzzese JL, Evans DB, Larry L, Cleary KR, Chiao PJ (1999) The nuclear factor-kappa B RelA transcription factor is constitutively activated in human pancreatic adenocarcinoma cells. Clin Cancer Res 5(1):119–127

Warburg O, Posener K, Negelein E (1924) On the metabolism of carcinoma cells. Biochem Z 152:309–344

Warne PH, Viciana PR, Downward J (1993) Direct interaction of Ras and the amino-terminal region of Raf-1 in vitro. Nature 364(6435):352–355. doi:10.1038/364352a0

Watanabe O, Baccino FM, Steer ML, Meldolesi J (1984) Supramaximal caerulein stimulation and ultrastructure of rat pancreatic acinar cell: early morphological changes during development of experimental pancreatitis. Am J Physiol 246(4 Pt 1):G457–G467

Whyte DB, Kirschmeier P, Hockenberry TN, Nunez-Oliva I, James L, Catino JJ, Bishop WR, Pai JK (1997) K- and N-Ras are geranylgeranylated in cells treated with farnesyl protein transferase inhibitors. J Biol Chem 272(22):14459–14464

Willemer S, Elsasser HP, Kern HF, Adler G (1987) Tubular complexes in cerulein- and oleic acid-induced pancreatitis in rats: glycoconjugate pattern, immunocytochemical, and ultrastructural findings. Pancreas 2(6):669–675

Willumsen BM, Norris K, Papageorge AG, Hubbert NL, Lowy DR (1984) Harvey murine sarcoma virus p21 ras protein: biological and biochemical significance of the cysteine nearest the carboxy terminus. EMBO J 3(11):2581–2585

Wolfman A, Macara IG (1990) A cytosolic protein catalyzes the release of GDP from p21ras. Science 248(4951):67–69

Wright LP, Philips MR (2006) Thematic review series: lipid posttranslational modifications. CAAX modification and membrane targeting of Ras. J Lipid Res 47(5):883–891, doi:R600004-JLR200 [pii]. 10.1194/jlr.R600004-JLR200

Yamamoto S, Tomita Y, Hoshida Y, Morooka T, Nagano H, Dono K, Umeshita K, Sakon M, Ishikawa O, Ohigashi H, Nakamori S, Monden M, Aozasa K (2004) Prognostic significance of activated Akt expression in pancreatic ductal adenocarcinoma. Clin Cancer Res 10(8): 2846–2850

Yang S, Wang X, Contino G, Liesa M, Sahin E, Ying H, Bause A, Li Y, Stommel JM, Dell'antonio G, Mautner J, Tonon G, Haigis M, Shirihai OS, Doglioni C, Bardeesy N, Kimmelman AC (2011) Pancreatic cancers require autophagy for tumor growth. Genes Dev 25(7):717–729, doi:gad.2016111 [pii]. 10.1101/gad.2016111

Yarden Y, Schlessinger J (1987a) Epidermal growth factor induces rapid, reversible aggregation of the purified epidermal growth factor receptor. Biochemistry 26(5):1443–1451

Yarden Y, Schlessinger J (1987b) Self-phosphorylation of epidermal growth factor receptor: evidence for a model of intermolecular allosteric activation. Biochemistry 26(5):1434–1442

Ying H, Kimmelman AC, Lyssiotis CA, Hua S, Chu GC, Fletcher-Sananikone E, Locasale JW, Son J, Zhang H, Coloff JL, Yan H, Wang W, Chen S, Viale A, Zheng H, Paik JH, Lim C, Guimaraes AR, Martin ES, Chang J, Hezel AF, Perry SR, Hu J, Gan B, Xiao Y, Asara JM, Weissleder R, Wang YA, Chin L, Cantley LC, DePinho RA (2012) Oncogenic Kras maintains pancreatic tumors through regulation of anabolic glucose metabolism. Cell 149(3):656–670, doi:S0092-8674(12)00352-2 [pii]. 10.1016/j.cell.2012.01.058

Yip-Schneider MT, Lin A, Barnard D, Sweeney CJ, Marshall MS (1999) Lack of elevated MAP kinase (Erk) activity in pancreatic carcinomas despite oncogenic K-ras expression. Int J Oncol 15(2):271–279

Yip-Schneider MT, Lin A, Marshall MS (2001) Pancreatic tumor cells with mutant K-ras suppress ERK activity by MEK-dependent induction of MAP kinase phosphatase-2. Biochem Biophys Res Commun 280(4):992–997, doi:10.1006/bbrc.2001.4243. S0006-291X(01)94243-3 [pii]

Yoshida T, Shiraki N, Baba H, Goto M, Fujiwara S, Kume K, Kume S (2008) Expression patterns of epiplakin1 in pancreas, pancreatic cancer and regenerating pancreas. Genes Cells 13(7):667–678, doi:GTC1196 [pii]. 10.1111/j.1365-2443.2008.01196.x

Yun J, Rago C, Cheong I, Pagliarini R, Angenendt P, Rajagopalan H, Schmidt K, Willson JK, Markowitz S, Zhou S, Diaz LA Jr, Velculescu VE, Lengauer C, Kinzler KW, Vogelstein B, Papadopoulos N (2009) Glucose deprivation contributes to the development of KRAS pathway mutations in tumor cells. Science 325(5947):1555–1559, doi:1174229 [pii]. 10.1126/science.1174229

Zhang XF, Settleman J, Kyriakis JM, Takeuchi-Suzuki E, Elledge SJ, Marshall MS, Bruder JT, Rapp UR, Avruch J (1993) Normal and oncogenic p21ras proteins bind to the amino-terminal regulatory domain of c-Raf-1. Nature 364(6435):308–313. doi:10.1038/364308a0

Zhang Z, Wang Y, Vikis HG, Johnson L, Liu G, Li J, Anderson MW, Sills RC, Hong HL, Devereux TR, Jacks T, Guan KL, You M (2001) Wildtype Kras2 can inhibit lung carcinogenesis in mice. Nat Genet 29(1):25–33, doi:10.1038/ng721. ng721 [pii]

Zhang X, Gureasko J, Shen K, Cole PA, Kuriyan J (2006) An allosteric mechanism for activation of the kinase domain of epidermal growth factor receptor. Cell 125(6):1137–1149, doi:S0092-8674(06)00584-8 [pii]. 10.1016/j.cell.2006.05.013

Zhou W, Capello M, Fredolini C, Piemonti L, Liotta LA, Novelli F, Petricoin EF (2011) Proteomic analysis of pancreatic ductal adenocarcinoma cells reveals metabolic alterations. J Proteome Res 10(4):1944–1952. doi:10.1021/pr101179t

Zhou W, Capello M, Fredolini C, Racanicchi L, Piemonti L, Liotta LA, Novelli F, Petricoin EF (2012) Proteomic analysis reveals Warburg effect and anomalous metabolism of glutamine in pancreatic cancer cells. J Proteome Res 11(2):554–563. doi:10.1021/pr2009274

Zhu Z, Kleeff J, Friess H, Wang L, Zimmermann A, Yarden Y, Buchler MW, Korc M (2000) Epiregulin is up-regulated in pancreatic cancer and stimulates pancreatic cancer cell growth. Biochem Biophys Res Commun 273(3):1019–1024, doi:10.1006/bbrc.2000.3033. S0006-291X(00)93033-X [pii]

Molecular Targeted Therapies in Pancreatic Cancer

Edward Kim, Ethan V. Abel, Arunima Ghosh, and Diane M. Simeone

Abstract Pancreatic cancer, one of the deadliest malignancies, is a complex disease consisting of heterogeneous cancer cells with deregulated signaling pathways and a myriad of microenvironment cells, including infiltrating immune cells and fibroblasts, that impact tumor growth and susceptibility to conventional chemotherapy. Understanding the signaling pathways that drive pancreatic cancer is crucial to the development of novel targeted therapies to combat the disease, which is largely refractory to conventional therapeutic options. Among these pathways are the Hedgehog, NOTCH, Wnt, MET, and TGF-β pathways that control not only bulk tumor growth, but also self-renewal of cancer stem cells and maintenance of the desmoplastic stroma characteristic of the disease. In addition to altered signaling pathways, many cells within the tumor microenvironment promote both tumor growth and serve as a barrier to chemotherapy. Here we will discuss how targeting these components of the disease may increase the efficacy with which it is treated.

E. Kim
Department of Internal Medicine, University of Michigan Medical Center, Ann Arbor, MI 48109, USA

E.V. Abel • A. Ghosh
Department of Surgery, University of Michigan Medical Center, Ann Arbor, MI 48109, USA

D.M. Simeone, M.D. (✉)
Department of Surgery, University of Michigan Medical Center, Ann Arbor, MI 48109, USA

Department of Molecular and Integrative Physiology, University of Michigan Medical Center, Ann Arbor, MI 48109, USA
e-mail: simeone@umich.edu

D.M. Simeone and A. Maitra (eds.), *Molecular Genetics of Pancreatic Cancer*,
DOI 10.1007/978-1-4614-6549-2_6,

Introduction

The lethality of pancreatic can be attributed to the absence of early detection, the inherent aggressive nature of the tumor, and resistance to currently available standard therapies. Gemcitabine has historically been the cornerstone of systemic chemotherapy, with limited improvement with the addition of other cytotoxic chemotherapies. Recently, a three-drug regimen, FOLFIRINOX, was shown to improve clinical outcomes in a clinically significant way but at a cost of toxicity that limits broad application (Conroy et al. 2011). Despite these therapeutic options, the duration of response to chemotherapy is limited in patients with pancreatic cancer, indicating a need to develop novel therapies against the disease. In this chapter we will focus on developmental signaling pathways that play a critical role in pancreatic cancer which may serve as promising therapeutic targets. We will also discuss how we may potentially improve therapeutic efficacy and clinical outcomes by targeting the desmoplastic stroma characteristic of pancreatic cancer as well as the particularly virulent pancreatic cancer stem cell (CSC) subpopulation.

Targeting Developmental Signaling Pathways

There is a distinct pattern of histologic changes in pancreatic tumorigenesis that begins with precursor pancreatic intraepithelial neoplasm (PanIN) lesions that eventually progress to invasive pancreatic ductal adenocarcinoma (PDAC). Specific mutations accompany these histologic changes, including KRAS mutations which can be found in the earliest PanIN lesions (i.e., PanIN-1) (Hezel et al. 2006). Ultimately, KRAS is mutated in greater than 95 % of pancreatic carcinomas. Other genetic changes commonly found in invasive pancreatic cancers such as mutations or deletions in the tumor suppressor genes p16/INK4A, p53, DPC4/SMAD4 also occur during PanIN progression to pancreatic cancer (Hezel et al. 2006). In addition to these and other genetic changes, there are molecular changes in pancreatic cancers that involve reactivation of developmental signaling pathways such as Hedgehog, NOTCH and Wnt. These developmental signaling pathways are included in a set of 12 core signaling pathways determined to be altered in pancreatic cancers through a comprehensive global genomic analysis (Jones et al. 2008). An average of 63 genetic alterations was found in pancreatic cancers, a majority of which were point mutations but also included deletions and amplifications. While distinct individual changes were seen within any given tumor, the specific alterations could be grouped into a set of 12 core signaling pathways including KRAS, TGF-β, Wnt/NOTCH/Hedgehog, cell cycle, and DNA repair genes (Jones et al. 2008). In the following section, we will discuss several of the developmental signaling pathways aberrantly activated in pancreatic cancer and describe their potential to serve as therapeutic targets.

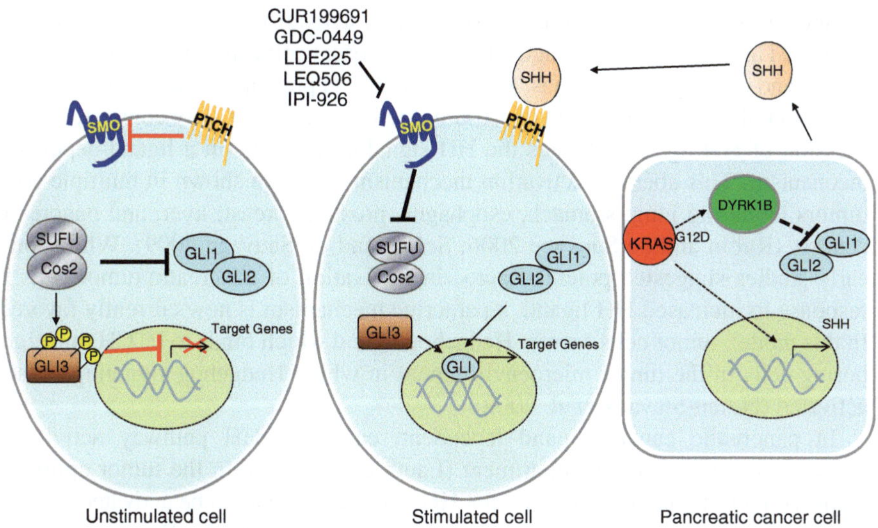

Fig. 1 The HH signaling pathway. The hedgehog signaling pathway is shown in three relevant scenarios: unstimulated cells (*left*), stimulated cells (*middle*), and pancreatic cancer cells (*right*). In unstimulated cells, PTCH inhibits the activity of SMO, resulting in inhibition of GLI-mediated transcription. In the presence of ligand (SHH), PTCH no longer inhibits SMO, which in turn inhibits SUFU and Cos2, resulting in GLI translocation to the nucleus and transcriptional activation of hedgehog signaling target genes. In pancreatic cancer cells, SHH is upregulated by oncogenic KRAS, which also blocks autonomous GLI activation through its effector, DYRK1B. Pancreatic cancer cell-secreted SHH stimulates neighboring cells, including pancreatic stellate cells

The Hedgehog Signaling Pathway

The Hedgehog signaling pathway is vital for spatial patterning during embryonic development (Ingham and McMahon 2001). Hedgehog signaling has been shown to regulate cell fate specification (heart, skin, eye), cell proliferation (lung, muscle, neural crest), and cell survival (gonad) in different target cells (Ingham and McMahon 2001; Ruiz i Altaba et al. 2002a, b; Berman et al. 2003). Canonical activation of the Hedgehog signaling pathway begins with binding of one of the three hedgehog (HH) ligands (Sonic, Indian, and Desert) to the 12-transmembrane protein Patched (PTCH). In the absence of HH ligand, PTCH actively represses the activity of Smoothened (SMO), a seven transmembrane receptor-like protein (Fig. 1). HH ligand binding to PTCH inhibits its repression of SMO, allowing SMO to then transduce the signal internally via the GLI family of transcription factors (Rubin and de Sauvage 2006; Gupta et al. 2010). This results in transcriptional activation of hedgehog transcriptional genes such *PTCH*, *GLI* and Hedgehog Interaction protein *HHIP*.

Aberrant Hedgehog signaling has been associated with cancer through several different mechanisms (Scales and de Sauvage 2009; Rubin and de Sauvage 2006). One mechanism of aberrant pathway activation is through mutation of a pathway component that allows for constitutive pathway activation in a ligand-independent manner. This is the pattern of activation seen in basal cell carcinomas, medulloblastomas, and rhabdomyosarcomas. Mutation of PTCH prevents it from its usual active inhibition of SMO, resulting in constitutive activation of SMO. By contrast, overexpression of HH ligand activates the HH signaling pathway in a ligand-dependent mechanism. This aberrant activation mechanism has been shown in multiple solid tumors including lung, stomach, esophagus, prostate, breast, liver, and pancreatic cancers (Rubin and de Sauvage 2006; Scales and de Sauvage 2009). While some early studies suggested potential autocrine activation of pancreatic tumor cells in response to increased HH ligand, a paracrine mechanism is now currently favored. In this model, tumor cells secrete Hedgehog ligand which binds to PTCH on neighboring cells in the tumor microenvironment in which Hedgehog signaling is then activated (Nolan-Stevaux et al. 2009).

In pancreatic cancer, ligand-dependent, canonical HH pathway activity is restricted to the stromal compartment (Lauth et al. 2010). In the tumor epithelial compartment of pancreatic cancer, the HH pathway appears to be activated by non-canonical upregulation of the effector transcription factor GLI1. Pancreatic cancer cells appear to be insensitive to HH ligand and in fact SMO is not required for pancreatic tumorigenesis (Nolan-Stevaux et al. 2009). Instead, GLI1 expression in pancreatic cancer epithelial cells is regulated by KRAS via its effector molecule DYRK1B, as well as by TGF-β signaling (Lauth et al. 2010; Nolan-Stevaux et al. 2009). While TGF-β signaling promotes the expression of GLI1 in pancreatic cancer cells, oncogenic KRAS and DYRK1B suppress its expression, limiting cell autonomous HH-signaling in pancreatic cancer cells (Lauth et al. 2010).

The normal adult pancreas does not normally express HH ligand, while expression of HH is a common feature of pancreatic cancer (Kim and Simeone 2011). Aberrant expression of HH ligand has been shown to occur as early as PanIN 1 lesions, with increasing levels expressed as these lesions progress to PDAC (Thayer et al. 2003). Sonic HH (SHH) is the dominant HH ligand expressed in pancreatic cancer and is aberrantly expressed in 70 % of patient tumors. A causal role for this aberrant SHH expression in pancreatic tumorigenesis is supported by evidence from a genetically engineered mouse model (Pdx-Shh) in which SHH is expressed in the pancreatic endoderm, resulting in development of abnormal tubular structures similar to human PanIN-1 and 2 lesions (Thayer et al. 2003). SHH has also been shown to play a critical role in formation and maintenance of the desmoplasia characteristic of pancreatic cancers (Bailey et al. 2008). Overexpression of SHH expression in a pancreatic epithelial cell line that forms xenograft tumors results in enhanced fibroblast infiltration. This fibrotic infiltration is accompanied by increased expression of the acellular components of the desmoplastic stroma, including collagen I and fibronectin (Bailey et al. 2008). HH signaling thus appears to play a role in the generation of the dense stroma that is seen in primary pancreatic tumors.

The formation of the dense stroma characteristic of pancreatic cancers appears to contribute to virulence of the cancer cells by promoting metastatic progression. It may also pose a physical barrier to drug delivery and contribute to the apparent resistance of pancreatic cancers to drug therapy. Given the above described role of the HH signaling pathway in the stroma of pancreatic cancers, blocking this pathway may facilitate improvement in drug efficacy by simply allowing greater penetration and drug delivery into tumor. Several novel agents that target the HH signaling pathway currently are in clinical development and share the general approach of inhibiting the SMO protein. HH pathway inhibition using a SMO antagonist has been studied in the KPC (KrasLSL.G12D/+; p53R172H/+; PdxCre) mouse model of pancreatic cancer, a well-studied model of pancreatic cancer that recapitulates human tumors, including formation of desmoplastic stroma (Olive et al. 2009; Hingorani et al. 2005). Treatment of KPC mice with the smoothened antagonist IPI-926, given in combination with the standard chemotherapy drug gemcitabine, produced a transient increase in tumor vascularity and intratumoral concentration of gemcitabine, leading to transient stabilization of disease (Olive et al. 2009). KPC mice treated with gemcitabine alone or IPI-926 alone showed no survival benefit in comparison with vehicle-treated controls; however, combination treatment with IPI-926 and gemcitabine extended the median survival of KPC mice from 11 to 25 days ($p=0.001$) (Olive et al. 2009). Although the effects were transient, these results provided preclinical evidence that targeting the Hedgehog signaling pathway may increase response to chemotherapy. There are a number of other HH pathway inhibitors in clinical development which all target SMO, including LDE225 (Novartis), LEQ506 (Novartis), GDC-0449 (Genentech), and IPI-926 (Infinity Pharmaceuticals). These novel agents are currently being studied in early phase trials for patients with advanced pancreatic cancer in combination with chemotherapy (www.clinicaltrials.gov).

The NOTCH Signaling Pathway

NOTCH signaling plays an important role in cell fate and differentiation through effects on cell proliferation, survival and apoptosis (Artavanis-Tsakonas et al. 1995; D'Souza et al. 2008; Fiuza and Arias 2007). This pathway also regulates adult stem cell homeostasis and maintenance (Gridley 1997, 2003). The NOTCH signaling pathway involves activation of the NOTCH receptor by ligand. Five NOTCH ligands have been identified to date, which include Dll-1 (Delta-like1), Dll-3 (Delta-like3), Dll-4 (Delta-like4) (Bettenhausen et al. 1995; Dunwoodie et al. 1997; Shutter et al. 2000), Jagged-1, and Jagged-2 (Lindsell et al. 1995; Shawber et al. 1996). Four members of the NOTCH family of receptors have been identified, NOTCH1-4. Upon activation by one of these ligands, the NOTCH receptor is cleaved by the metalloprotease tumor necrosis factor α convertase enzyme (TACE) and γ-secretase, releasing the intracellular domain of NOTCH (ICD) (Fig. 2). ICD translocates from the cell surface to the nucleus and binds the transcription factor CSL. In the absence of NOTCH pathway activation, CSL is free to bind to co-repressors which inhibit transcription (Kao et al. 1998; Hsieh et al. 1999; Morel et al. 2001). NOTCH pathway activation allows ICD to compete with inhibitory proteins to bind to CSL and to recruit co-activators, including p300, mastermind-like 1–3 (MAML1-3), and

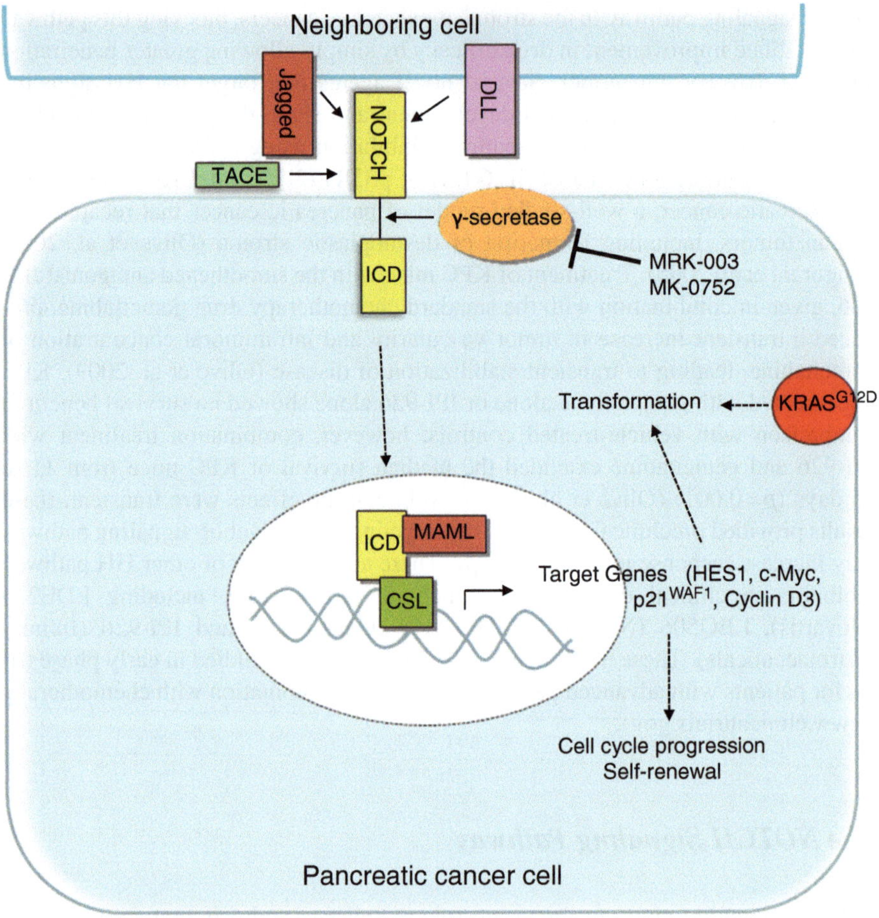

Fig. 2 The NOTCH signaling pathway. In NOTCH signaling, ligand-presenting cells stimulate the extracellular domain of NOTCH proteins (here on a pancreatic cancer cell) with either Delta-like (DLL) or Jagged family ligands. Upon stimulation, TACE and γ-secretase cleave NOTCH proteins, and the intracellular domain (ICD) translocates to the nucleus where it binds the transcription factor CSL and recruits coactivators like MAML. This promotes the transcription of NOTCH target genes

histone acetyltransferases. This process converts CSL from a transcriptional repressor to transcriptional activator (Zhou et al. 2000; Kurooka and Honjo 2000; Fryer et al. 2002, 2004). Several NOTCH target genes have been identified, including HES1 (hairy/enhancer of Split), c-Myc, cyclin D3, and $p21^{WAF1}$ (Blaumueller et al. 1997).

NOTCH1 was first characterized as an oncogene in human T-cell acute lymphoblastic leukemia (Reynolds et al. 1987) and subsequently in several epithelial tumors (Gallahan and Callahan 1997; Gallahan et al. 1996; Jhappan et al. 1992), including head and neck, breast, renal, lung, and colon cancers (Radtke and Raj 2003). In the pancreas, NOTCH signaling is normally suppressed in early development (Apelqvist et al. 1999; Jensen et al. 2000) but has been found to be upregulated in pancreatic cancer (Miyamoto et al. 2003). Further evidence supporting a causal role for aberrant NOTCH pathway activation in pancreatic cancer can be found from genetically engineered mouse models of pancreatic cancer. In the KRAS mouse model of pancreas cancer, NOTCH pathway activation can be seen in PanIN lesions (Hingorani et al. 2003). Co-expression of NOTCH1 with oncogenic KRAS in pancreatic acinar cells results in rapid and widespread transformation of acinar cells to duct-like cells and progression to aggressive, high-grade lesions (De La O et al. 2008).

Based on evidence of upregulation of NOTCH signaling in pancreatic cancer, targeting this pathway is of clinical interest for therapeutic application to patients with pancreatic cancer. The primary target for therapeutic intervention in the NOTCH signaling pathway thus far has been the enzyme γ-secretase, responsible for the last cleavage step of the NOTCH receptor that releases ICD. In preclinical studies, inhibition of NOTCH signaling by down regulation of NOTCH1 receptors using specific siRNA or γ-secretase inhibitors (GSI) reduced proliferation, increased apoptosis and decreased invasion of pancreatic cancer cells (Plentz et al. 2009; Mullendore et al. 2009; Wang et al. 2006). In KPC mice, treatment with the GSI, MRK-003 (Merck), attenuated the progression of PanIN lesions to PDAC (Plentz et al. 2009). A recent study exploring the effects of MRK-003 and gemcitabine in the same KPC mouse model of pancreatic cancer found the combined treatment reduced the proliferation of neoplastic cells, significantly induced endothelial cell death and reduced the density of intratumoral vessels (Cook et al. 2012). In this study it was proposed that the hypoxia caused by endothelial cell death sensitized the tumor cells to the effects of GSI by activating target genes such as survivin and NOTCH3 (Cook et al. 2012). This combination of MRK003/gemcitabine is currently being tested in an ongoing clinical trial in the United Kingdom. Another GSI, MK-0752 (Merck), is being tested in combination with gemcitabine in patients with advanced pancreatic cancer (www.clinicaltrials.gov).

The Wnt Signaling Pathway

Wnt-β-catenin signaling is required for morphogenesis, proliferation and differentiation of many organs. *Wnt* genes encode small, secreted proteins that are involved in many aspects of embryonic development and also control homeostatic self-renewal in a number of adult tissues (Clevers 2006; Willert and Jones 2006). To initiate pathway signaling, Wnt ligands (19 family members) bind to receptors of the Frizzled (Fzd) family (10 members), which in turn interact with transmembrane co-receptors LRP5/6 (Fig. 3). Activated LRP5/6 then recruits the protein, Dishevelled (Dsh), at which point Wnt signaling can branch into two different pathways, a canonical and noncanonical pathway. In the canonical pathway (Fig. 3), in the absence of Wnt, unstimulated cells regulate β-catenin levels by a multiprotein complex which phosphorylates β-catenin, leading to its subsequent ubiquitination and degradation. This β-catenin degradation complex consists of the adenomatous

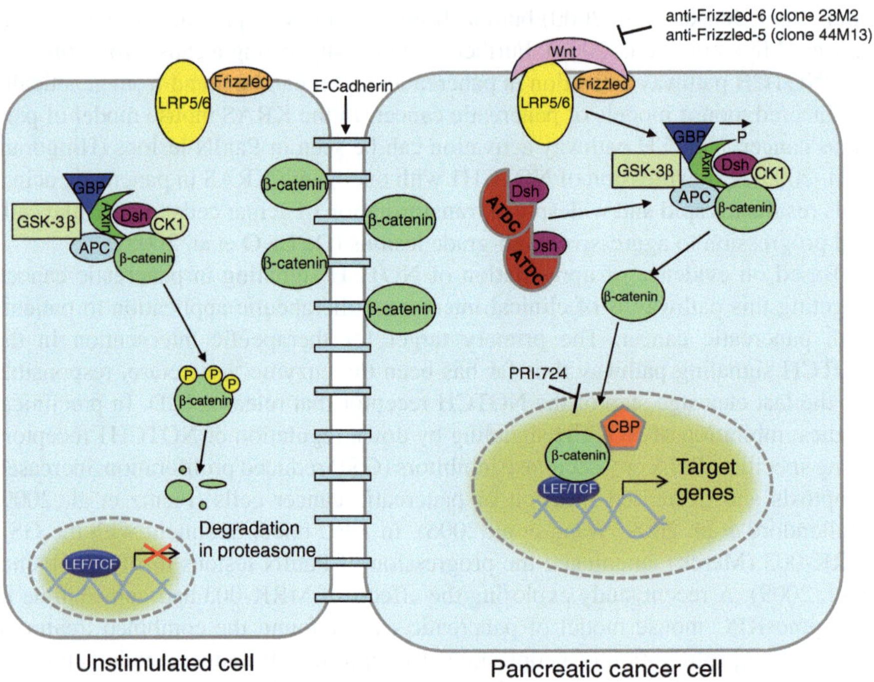

Fig. 3 The Wnt signaling pathway. In canonical Wnt signaling, cells exist in either an unstimulated (*left*) or stimulated (*right*) state. In unstimulated cells, β-catenin is mostly complexed with E-cadherin, while free β-catenin is phosphorylated and degraded by a complex consisting of APC, Axin, and GSK3β. Upon stimulation by Wnt proteins, the Frizzled receptors and LRP5/6 co-receptors activate disheveled (Dsh), which inhibits the degradation complex, allowing β-catenin to translocate to the nucleus, bind to LEF/TCF, and promote target gene transcription. In pancreatic cancer, upregulation of ATDC promotes β-catenin stability through binding and stabilization of Dsh

polyposis coli (APC) tumor suppressor protein, Axin, and the glycogen synthase kinase, GSK3β. Binding of Wnt to Fzd leads to inactivation of the degradation complex and accumulation of unphosphorylated β-catenin, which localizes to the nucleus. In the nucleus, β-catenin binds to TCF/LEF (T-cell factor/lymphoid enhancing factor) to activate downstream target genes (Willert and Jones 2006; Clevers 2006) (Fig. 3).

Wnt also is activated by the "noncanonical" pathway which is independent of TCF/LEF and β-catenin. The "noncanonical" pathway is divided into two types: the Planar Cell Polarity (PCP) pathway and Wnt-Calcium pathway. In the PCP pathway, which has mostly been studied in Drosophila, Wnt signaling is transduced through Fzd independent of the co-receptors LRP5/6, leading to the activation of Dsh (Nishimura et al. 2012). Dsh, through Daam1 (Dishevelled associated activator of morphogenesis), mediates activation of Rho, Rock and JNK, inducing cytoskeletal changes important for cell polarization and motility during gastrulation (Nishimura et al. 2012; Kohn and Moon 2005). In the Wnt-Ca pathway, Wnt 5a and Wnt11, through activation of Fzd receptors, can stimulate intracellular Ca^{2+} release from the endoplasmic reticulum, which activates G-proteins without affecting β-catenin stabilization (Kohn and Moon 2005).

Several studies have proposed a role for the canonical Wnt pathway in pancreatic organogenesis. Evidence that Wnt/β-catenin signaling is important for the developing pancreas came from Heller et al. and others who have demonstrated expression of Wnt2b, Wnt 4, Wnt5a, Wnt7b and Frizzled receptors in the developing pancreas (Heller et al. 2002; Murtaugh et al. 2005). Later in development, the Wnt signaling pathway appears to promote proliferation and/or differentiation of acinar cells (Murtaugh et al. 2005; Wells et al. 2007; Morris et al. 2010). Wnt/β-catenin signaling may also be involved in maintaining normal islet cell development (Dessimoz et al. 2005).

The Wnt β-catenin pathway has been implicated as playing a key role in initiation and progression of cancer in many tissue types. The best studied pathway mutations are the inherited and sporadic mutations in the tumor suppressor APC. Monoallelic inactivating mutations in APC result in familial adenomatous polyposis (FAP), an inherited autosomal dominant condition leading to the development of multiple adenomas in the colorectum (Groden et al. 1991; Nishisho et al. 1991). Additionally, mutations in the gene encoding β-catenin (*CTNNB1*) are present in approximately 10 % of the remaining CRC tumors, mostly in early or smaller, less aggressive tumors (Samowitz et al. 1999). Loss of function mutations in APC or gain of function mutations in β-catenin are both rare in pancreatic cancer, except in the setting of pseudopapillary tumors in the pancreas, where mutations in β-catenin are driver mutations for the disease (Abraham et al. 2002). The contribution of aberrant Wnt signaling to pancreatic tumorigenesis was first demonstrated by Pasca di Magliano and colleagues, where they showed that the canonical arm of the Wnt pathway is induced in human PDA as well as in mouse models of pancreatic cancer. Wnt inhibition could block proliferation and apoptosis in cultured pancreatic adenocarcinoma cells (Pasca di Magliano et al. 2007).

In addition to the core components of canonical and noncanonical Wnt signaling, other novel regulators of Wnt signaling have been identified in pancreatic cancer. A recently identified oncogene in pancreatic cancer, the ataxia telangiectasia Group D associated gene (ATDC), has been shown to promote pancreatic tumor growth and metastasis, at least in part, through upregulation of the β-catenin signaling pathway (Wang et al. 2009a). ATDC was shown to bind and stabilize Disheveled-2, bringing it to the β-catenin degradation complex. Binding of ATDC and Disheveled-2 to the degradation complex results in inhibition of degradation complex, release of β catenin from the complex, and subsequent activation of the downstream target genes (Wang et al. 2009a). Another mechanism of activating the Wnt signaling pathway in pancreatic cancer involves Sulfs. The extracellular sulfatases, Sulf1 and Sulf2, act on internal glucosamine-6-sulfate (6S) modifications within heparan sulfate proteoglycans (HSPGs) and modulate HSPG interactions with various signaling molecules, including Wnt ligands (Nawroth et al. 2007).

The Wnt pathway can be potentially targeted at multiple levels, either by antibodies against Fzd or by the use of Wnt inhibitors. Antibodies directed against

Fzd6 (clone 23M2) and Fzd5 (clone 44M13) have been shown to have antitumor properties (Deonarain et al. 2009). The inhibitor PRI-724 (Prism Biolabs), which blocks the interaction of β-catenin with CBP and is being tested in a phase 1 clinical trial in patients with advanced solid tumors, including pancreatic cancer (www.clinicaltrials.gov).

The MET Signaling Pathway

Embryogenesis, tissue repair, organ regeneration, and cancer invasion involves epithelial mesenchymal transition (EMT) (Kalluri 2009). This is stimulated by extracellular signaling which leads to modification of cellular proteins, intercellular junctional molecules and the cell cytoskeleton, leading to ordered cell migration and morphogenesis of new structures. One of the key signaling pathway that participates in these events is the hepatocyte growth factor (HGF) ligand and its receptor MET.

MET (also known as c-Met) is an integral plasma membrane protein that relays signals from the extracellular environment into the cytoplasm. MET, which is expressed by progenitors as well as epithelial and endothelial cells, is activated when its extracellular domain binds to HGF, also known as scatter factor (Sonnenberg et al. 1993). HGF is secreted predominantly by mesenchymal cells and bound in an inactive form to heparin proteoglycans within the extracellular matrix (Kobayashi et al. 1994; Lyon et al. 1994). HGF mRNA is also found in fibroblasts, smooth muscle cells, mast cells, macrophages, endothelial cells, leukocytes, and megakaryocytes (Zarnegar and Michalopoulos 1995). The HGF polypeptide is inactive in its initial form and must be cleaved into a disulfide-linked α-β heterodimer by an extracellular protease to acquire MET-binding activity (Zarnegar and Michalopoulos 1995).

Once HGF binds MET, its kinase activity is switched on by receptor dimerization and trans-phosphorylation of two catalytic tyrosine residues (Tyr1234 and Tyr1235) within the kinase activation loop (Trusolino et al. 2010). This leads to phosphorylation of two additional docking tyrosines in the carboxyl terminal tail; this site acts as a harbor for recruitment of several other signaling molecules. MET is negatively regulated by several protein-tyrosine phosphatases (PTP) which dephosphorylate either the catalytic or the docking tyrosines (PTP1, 2, 3) which prevents engagement of binding partners as well as downstream signaling (Trusolino et al. 2010) (Fig. 4).

MET signaling is augmented by a few other scaffolding partners, including GRB2-associated protein (GAB1) and CD44. GAB1 has a unique binding site for MET; upon binding and phosphorylation by MET receptor, GAB1 provides extra adapter sites for PI3K, SHP2, CRK, PLCγ1, and p120 Ras-GAP (Maroun et al. 2003; Maroun et al. 2000; Weidner et al. 1996). CD44 is a transmembrane cell adhesion molecule that activates MET in two ways; the extracellular domain tethers MET, HGF, and CD44, while the cytoplasmic tail helps to transduce signal from MET to Ras (Orian-Rousseau et al. 2002). Recently, ICAM-1 was identified

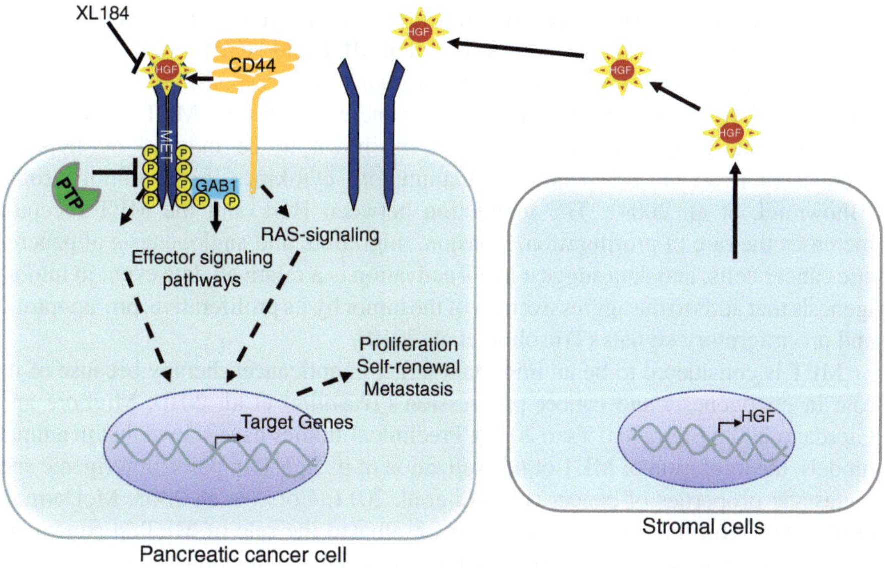

Fig. 4 The MET signaling pathway. In MET signaling, stromal (*right*) cell-secreted HGF binds to the MET receptor on pancreatic cancer cells (*left*). Upon ligand binding, MET dimerizes, autophosphorylates itself, and promotes signal transduction. Docking of GAB1 to phosphorylated MET promotes additional signal transduction. The cell surface molecule CD44 can promote MET signaling by interacting with MET and HGF outside the cell, as well as promoting RAS-signaling inside the cell. MET signaling ultimately drives metastasis, proliferation, and self-renewal of pancreatic cancer cells

as a new co-receptor for MET (Olaku et al. 2011), although the exact signaling mechanism has not yet been elucidated. Thus the basic signaling machinery of MET is regulated by a complex group of signal modifiers.

MET activates a cascade of downstream signaling pathways that include the MAP kinase PI3K-AKT, STAT, and NF-κB pathways (Trusolino et al. 2010) which function to modulate downstream gene expression. The mesenchymal-epithelial communication mediated by HGF-MET signaling integrates several pathways that control cell proliferation essential for normal processes such as embryogenesis, organ regeneration, and wound healing (Bhowmick et al. 2004; Boccaccio and Comoglio 2006). A role of MET in cancer was first noted in 1984, when it was cloned as a fusion oncogene from a human osteosarcoma cell line (Cooper et al. 1984). Germ line mutations in MET were observed in hereditary kidney cancer (Schmidt et al. 1997) and MET-activating mutations have also been observed in sporadic papillary renal cancer (Schmidt et al. 1997), childhood hepatocellular cancer (Park et al. 1999), and gastric cancer (Soman et al. 1991). More frequently, MET is overexpressed rather than mutated in cancer, as in colorectal (Takeuchi et al. 2003; Di Renzo et al. 1995a), hepatocellular (Suzuki et al. 1994), gastric (Amemiya et al. 2002), prostate (Humphrey et al. 1995), breast (Beviglia et al. 1997; Ghoussoub et al. 1998; Lee et al. 2005), and pancreatic cancers (Di Renzo et al. 1995b).

Evidence of MET/HGF upregulation in pancreatic cancer came from work in pancreatic cancer cell lines which showed that MET and HGF were overexpressed in a panel of 31 pancreatic cancer cell lines and were responsible for a "ductal" phenotype (Di Renzo et al. 1995b).In most pancreatic cancers, MET expression is transcriptionally upregulated and has been shown to be induced by hypoxia (Pennacchietti et al. 2003) and/or inflammatory cytokines in the tumor stroma (Bhowmick et al. 2004). The interaction between HGF and the MET receptor increases the rate of proliferation, invasion, migration, and angiogenesis of pancreatic cancer cells, and data suggest MET activation is a relatively late event in tumorigenesis that adds to the aggressiveness of the tumor by its proliferative, pro-apoptotic and pro-migratory signals (Trusolino et al. 2010).

MET is considered to be an important target in anticancer therapy because of its role in oncogenesis and cancer progression (Trusolino et al. 2010; Migliore and Giordano 2008; Sierra and Tsao 2011). Preclinical studies have shown that in animal models, the inhibition of MET or neutralization of its ligand impairs tumorigenic and metastatic properties of cancer cells (Li et al. 2011; Corso et al. 2008; McDermott et al. 2007). Recently, Li, and colleagues evaluated the role of MET in pancreatic cancer stem cell (CSC, reviewed later in the chapter) function (Li et al. 2011). Pancreatic cancer cells expressing high levels of MET cells had increased tumorigenic potential in mice, and cells that expressed MET and CD44 (0.5–5 % of the pancreatic cancer cells) had the capability for self-renewal and the highest tumorigenic potential of all cell populations studied. MET inhibition using the pharmacologic inhibitor XL184 or knockdown by shRNA slowed tumor growth and reduced the population of CSCs, either alone or in combination with gemcitabine. Additionally, targeting of MET prevented the development of metastases (Li et al. 2011). Based on this data, clinical trials targeting MET are currently in development.

The TGF-β Signaling Pathway

TGF-β is a multifunctional cytokine that controls cell growth, differentiation, proliferation, and angiogenesis, both during embryonic development and in adult tissues (Massague 1998). The TGF-β family contains two subfamilies, the TGF-β/Activin/Nodal subfamily and the bone morphogenetic protein (BMP)/growth and differentiation factor (GDF)/Muellerian inhibiting substance (MIS) subfamily, as defined by sequence similarity and the specific signaling pathways that they activate. The ligand family is comprised of three isoforms, TGF-β1, TGF-β2, and TGF-β3 (Massague 1998). TGF-β1 is expressed in epithelial, endothelial, hematopoietic and connective tissue cells; TGF-β2 is expressed in epithelial and neuronal cells and TGF-β3 is expressed in mesenchymal cells (Pasche 2001). There is 70–80 % homology among TGF-β isoforms which have different binding affinities to their tissue-specific receptors (Massague 1998). In general, they exhibit similar functions in vitro on cell growth regulation, ECM production and immune modulation. However, each ligand has distinct activities in vivo (Pasche 2001; Massague 1998).

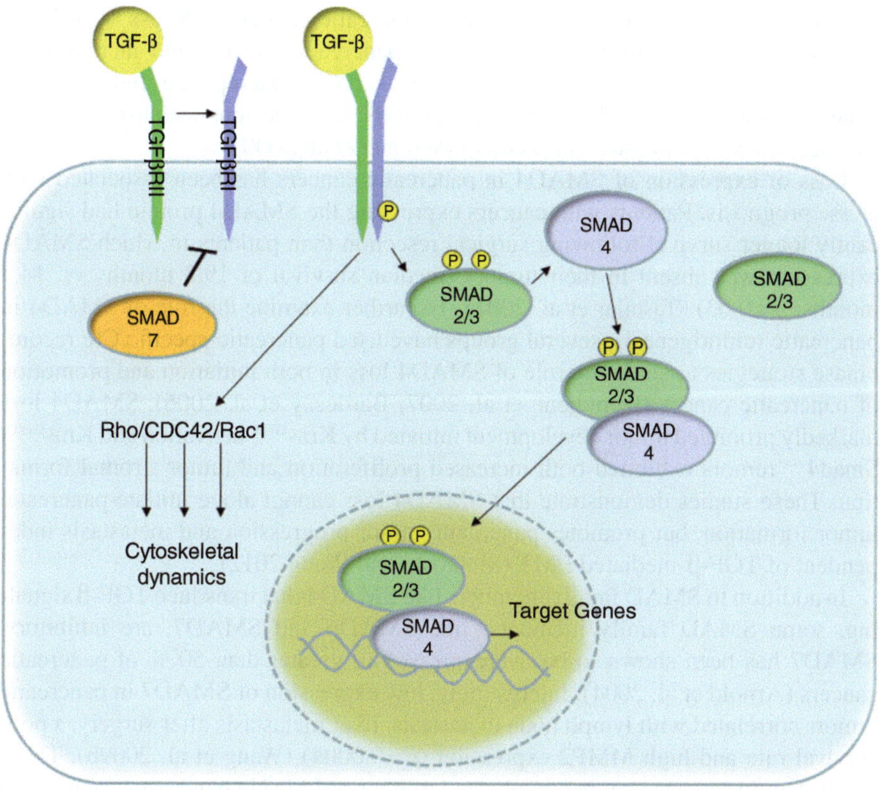

Fig. 5 The TGF-β signaling pathway. Binding of TGF-β to TGFβRII promotes dimerization with and phosphorylation of TGFβRI. This leads to recruitment and phosphorylation of SMAD2/3, which in turn bind to SMAD4 and translocate to the nucleus to promote transcription of target genes. Additionally, repressor SMADs, like SMAD7, inhibit TGF-β signaling. In addition to driving transcription through SMAD2/3/4, TGF-β signaling also activates small GTPases like Rho, Rac1, and CDC42, which regulate cytoskeletal dynamics

To initiate signaling, TGF-β ligands interact with two receptors, TGFβRI and TGFβRII. TGF-β binds to TGFβRII, which then recruits and phosphorylates TGFβRI (Fig. 5). This allows activation and phosphorylation of SMAD2 and SMAD3. Phosphorylated SMAD2 and 3 then combine with SMAD4 to translocate into the nucleus (Massague 1998). Once in the nucleus, the SMAD complex can associate with cofactors to transcriptionally regulate target genes. In addition to SMAD dependent signaling pathways, TGF-β also activates many other signaling pathways such as PI3K (Krymskaya et al. 1997), MAPK (Hartsough and Mulder 1995), and the small GTPases Rho (Bhowmick et al. 2001), Cdc42 (Edlund et al. 2002), and Rac1 (Mucsi et al. 1996).

SMAD4 (or Deleted in Pancreatic Cancer, locus 4/DPC4) inactivation through homozygous deletion or intragenic mutations are found in more than half of pancreatic cancers, (Jaffee et al. 2002). It is thought that loss of the SMAD4 expression is a rather late event in the pathogenesis of pancreatic cancer, with loss of SMAD4 expression occurring in 14.3 % of stage I pancreatic cancers and increasing to 60.0 % of stage IV pancreatic cancers (Hua et al. 2003). In a separate study, SMAD4 gene expression was found to be normal in PanIN1 and 2 lesions with loss of expression seen in 31 % of cases with PanIN3 (Wilentz et al. 2000).

Loss of expression of SMAD4 in pancreatic cancers has been associated with worse prognosis. Patients with cancers expressing the SMAD4 protein had significantly longer survival following surgical resection than patients in which SMAD4 expression was absent in their tumors (median survival of 19.2 months vs. 14.7 months; $p=0.03$) (Tascilar et al. 2001). To further examine the role of SMAD4 in pancreatic tumorigenesis, several groups have used pancreatic-specific Cre recombinase strategies to study the role of SMAD4 loss in both initiation and promotion of pancreatic cancer (Izeradjene et al. 2007; Bardeesy et al. 2006). SMAD4 loss markedly promoted tumor development initiated by $Kras^{G12D}$ activation and $Kras^{G12D}$/$Smad4^{-/-}$ tumors exhibited both increased proliferation and tumor stromal formation. These studies demonstrate that SMAD4 loss cannot alone initiate pancreatic tumor formation, but promotes pancreatic tumor progression and metastasis independent of TGF-β-mediated EMT (Malkoski and Wang 2012).

In addition to SMAD family members like SMAD4 that transduce TGF-β signaling, some SMAD family members, like SMAD6 and SMAD7, are inhibitory. SMAD7 has been shown to be overexpressed in greater than 50 % of pancreatic cancers (Arnold et al. 2004). Interestingly, low expression of SMAD7 in pancreatic tumors correlated with lymph node metastasis, liver metastasis after surgery, a poor survival rate and high MMP2 expression ($p=0.0004$) (Wang et al. 2009b). These results would suggest a more complicated role for SMAD7 in pancreatic cancer, and not simply one of an oncogene. Several other molecules, like KLF11, retinoblastoma, and thioredoxin have been associated with SMAD7-dependent aggressiveness of pancreatic cancer (Ellenrieder et al. 2004; Arnold et al. 2004).

TGF-β signaling is complex in tumor development as it appears to have dual roles, with growth inhibitory function in early tumor development but apparent promotion of invasion and metastasis later in tumorigenesis. This latter role of TGF-β is the basis for interest in targeting this pathway in pancreatic cancer. Several inhibitory approaches have shown efficacy in preclinical and clinical studies. These include blocking production of TGF-β ligands with antisense molecules, small-molecule inhibitors of the kinase activity of TGFβRI and TGFβRII, monoclonal antibodies that block TGF-β signaling and soluble forms of TGFβRII and TGFβRIII that function as ligand traps (Flavell et al. 2010; Rowland-Goldsmith et al. 2001, 2002; Kelly and Morris 2010). In addition, combined therapies of small-molecule inhibitors with immune-stimulating vaccines represents an additional therapeutic approach that is being tested (Terabe et al. 2009). Another agent being utilized to target the TGF-β pathway in pancreatic cancer is trabedersen (AP 12009), a phosphorothioate antisense mRNA targeting TGF-β2 (Schlingensiepen et al. 2011). Using an orthotopic xenograft model, trabedersen was effective at inhibiting tumor cell growth and cell migration, while reversing TGF-β2-mediated immunosuppression of lymphokine activated killer (LAK) cells (Schlingensiepen et al. 2011).

These data support the idea that the TGF-β is a desirable target in pancreatic cancer; however, further evaluation of these TGF-β inhibitory agents is necessary to assess actual efficacy in controlled clinical trials.

Stromal Biology and Therapeutic Targets

Pancreatic cancer characteristically has an abundantly dense stroma composed of a mixture of both cellular and acellular components including extracellular matrix proteins (ECM), growth factors, cytokines. The different cellular components include cells of mesenchymal and immune origin. In the following section, we will describe how these stromal cells contribute to pancreatic cancer growth and how they may be targeted.

Cancer-Associated Fibroblasts

The dense stroma found in pancreatic cancer appears to be formed through the actions of cancer-associated fibroblasts (CAFs) (Apte et al. 2004; Hwang et al. 2008). Current understanding of the actual cell of origin for CAFs is incomplete and although the term CAF is often used interchangeably with activated pancreatic stellate cells (PSCs), CAFs may also be derived from other cell types including infiltrating cells from the bone marrow (Direkze et al. 2004). Further demonstrating the complexity of CAFs is a recent study in human pancreatic tumors that identified a subpopulation of CAFs that are CD10+ which more robustly support tumor growth, highlighting the fact that CAFS represent a heterogeneous population of cells (Ikenaga et al. 2010). In addition to supporting enhanced tumorigenicity, CAFs appear to also contribute to resistance of pancreatic cancer cells to chemotherapy and radiation and promote metastatic spread (Hwang et al. 2008).

There have been multiple mechanisms proposed by which CAFS contribute to the tumor progression, including signaling pathways such as SDF-1/CXCR4 axis, the Hedgehog pathway (discussed previously), hypoxia-mediated signaling, and innate immunity. Stromal cell-derived factor-1 (SDF-1) is a member of the CXC subfamily of chemokines and interacts with its receptor CXCR4. SDF1-CXCR4 signaling has been implicated in the process of local invasion and distant metastasis of pancreatic cancer (Hermann et al. 2007). CAFs have been shown to express SDF-1, whereas CXCR4 is expressed by pancreatic cancer cells (Koshiba et al. 2000). Increased proliferation and metastatic spread of pancreatic cancer cells expressing CXCR4 can be abrogated by anti-SDF-1 neutralizing antibodies or the CXCR4 inhibitor AMD3100/plerixafor (Johnson Matthey), suggesting that the SDF-1/CXCR4 axis contributes to CAF stimulation of pancreatic cancer cells (Gao et al. 2010). In one study, pancreatic cancer cell lines treated with recombinant SDF-1 were resistant to gemcitabine, and this effect was reversed by blocking CXCR4 with AMD3100 (Singh et al. 2010). In addition, a potential important role of CXCR4 has been described in pancreatic CSCs, in which a subpopulation of CSCs expressing CD133 and CXCR4 were found to be highly invasive and responsible for metastasis (Hermann et al. 2007), further supporting the rationale for exploring SDF-1/CXCR4 for therapeutic targeting.

In addition to provided growth factors and chemoattractants that promote pancreatic cancer cell growth, the extremely dense stroma of pancreatic cancer serves to protect cells from chemotherapy by "crushing" blood vessels. Using a murine pancreatic cancer model, Olive and colleagues showed that the dense tumor stroma was driven by tumor cell-derived SHH which activated the Hedgehog pathway in stromal cells (Olive et al. 2009). By inhibiting SMO with IPI-926, blood vessels could be transiently reopened by decreasing the stroma, which allowed for enhanced efficacy of gemcitabine treatment (Olive et al. 2009). In addition to being driven by paracrine SHH signaling, the desmoplastic tumor stroma has also been shown to be sustained by excessive amounts of the extracellular matrix component, hyaluronic acid (hyaluronan) (Provenzano et al. 2012; Jacobetz et al. 2013). Provenzano and Jacobetz and their colleagues simultaneously reported that in murine pancreatic cancer models, hyaluronic acid in the stroma led to the collapse of tumor vasculature, which impeded drug delivery. Using a PEGylated form of the hyaluronic acid-degrading enzyme, PH20 hyaluronidase (PEGPH20), the authors were able to restore a normalized stroma and tumor vasculature. When combined with gemcitabine, PEGPH20 was able to substantially reduce tumor burden and extend animal survival (Provenzano et al. 2012; Jacobetz et al. 2013). Based on these findings, the tumor stroma can be thought of as both nurturing and protecting pancreatic cancer cells and a valuable target in pancreatic cancer therapy.

Hypoxia-Driven Signaling Pathways

Hypoxia is a common condition in zones of rapidly proliferating tumors which influences signaling pathways that control cell proliferation, angiogenesis, and apoptosis (Harris 2002). Hypoxia is also believed to be a prevalent state in pancreatic tumors due to hypovascularity that is concomitantly found within the dense stroma. Hypoxic conditions are also associated with resistance to chemotherapy and radiation therapy (Harris 2002; Yokoi and Fidler 2004). In pancreatic cancer, hypoxia confers multidrug resistance primarily through the PI3K/AKT/NF-κB pathway and partially through the MAPK signaling pathway (Yokoi and Fidler 2004). Inhibition of PI3K with the inhibitor LY294002 (Eli Lilly), in combination with a Chk1 inhibitor, UCN-01 (Tokyo Research Laboratories), has been shown to partially sensitize pancreatic cancer cells to cytotoxic chemotherapy under hypoxic conditions (Onozuka et al. 2011).

Immune Cells

Immune cells form an integral part of the tumor stroma and various types of immune cells have either tumor-promoting or tumor-antagonistic properties. The balance between these two properties contributes to tumor growth. Tumor-promoting cells include macrophages, mast cells, neutrophils, T and B lymphocytes (Ruffell et al. 2010; DeNardo et al. 2010). These cells are activated by a number of signaling molecules that have been extensively studied in different cell systems (Ruffell et al. 2010; Murdoch et al. 2008; Qian and Pollard 2010). Infiltration with immune cells has been observed in all stages of pancreatic cancer, from PanIN lesions to invasive cancer (Clark et al. 2007). These immune cells secrete a number of molecules that modulate tumor and stromal growth, including VEGF, FGF2, chemokines and cytokines, pro-angiogenic factors such as MMP-9 and other matrix metalloproteases, and heparinase (Murdoch et al. 2008; Qian and Pollard 2010; Hanahan and Weinberg 2011). Kraman and colleagues identified a specific subpopulation of stromal cells expressing fibroblast activation protein (FAP) that play a role in suppressing antitumor immunity (Kraman et al. 2010). Depletion of this subpopulation led to IFNγ and TNFα mediated modulation of tumor growth (Kraman et al. 2010). Pancreatic cancer cells are also responsible for recruiting immune cells to suppress the antitumor activity of CD8(+) T cells. Using murine models for pancreatic cancer, two groups simultaneously found that oncogenic KRAS results in the secretion of GM-CSF by pancreatic cancer cells, which in turn attracts Gr-1(+) CD11b(+) cells that can inhibit the activity of CD8(+) T cells in the tumor (Bayne et al. 2012; Pylayeva-Gupta et al. 2012). These data emphasize the complex nature of the immune system and tumor development and represent a venue to target to alter the immune suppressive environment that exists in pancreatic cancer.

Cancer Stem Cells

A subset of cancer cells has been identified in many solid tumors which has the capacity to efficiently propagate a new tumor with the heterogeneity and pathologic features of the original cancer. These cells are called CSCs because they share normal stem cell features such as self-renewal and the ability to undergo both symmetric and asymmetric cell division (Reya et al. 2001). Conventional therapies are directed at eliminating bulk tumor cells; however, these therapies are usually short-lived, and tumors eventually reestablish themselves. One reason for this phenomenon is that the CSCs are intrinsically resistant to cytotoxic chemotherapy and persist despite apparent response in bulk tumor (Kim and Simeone 2011). Therefore, understanding differences between CSCs and bulk tumor cells is relevant to improving overall efficacy of treatment.

CSCs were first described in acute myeloid leukemia (AML) as a distinct CD34+/CD38− population capable of both self-renewal and distinct progeny (Bonnet and

Dick 1997). Subsequently, CSCs have been identified by surface marker analysis in solid tumors, including pancreatic cancer, with a first report demonstrating a subset of CD44+/CD24+/ESA+pancreatic CSCs (Li et al. 2007). Additionally, both CD133 and ALDH have been identified as potential independent markers for pancreatic CSCs (Hermann et al. 2007; Jimeno et al. 2009). Most recently, c-Met+/CD44+ pancreatic cancer cells have been described to potently enrich for a population of pancreatic CSCs (Li et al. 2011).

Although tumorigenesis is generally considered a clonal process, there is ultimately genetic diversity within an individual tumor (Marusyk and Polyak 2010). Recent detailed analysis of tumor cells from different metastatic sites within an individual patient confirmed that there are subclonal populations due to genomic instability (Campbell et al. 2010). We currently lack a detailed understanding of how genetic heterogeneity of pancreatic cancer correlates with the hierarchy of CSCs. Although one could view the complexity of genetic heterogeneity as an insurmountable barrier to the development of targeted therapies, an alternate view in the context of CSCs is that identifying the dominant signaling pathways in the CSC subpopulation is the key to eliminating the subpopulation of cells that may be most important for clinical progression and recurrence of disease. Based on this latter view, we describe below the results of increased attention focused on CSC biology, including the role of developmental signaling pathways and commonalities between CSC and cancer cells that have undergone EMT.

There are multiple signaling pathways that are upregulated in pancreatic CSCs that represent possible therapeutic targets. MET has recently been described as a potent marker for identifying pancreatic CSCs when studied in combination with CD44 expression (Li et al. 2011). This selective feature of CSC is now being targeted therapeutically with agents that inhibit MET. Treatment with the MET inhibitor XL184 has been shown in preclinical studies to reduce the percentage of pancreatic CSCs, decrease tumorsphere-forming capacity, and decrease in vivo tumorigenicity (Li et al. 2011). In addition to high levels of MET expression, Lonardo and colleagues found Nodal/Activin signaling to be elevated in pancreatic CSCs (Lonardo et al. 2011). By inhibiting the Nodal/Activin receptors Alk4/7 with the chemical inhibitor SB431542 or targeted siRNA, CSCs could be ablated in vitro. Additionally, the combination of SB431542 with gemcitabine and the SMO inhibitor CUR199691 could effectively ablate tumor growth in vivo (Lonardo et al. 2011). Together, HGF/MET and Nodal/Activin/Alk4 represent signaling pathways that may allow for the development of CSC-targeted therapies that can potentially be used in combination with standard chemotherapeutic regimens to reduce disease recurrence by specifically eliminating CSCs.

CSCs have also found to be resistant to chemotherapy and radiotherapy as evidenced by the increased percentage of CSC isolated following treatment. Cells that have undergone EMT share some of the same characteristics. Induction of EMT has been shown in breast cancer to cause transition to development of a CSC marker profile with associated phenotypic changes such as increased ability to form tumorspheres (Mani et al. 2008). In pancreatic cancer, recent gene expression profiling analysis of human and murine pancreatic cancer cell samples revealed three distinct tumor types: classical epithelial, quasimesenchymal, and endocrine-like type (Collisson et al. 2011). The most EMT-like quasimesenchymal tumors were associated with poor patient prognosis (Collisson et al. 2011). It has also been reported in preclinical studies that gemcitabine-resistant cells appear to undergo EMT with associated phenotypic changes of increased invasiveness and migration (Wang et al. 2009c). These resistant cells correspondingly had an increased population of pancreatic CSCs (Wang et al. 2009c). Although these data do not prove equivalence of CSC with cancer cells that have undergone EMT, the correlation between EMT and CSCs may provide insight into shared phenotypes of chemoresistance and allow for identification of new targets for therapy.

Conclusions

From the analyses of the complex pathways present in pancreatic cancer and evolving evidence of tumor heterogeneity, it is evident that in pancreatic tumors, changes are not often due to a single driver mutation, but more often a combination of many mutations collaborating together. Therefore, targeting a single pathway or molecule is unlikely to be successful. Adding to this complexity is the genetic clonal variation within the tumor itself. It is also becoming more apparent that in order to effectively treat pancreatic cancer, other cells in the tumor microenvironment must also be targeted, such as CAFs and immune cells that nourish and protect pancreatic cancer cells from chemotherapies. Ultimately, effective treatment of pancreatic cancer will need to combine means of normalizing the tumor stroma, removing bulk tumor cells, and eliminating elusive pancreatic CSCs that might drive tumor reestablishment and disease relapse. By achieving these goals, pancreatic cancer may someday become a manageable condition rather than a certain death sentence.

References

Abraham SC, Klimstra DS, Wilentz RE, Yeo CJ, Conlon K, Brennan M, Cameron JL, Wu TT, Hruban RH (2002) Solid-pseudopapillary tumors of the pancreas are genetically distinct from pancreatic ductal adenocarcinomas and almost always harbor beta-catenin mutations. Am J Pathol 160(4):1361–1369

Amemiya H, Kono K, Itakura J, Tang RF, Takahashi A, An FQ, Kamei S, Iizuka H, Fujii H, Matsumoto Y (2002) c-Met expression in gastric cancer with liver metastasis. Oncology 63(3):286–296. doi:ocl63286 [pii]

Apelqvist A, Li H, Sommer L, Beatus P, Anderson DJ, Honjo T, Hrabe de Angelis M, Lendahl U, Edlund H (1999) Notch signalling controls pancreatic cell differentiation. Nature 400(6747):877–881

Apte MV, Park S, Phillips PA, Santucci N, Goldstein D, Kumar RK, Ramm GA, Buchler M, Friess H, McCarroll JA, Keogh G, Merrett N, Pirola R, Wilson JS (2004) Desmoplastic reaction in pancreatic cancer: role of pancreatic stellate cells. Pancreas 29(3):179–187

Arnold NB, Ketterer K, Kleeff J, Friess H, Buchler MW, Korc M (2004) Thioredoxin is downstream of Smad7 in a pathway that promotes growth and suppresses cisplatin-induced apoptosis in pancreatic cancer. Cancer Res 64(10):3599–3606. doi:10.1158/0008-5472.CAN-03-2999 64/10/3599 [pii]

Artavanis-Tsakonas S, Matsuno K, Fortini ME (1995) Notch signaling. Science 268(5208): 225–232

Bailey JM, Swanson BJ, Hamada T, Eggers JP, Singh PK, Caffery T, Ouellette MM, Hollingsworth MA (2008) Sonic hedgehog promotes desmoplasia in pancreatic cancer. Clin Cancer Res 14(19):5995–6004. doi:14/19/5995 [pii] 10.1158/1078-0432.CCR-08-0291

Bardeesy N, Cheng KH, Berger JH, Chu GC, Pahler J, Olson P, Hezel AF, Horner J, Lauwers GY, Hanahan D, DePinho RA (2006) Smad4 is dispensable for normal pancreas development yet critical in progression and tumor biology of pancreas cancer. Genes Dev 20(22):3130–3146. doi:20/22/3130 [pii] 10.1101/gad.1478706

Bayne LJ, Beatty GL, Jhala N, Clark CE, Rhim AD, Stanger BZ, Vonderheide RH (2012) Tumor-derived granulocyte-macrophage colony-stimulating factor regulates myeloid inflammation and T cell immunity in pancreatic cancer. Cancer Cell 21(6):822–835. doi:10.1016/j.ccr.2012.04.025

Berman DM, Karhadkar SS, Maitra A, Montes De Oca R, Gerstenblith MR, Briggs K, Parker AR, Shimada Y, Eshleman JR, Watkins DN, Beachy PA (2003) Widespread requirement for Hedgehog ligand stimulation in growth of digestive tract tumours. Nature 425(6960):846–851. doi:10.1038/nature01972 nature01972 [pii]

Bettenhausen B, Hrabe de Angelis M, Simon D, Guenet JL, Gossler A (1995) Transient and restricted expression during mouse embryogenesis of Dll1, a murine gene closely related to Drosophila Delta. Development 121(8):2407–2418

Beviglia L, Matsumoto K, Lin CS, Ziober BL, Kramer RH (1997) Expression of the c-Met/HGF receptor in human breast carcinoma: correlation with tumor progression. Int J Cancer 74(3): 301–309. doi:10.1002/(SICI)1097-0215(19970620)74:3<301::AID-IJC12>3.0.CO;2-E [pii]

Bhowmick NA, Ghiassi M, Bakin A, Aakre M, Lundquist CA, Engel ME, Arteaga CL, Moses HL (2001) Transforming growth factor-beta1 mediates epithelial to mesenchymal transdifferentiation through a RhoA-dependent mechanism. Mol Biol Cell 12(1):27–36

Bhowmick NA, Neilson EG, Moses HL (2004) Stromal fibroblasts in cancer initiation and progression. Nature 432(7015):332–337. doi:nature03096 [pii] 10.1038/nature03096

Blaumueller CM, Qi H, Zagouras P, Artavanis-Tsakonas S (1997) Intracellular cleavage of Notch leads to a heterodimeric receptor on the plasma membrane. Cell 90(2):281–291. doi:S0092-8674(00)80336-0 [pii]

Boccaccio C, Comoglio PM (2006) Invasive growth: a MET-driven genetic programme for cancer and stem cells. Nat Rev Cancer 6(8):637–645. doi:nrc1912 [pii] 10.1038/nrc1912

Bonnet D, Dick JE (1997) Human acute myeloid leukemia is organized as a hierarchy that originates from a primitive hematopoietic cell. Nat Med 3(7):730–737

Campbell PJ, Yachida S, Mudie LJ, Stephens PJ, Pleasance ED, Stebbings LA, Morsberger LA, Latimer C, McLaren S, Lin ML, McBride DJ, Varela I, Nik-Zainal SA, Leroy C, Jia M, Menzies A, Butler AP, Teague JW, Griffin CA, Burton J, Swerdlow H, Quail MA, Stratton MR, Iacobuzio-Donahue C, Futreal PA (2010) The patterns and dynamics of genomic instability in metastatic pancreatic cancer. Nature 467(7319):1109–1113. doi:nature09460 [pii] 10.1038/nature09460

Clark CE, Hingorani SR, Mick R, Combs C, Tuveson DA, Vonderheide RH (2007) Dynamics of the immune reaction to pancreatic cancer from inception to invasion. Cancer Res 67(19):9518–9527. doi:67/19/9518 [pii] 10.1158/0008-5472.CAN-07-0175

Clevers H (2006) Wnt/beta-catenin signaling in development and disease. Cell 127(3):469–480. doi:S0092-8674(06)01344-4 [pii] 10.1016/j.cell.2006.10.018

Collisson EA, Sadanandam A, Olson P, Gibb WJ, Truitt M, Gu S, Cooc J, Weinkle J, Kim GE, Jakkula L, Feiler HS, Ko AH, Olshen AB, Danenberg KL, Tempero MA, Spellman PT, Hanahan D, Gray JW (2011) Subtypes of pancreatic ductal adenocarcinoma and their differing responses to therapy. Nat Med 17(4):500–503. doi:nm.2344 [pii] 10.1038/nm.2344

Conroy T, Desseigne F, Ychou M, Bouché O, Guimbaud R, Bécouarn Y, Adenis A, Raoul JL, Gourgou-Bourgade S, de la Fouchardière C, Bennouna J, Bachet JB, Khemissa-Akouz F, Péré-Vergé D, Delbaldo C, Assenat E, Chauffert B, Michel P, Montoto-Grillot C, Ducreux M (2011) FOLFIRINOX versus gemcitabine for metastatic pancreatic cancer. N Engl J Med 364(19):1817–1825. doi:doi:10.1056/NEJMoa1011923

Cook N, Frese KK, Bapiro TE, Jacobetz MA, Gopinathan A, Miller JL, Rao SS, Demuth T, Howat WJ, Jodrell DI, Tuveson DA (2012) Gamma secretase inhibition promotes hypoxic necrosis in mouse pancreatic ductal adenocarcinoma. J Exp Med 209(3):437–444. doi:jem.20111923 [pii] 10.1084/jem.20111923

Cooper CS, Park M, Blair DG, Tainsky MA, Huebner K, Croce CM, Vande Woude GF (1984) Molecular cloning of a new transforming gene from a chemically transformed human cell line. Nature 311(5981):29–33

Corso S, Migliore C, Ghiso E, De Rosa G, Comoglio PM, Giordano S (2008) Silencing the MET oncogene leads to regression of experimental tumors and metastases. Oncogene 27(5): 684–693. doi:1210697 [pii] 10.1038/sj.onc.1210697

D'Souza B, Miyamoto A, Weinmaster G (2008) The many facets of Notch ligands. Oncogene 27(38):5148–5167. doi:onc2008229 [pii] 10.1038/onc.2008.229

De La O JP, Emerson LL, Goodman JL, Froebe SC, Illum BE, Curtis AB, Murtaugh LC (2008) Notch and Kras reprogram pancreatic acinar cells to ductal intraepithelial neoplasia. Proc Natl Acad Sci U S A 105(48):18907–18912. doi:0810111105 [pii] 10.1073/pnas.0810111105

DeNardo DG, Andreu P, Coussens LM (2010) Interactions between lymphocytes and myeloid cells regulate pro- versus anti-tumor immunity. Cancer Metastasis Rev 29(2):309–316. doi:10.1007/s10555-010-9223-6

Deonarain MP, Kousparou CA, Epenetos AA (2009) Antibodies targeting cancer stem cells: a new paradigm in immunotherapy? MAbs 1(1):12–25

Dessimoz J, Bonnard C, Huelsken J, Grapin-Botton A (2005) Pancreas-specific deletion of beta-catenin reveals Wnt-dependent and Wnt-independent functions during development. Curr Biol 15(18):1677–1683. doi:10.1016/j.cub.2005.08.037

Di Renzo MF, Olivero M, Giacomini A, Porte H, Chastre E, Mirossay L, Nordlinger B, Bretti S, Bottardi S, Giordano S et al (1995a) Overexpression and amplification of the met/HGF receptor gene during the progression of colorectal cancer. Clin Cancer Res 1(2):147–154

Di Renzo MF, Poulsom R, Olivero M, Comoglio PM, Lemoine NR (1995b) Expression of the Met/hepatocyte growth factor receptor in human pancreatic cancer. Cancer Res 55(5):1129–1138

Direkze NC, Hodivala-Dilke K, Jeffery R, Hunt T, Poulsom R, Oukrif D, Alison MR, Wright NA (2004) Bone marrow contribution to tumor-associated myofibroblasts and fibroblasts. Cancer Res 64(23):8492–8495. doi:10.1158/0008-5472.can-04-1708

Dunwoodie SL, Henrique D, Harrison SM, Beddington RS (1997) Mouse Dll3: a novel divergent Delta gene which may complement the function of other Delta homologues during early pattern formation in the mouse embryo. Development 124(16):3065–3076

Edlund S, Landström M, Heldin CH, Aspenström P (2002) Transforming growth factor-beta-induced mobilization of actin cytoskeleton requires signaling by small GTPases Cdc42 and RhoA. Mol Biol Cell 13(3):902–914. doi:10.1091/mbc.01-08-0398

Ellenrieder V, Buck A, Harth A, Jungert K, Buchholz M, Adler G, Urrutia R, Gress TM (2004) KLF11 mediates a critical mechanism in TGF-beta signaling that is inactivated by Erk-MAPK in pancreatic cancer cells. Gastroenterology 127(2):607–620. doi:S0016508504008649 [pii]

Fiuza UM, Arias AM (2007) Cell and molecular biology of Notch. J Endocrinol 194(3):459–474. doi:194/3/459 [pii] 10.1677/JOE-07-0242

Flavell RA, Sanjabi S, Wrzesinski SH, Licona-Limon P (2010) The polarization of immune cells in the tumour environment by TGFbeta. Nat Rev Immunol 10(8):554–567. doi:nri2808 [pii] 10.1038/nri2808

Fryer CJ, Lamar E, Turbachova I, Kintner C, Jones KA (2002) Mastermind mediates chromatin-specific transcription and turnover of the Notch enhancer complex. Genes Dev 16(11):1397–1411. doi:10.1101/gad.991602

Fryer CJ, White JB, Jones KA (2004) Mastermind recruits CycC:CDK8 to phosphorylate the Notch ICD and coordinate activation with turnover. Mol Cell 16(4):509–520. doi:S1097276504006409 [pii] 10.1016/j.molcel.2004.10.014

Gallahan D, Callahan R (1997) The mouse mammary tumor associated gene INT3 is a unique member of the NOTCH gene family (NOTCH4). Oncogene 14(16):1883–1890. doi:10.1038/sj.onc.1201035

Gallahan D, Jhappan C, Robinson G, Hennighausen L, Sharp R, Kordon E, Callahan R, Merlino G, Smith GH (1996) Expression of a truncated Int3 gene in developing secretory mammary epithelium specifically retards lobular differentiation resulting in tumorigenesis. Cancer Res 56(8):1775–1785

Gao Z, Wang X, Wu K, Zhao Y, Hu G (2010) Pancreatic stellate cells increase the invasion of human pancreatic cancer cells through the stromal cell-derived factor-1/CXCR4 axis. Pancreatology 10(2–3):186–193. doi:10.1159/000236012

Ghoussoub RA, Dillon DA, D'Aquila T, Rimm EB, Fearon ER, Rimm DL (1998) Expression of c-met is a strong independent prognostic factor in breast carcinoma. Cancer 82(8):1513–1520. doi:10.1002/(SICI)1097-0142(19980415)82:8<1513::AID-CNCR13>3.0.CO;2-7 [pii]

Gridley T (1997) Notch signaling in vertebrate development and disease. Mol Cell Neurosci 9(2):103–108. doi:S1044-7431(97)90610-2 [pii] 10.1006/mcne.1997.0610

Gridley T (2003) Notch signaling and inherited disease syndromes. Hum Mol Genet 12 Spec No. 1:R9–R13

Groden J, Thliveris A, Samowitz W, Carlson M, Gelbert L, Albertsen H, Joslyn G, Stevens J, Spirio L, Robertson M et al (1991) Identification and characterization of the familial adenomatous polyposis coli gene. Cell 66(3):589–600

Gupta S, Takebe N, Lorusso P (2010) Targeting the Hedgehog pathway in cancer. Ther Adv Med Oncol 2(4):237–250. doi:10.1177/1758834010366430 10.1177_1758834010366430 [pii]

Hanahan D, Weinberg RA (2011) Hallmarks of cancer: the next generation. Cell 144(5):646–674. doi:S0092-8674(11)00127-9 [pii] 10.1016/j.cell.2011.02.013

Harris AL (2002) Hypoxia—a key regulatory factor in tumour growth. Nat Rev Cancer 2(1): 38–47. doi: http://www.nature.com/nrc/journal/v2/n1/suppinfo/nrc704_S1.html

Hartsough MT, Mulder KM (1995) Transforming growth factor beta activation of p44mapk in proliferating cultures of epithelial cells. J Biol Chem 270(13):7117–7124. doi:10.1074/jbc.270.13.7117

Heller RS, Dichmann DS, Jensen J, Miller C, Wong G, Madsen OD, Serup P (2002) Expression patterns of Wnts, Frizzleds, sFRPs, and misexpression in transgenic mice suggesting a role for Wnts in pancreas and foregut pattern formation. Dev Dyn 225(3):260–270. doi:10.1002/dvdy.10157

Hermann PC, Huber SL, Herrler T, Aicher A, Ellwart JW, Guba M, Bruns CJ, Heeschen C (2007) Distinct populations of cancer stem cells determine tumor growth and metastatic activity in human pancreatic cancer. Cell Stem Cell 1(3):313–323. doi:S1934-5909(07)00066-5 [pii] 10.1016/j.stem.2007.06.002

Hezel AF, Kimmelman AC, Stanger BZ, Bardeesy N, Depinho RA (2006) Genetics and biology of pancreatic ductal adenocarcinoma. Genes Dev 20(10):1218–1249. doi:20/10/1218 [pii] 10.1101/gad.1415606

Hingorani SR, Petricoin EF, Maitra A, Rajapakse V, King C, Jacobetz MA, Ross S, Conrads TP, Veenstra TD, Hitt BA, Kawaguchi Y, Johann D, Liotta LA, Crawford HC, Putt ME, Jacks T, Wright CV, Hruban RH, Lowy AM, Tuveson DA (2003) Preinvasive and invasive ductal pancreatic cancer and its early detection in the mouse. Cancer Cell 4(6):437–450. doi:S153561080300309X [pii]

Hingorani SR, Wang L, Multani AS, Combs C, Deramaudt TB, Hruban RH, Rustgi AK, Chang S, Tuveson DA (2005) Trp53R172H and KrasG12D cooperate to promote chromosomal instability and widely metastatic pancreatic ductal adenocarcinoma in mice. Cancer Cell 7(5): 469–483. doi:S1535-6108(05)00128-5 [pii] 10.1016/j.ccr.2005.04.023

Hsieh JJ, Zhou S, Chen L, Young DB, Hayward SD (1999) CIR, a corepressor linking the DNA binding factor CBF1 to the histone deacetylase complex. Proc Natl Acad Sci USA 96(1):23–28

Hua Z, Zhang YC, Hu XM, Jia ZG (2003) Loss of DPC4 expression and its correlation with clinicopathological parameters in pancreatic carcinoma. World J Gastroenterol 9(12):2764–2767

Humphrey PA, Zhu X, Zarnegar R, Swanson PE, Ratliff TL, Vollmer RT, Day ML (1995) Hepatocyte growth factor and its receptor (c-MET) in prostatic carcinoma. Am J Pathol 147(2):386–396

Hwang RF, Moore T, Arumugam T, Ramachandran V, Amos KD, Rivera A, Ji B, Evans DB, Logsdon CD (2008) Cancer-associated stromal fibroblasts promote pancreatic tumor progression. Cancer Res 68(3):918–926. doi:68/3/918 [pii] 10.1158/0008-5472.CAN-07-5714

Ikenaga N, Ohuchida K, Mizumoto K, Cui L, Kayashima T, Morimatsu K, Moriyama T, Nakata K, Fujita H, Tanaka M (2010) CD10+ pancreatic stellate cells enhance the progression of pancreatic cancer. Gastroenterology 139(3):1041–1051. doi:S0016-5085(10)00849-8 [pii] 10.1053/j.gastro.2010.05.084, 1051.e1–8

Ingham PW, McMahon AP (2001) Hedgehog signaling in animal development: paradigms and principles. Genes Dev 15(23):3059–3087. doi:10.1101/gad.938601

Izeradjene K, Combs C, Best M, Gopinathan A, Wagner A, Grady WM, Deng CX, Hruban RH, Adsay NV, Tuveson DA, Hingorani SR (2007) Kras(G12D) and Smad4/Dpc4 haploinsufficiency cooperate to induce mucinous cystic neoplasms and invasive adenocarcinoma of the pancreas. Cancer Cell 11(3):229–243. doi:S1535-6108(07)00057-8 [pii] 10.1016/j.ccr.2007.01.017

Jacobetz MA, Chan DS, Neesse A, Bapiro TE, Cook N, Frese KK, Feig C, Nakagawa T, Caldwell ME, Zecchini HI, Lolkema MP, Jiang P, Kultti A, Thompson CB, Maneval DC, Jodrell DI, Frost GI, Shepard HM, Skepper JN, Tuveson DA (2013) Hyaluronan impairs vascular function and drug delivery in a mouse model of pancreatic cancer. Gut 62(1):112–120. doi:10.1136/gutjnl-2012-302529

Jaffee EM, Hruban RH, Canto M, Kern SE (2002) Focus on pancreas cancer. Cancer Cell 2(1):25–28. doi:10.1016/s1535-6108(02)00093-4

Jensen J, Heller RS, Funder-Nielsen T, Pedersen EE, Lindsell C, Weinmaster G, Madsen OD, Serup P (2000) Independent development of pancreatic alpha- and beta-cells from neurogenin3-expressing precursors: a role for the notch pathway in repression of premature differentiation. Diabetes 49(2):163–176

Jhappan C, Gallahan D, Stahle C, Chu E, Smith GH, Merlino G, Callahan R (1992) Expression of an activated Notch-related int-3 transgene interferes with cell differentiation and induces neoplastic transformation in mammary and salivary glands. Genes Dev 6(3):345–355

Jimeno A, Feldmann G, Suárez-Gauthier A, Rasheed Z, Solomon A, Zou GM, Rubio-Viqueira B, García-García E, López-Ríos F, Matsui W, Maitra A, Hidalgo M (2009) A direct pancreatic cancer xenograft model as a platform for cancer stem cell therapeutic development. Mol Cancer Ther 8(2):310–314

Jones S, Zhang X, Parsons DW, Lin JC, Leary RJ, Angenendt P, Mankoo P, Carter H, Kamiyama H, Jimeno A, Hong SM, Fu B, Lin MT, Calhoun ES, Kamiyama M, Walter K, Nikolskaya T, Nikolsky Y, Hartigan J, Smith DR, Hidalgo M, Leach SD, Klein AP, Jaffee EM, Goggins M, Maitra A, Iacobuzio-Donahue C, Eshleman JR, Kern SE, Hruban RH, Karchin R, Papadopoulos N, Parmigiani G, Vogelstein B, Velculescu VE, Kinzler KW (2008) Core signaling pathways in human pancreatic cancers revealed by global genomic analyses. Science 321(5897):1801–1806. doi:10.1126/science.1164368

Kalluri R (2009) EMT: when epithelial cells decide to become mesenchymal-like cells. J Clin Invest 119(6):1417–1419. doi:10.1172/jci39675

Kao HY, Ordentlich P, Koyano-Nakagawa N, Tang Z, Downes M, Kintner CR, Evans RM, Kadesch T (1998) A histone deacetylase corepressor complex regulates the Notch signal transduction pathway. Genes Dev 12(15):2269–2277

Kelly RJ, Morris JC (2010) Transforming growth factor-beta: a target for cancer therapy. J Immunotoxicol 7(1):15–26. doi:10.3109/15476910903389920

Kim EJ, Simeone DM (2011) Advances in pancreatic cancer. Curr Opin Gastroenterol 27(5):460–466. doi:10.1097/MOG.0b013e328349e31f

Kobayashi T, Honke K, Gasa S, Miyazaki T, Tajima H, Matsumoto K, Nakamura T, Makita A (1994) Hepatocyte growth factor elevates the activity levels of glycolipid sulfotransferases in renal cell carcinoma cells. Eur J Biochem 219(1–2):407–413

Kohn AD, Moon RT (2005) Wnt and calcium signaling: beta-catenin-independent pathways. Cell Calcium 38(3–4):439–446. doi:10.1016/j.ceca.2005.06.022

Koshiba T, Hosotani R, Miyamoto Y, Ida J, Tsuji S, Nakajima S, Kawaguchi M, Kobayashi H, Doi R, Hori T, Fujii N, Imamura M (2000) Expression of stromal cell-derived factor 1 and CXCR4 ligand receptor system in pancreatic cancer: a possible role for tumor progression. Clin Cancer Res 6(9):3530–3535

Kraman M, Bambrough PJ, Arnold JN, Roberts EW, Magiera L, Jones JO, Gopinathan A, Tuveson DA, Fearon DT (2010) Suppression of antitumor immunity by stromal cells expressing fibroblast activation protein-alpha. Science 330(6005):827–830. doi:330/6005/827 [pii] 10.1126/science.1195300

Krymskaya VP, Hoffman R, Eszterhas A, Ciocca V, Panettieri RA (1997) TGF-β1 modulates EGF-stimulated phosphatidylinositol 3-kinase activity in human airway smooth muscle cells. Am J Physiol 273(6):L1220–L1227

Kurooka H, Honjo T (2000) Functional interaction between the mouse notch1 intracellular region and histone acetyltransferases PCAF and GCN5. J Biol Chem 275(22):17211–17220. doi:10.1074/jbc.M000909200 M000909200 [pii]

Lauth M, Bergstrom A, Shimokawa T, Tostar U, Jin Q, Fendrich V, Guerra C, Barbacid M, Toftgard R (2010) DYRK1B-dependent autocrine-to-paracrine shift of Hedgehog signaling by mutant RAS. Nat Struct Mol Biol 17(6):718–725. doi:nsmb.1833 [pii] 10.1038/nsmb.1833

Lee WY, Chen HH, Chow NH, Su WC, Lin PW, Guo HR (2005) Prognostic significance of co-expression of RON and MET receptors in node-negative breast cancer patients. Clin Cancer Res 11(6):2222–2228. doi:11/6/2222 [pii] 10.1158/1078-0432.CCR-04-1761

Li C, Heidt DG, Dalerba P, Burant CF, Zhang L, Adsay V, Wicha M, Clarke MF, Simeone DM (2007) Identification of pancreatic cancer stem cells. Cancer Res 67(3):1030–1037. doi:67/3/1030 [pii] 10.1158/0008-5472.CAN-06-2030

Li C, Wu JJ, Hynes M, Dosch J, Sarkar B, Welling TH, Pasca di Magliano M, Simeone DM (2011) c-Met is a marker of pancreatic cancer stem cells and therapeutic target. Gastroenterology 141(6):2218.e5–2227.e5. doi:S0016-5085(11)01157-7 [pii] 10.1053/j.gastro.2011.08.009

Lindsell CE, Shawber CJ, Boulter J, Weinmaster G (1995) Jagged: a mammalian ligand that activates Notch1. Cell 80(6):909–917. doi:0092-8674(95)90294-5 [pii]

Lonardo E, Hermann PC, Mueller MT, Huber S, Balic A, Miranda-Lorenzo I, Zagorac S, Alcala S, Rodriguez-Arabaolaza I, Ramirez JC, Torres-Ruíz R, Garcia E, Hidalgo M, Cebrián DÁ, Heuchel R, Löhr M, Berger F, Bartenstein P, Aicher A, Heeschen C (2011) Nodal/Activin signaling drives self-renewal and tumorigenicity of pancreatic cancer stem cells and provides a target for combined drug therapy. Cell Stem Cell 9(5):433–446

Lyon M, Deakin JA, Mizuno K, Nakamura T, Gallagher JT (1994) Interaction of hepatocyte growth factor with heparan sulfate. Elucidation of the major heparan sulfate structural determinants. J Biol Chem 269(15):11216–11223

Malkoski SP, Wang XJ (2012) Two sides of the story? Smad4 loss in pancreatic cancer versus head-and-neck cancer. FEBS Lett. doi:S0014-5793(12)00101-9 [pii] 10.1016/j.febslet.2012.01.054

Mani SA, Guo W, Liao MJ, Eaton EN, Ayyanan A, Zhou AY, Brooks M, Reinhard F, Zhang CC, Shipitsin M, Campbell LL, Polyak K, Brisken C, Yang J, Weinberg RA (2008) The epithelial-mesenchymal transition generates cells with properties of stem cells. Cell 133(4):704–715. doi:S0092-8674(08)00444-3 [pii] 10.1016/j.cell.2008.03.027

Maroun CR, Naujokas MA, Holgado-Madruga M, Wong AJ, Park M (2000) The tyrosine phosphatase SHP-2 is required for sustained activation of extracellular signal-regulated kinase and epithelial morphogenesis downstream from the met receptor tyrosine kinase. Mol Cell Biol 20(22):8513–8525

Maroun CR, Naujokas MA, Park M (2003) Membrane targeting of Grb2-associated binder-1 (Gab1) scaffolding protein through Src myristoylation sequence substitutes for Gab1 pleckstrin homology domain and switches an epidermal growth factor response to an invasive morphogenic program. Mol Biol Cell 14(4):1691–1708. doi:10.1091/mbc.E02-06-0352

Marusyk A, Polyak K (2010) Tumor heterogeneity: causes and consequences. Biochim Biophys Acta 1805(1):105–117. doi:S0304-419X(09)00074-2 [pii] 10.1016/j.bbcan.2009.11.002

Massague J (1998) TGF-beta signal transduction. Annu Rev Biochem 67:753–791. doi:10.1146/annurev.biochem.67.1.753

McDermott U, Sharma SV, Dowell L, Greninger P, Montagut C, Lamb J, Archibald H, Raudales R, Tam A, Lee D, Rothenberg SM, Supko JG, Sordella R, Ulkus LE, Iafrate AJ, Maheswaran S, Njauw CN, Tsao H, Drew L, Hanke JH, Ma XJ, Erlander MG, Gray NS, Haber DA, Settleman J (2007) Identification of genotype-correlated sensitivity to selective kinase inhibitors by using high-throughput tumor cell line profiling. Proc Natl Acad Sci USA 104(50): 19936–19941. doi:0707498104 [pii] 10.1073/pnas.0707498104

Migliore C, Giordano S (2008) Molecular cancer therapy: can our expectation be MET? Eur J Cancer 44(5):641–651. doi:S0959-8049(08)00067-1 [pii] 10.1016/j.ejca.2008.01.022

Miyamoto Y, Maitra A, Ghosh B, Zechner U, Argani P, Iacobuzio-Donahue CA, Sriuranpong V, Iso T, Meszoely IM, Wolfe MS, Hruban RH, Ball DW, Schmid RM, Leach SD (2003) Notch mediates TGF alpha-induced changes in epithelial differentiation during pancreatic tumorigenesis. Cancer Cell 3(6):565–576. doi:S1535610803001405 [pii]

Morel V, Lecourtois M, Massiani O, Maier D, Preiss A, Schweisguth F (2001) Transcriptional repression by suppressor of hairless involves the binding of a hairless-dCtBP complex in Drosophila. Curr Biol 11(10):789–792. doi:S0960-9822(01)00224-X [pii]

Morris JP 4th, Cano DA, Sekine S, Wang SC, Hebrok M (2010) Beta-catenin blocks Kras-dependent reprogramming of acini into pancreatic cancer precursor lesions in mice. J Clin Invest 120(2):508–520. doi:10.1172/JCI40045 40045 [pii]

Mucsi I, Skorecki KL, Goldberg HJ (1996) Extracellular signal-regulated kinase and the small GTP-binding protein, Rac, contribute to the effects of transforming growth factor-beta1 on gene expression. J Biol Chem 271(28):16567–16572. doi:10.1074/jbc.271.28.16567

Mullendore ME, Koorstra JB, Li YM, Offerhaus GJ, Fan X, Henderson CM, Matsui W, Eberhart CG, Maitra A, Feldmann G (2009) Ligand-dependent Notch signaling is involved in tumor initiation and tumor maintenance in pancreatic cancer. Clin Cancer Res 15(7):2291–2301. doi:1078-0432.CCR-08-2004 [pii] 10.1158/1078-0432.CCR-08-2004

Murdoch C, Muthana M, Coffelt SB, Lewis CE (2008) The role of myeloid cells in the promotion of tumour angiogenesis. Nat Rev Cancer 8(8):618–631. doi:nrc2444 [pii] 10.1038/nrc2444

Murtaugh LC, Law AC, Dor Y, Melton DA (2005) Beta-catenin is essential for pancreatic acinar but not islet development. Development 132(21):4663–4674. doi:10.1242/dev.02063

Nawroth R, van Zante A, Cervantes S, McManus M, Hebrok M, Rosen SD (2007) Extracellular sulfatases, elements of the Wnt signaling pathway, positively regulate growth and tumorigenicity of human pancreatic cancer cells. PLoS One 2(4):e392. doi:10.1371/journal.pone.0000392

Nishimura T, Honda H, Takeichi M (2012) Planar cell polarity links axes of spatial dynamics in neural-tube closure. Cell 149(5):1084–1097

Nishisho I, Nakamura Y, Miyoshi Y, Miki Y, Ando H, Horii A, Koyama K, Utsunomiya J, Baba S, Hedge P (1991) Mutations of chromosome 5q21 genes in FAP and colorectal cancer patients. Science 253(5020):665–669

Nolan-Stevaux O, Lau J, Truitt ML, Chu GC, Hebrok M, Fernandez-Zapico ME, Hanahan D (2009) GLI1 is regulated through Smoothened-independent mechanisms in neoplastic pancreatic ducts and mediates PDAC cell survival and transformation. Genes Dev 23(1):24–36. doi:23/1/24 [pii] 10.1101/gad.1753809

Olaku V, Matzke A, Mitchell C, Hasenauer S, Sakkaravarthi A, Pace G, Ponta H, Orian-Rousseau V (2011) c Met recruits ICAM-1 as a coreceptor to compensate for the loss of CD44 in Cd44 null mice. Mol Biol Cell 22(15):2777–2786. doi:mbc.E11-02-0134 [pii] 10.1091/mbc.E11-02-0134

Olive KP, Jacobetz MA, Davidson CJ, Gopinathan A, McIntyre D, Honess D, Madhu B, Goldgraben MA, Caldwell ME, Allard D, Frese KK, DeNicola G, Feig C, Combs C, Winter SP, Ireland-Zecchini H, Reichelt S, Howat WJ, Chang A, Dhara M, Wang L, Rückert F, Grützmann R, Pilarsky C, Izeradjene K, Hingorani SR, Huang P, Davies SE, Plunkett W, Egorin M, Hruban RH, Whitebread N, McGovern K, Adams J, Iacobuzio-Donahue C, Griffiths J, Tuveson DA (2009) Inhibition of Hedgehog signaling enhances delivery of chemotherapy in a mouse model of pancreatic cancer. Science 324(5933):1457–1461. doi:10.1126/science.1171362

Onozuka H, Tsuchihara K, Esumi H (2011) Hypoglycemic/hypoxic condition in vitro mimicking the tumor microenvironment markedly reduced the efficacy of anticancer drugs. Cancer Sci 102(5):975–982. doi:10.1111/j.1349-7006.2011.01880.x

Orian-Rousseau V, Chen L, Sleeman JP, Herrlich P, Ponta H (2002) CD44 is required for two consecutive steps in HGF/c-Met signaling. Genes Dev 16(23):3074–3086. doi:10.1101/gad.242602

Park WS, Dong SM, Kim SY, Na EY, Shin MS, Pi JH, Kim BJ, Bae JH, Hong YK, Lee KS, Lee SH, Yoo NJ, Jang JJ, Pack S, Zhuang Z, Schmidt L, Zbar B, Lee JY (1999) Somatic mutations in the kinase domain of the Met/hepatocyte growth factor receptor gene in childhood hepatocellular carcinomas. Cancer Res 59(2):307–310

Pasca di Magliano M, Biankin AV, Heiser PW, Cano DA, Gutierrez PJ, Deramaudt T, Segara D, Dawson AC, Kench JG, Henshall SM, Sutherland RL, Dlugosz A, Rustgi AK, Hebrok M (2007) Common activation of canonical Wnt signaling in pancreatic adenocarcinoma. PLoS One 2(11):e1155. doi:10.1371/journal.pone.0001155

Pasche B (2001) Role of transforming growth factor beta in cancer. J Cell Physiol 186(2):153–168. doi:10.1002/1097-4652(200002)186:2<153::aid-jcp1016>3.0.co;2-j

Pennacchietti S, Michieli P, Galluzzo M, Mazzone M, Giordano S, Comoglio PM (2003) Hypoxia promotes invasive growth by transcriptional activation of the met protooncogene. Cancer Cell 3(4):347–361. doi:S1535610803000850 [pii]

Plentz R, Park JS, Rhim AD, Abravanel D, Hezel AF, Sharma SV, Gurumurthy S, Deshpande V, Kenific C, Settleman J, Majumder PK, Stanger BZ, Bardeesy N (2009) Inhibition of gamma-secretase activity inhibits tumor progression in a mouse model of pancreatic ductal adenocarcinoma. Gastroenterology 136(5):1741.e6–1749.e6. doi:S0016-5085(09)00014-6 [pii] 10.1053/j.gastro.2009.01.008

Provenzano PP, Cuevas C, Chang AE, Goel VK, Von Hoff DD, Hingorani SR (2012) Enzymatic targeting of the stroma ablates physical barriers to treatment of pancreatic ductal adenocarcinoma. Cancer Cell 21(3):418–429

Pylayeva-Gupta Y, Lee KE, Hajdu CH, Miller G, Bar-Sagi D (2012) Oncogenic Kras-induced GM-CSF production promotes the development of pancreatic neoplasia. Cancer Cell 21(6):836–847. doi:10.1016/j.ccr.2012.04.024

Qian BZ, Pollard JW (2010) Macrophage diversity enhances tumor progression and metastasis. Cell 141(1):39–51. doi:S0092-8674(10)00287-4 [pii] 10.1016/j.cell.2010.03.014

Radtke F, Raj K (2003) The role of Notch in tumorigenesis: oncogene or tumour suppressor? Nat Rev Cancer 3(10):756–767. doi:10.1038/nrc1186

Reya T, Morrison SJ, Clarke MF, Weissman IL (2001) Stem cells, cancer, and cancer stem cells. Nature 414(6859):105–111. doi:10.1038/35102167 35102167 [pii]

Reynolds TC, Smith SD, Sklar J (1987) Analysis of DNA surrounding the breakpoints of chromosomal translocations involving the beta T cell receptor gene in human lymphoblastic neoplasms. Cell 50(1):107–117. doi:0092-8674(87)90667-2 [pii]

Rowland-Goldsmith MA, Maruyama H, Kusama T, Ralli S, Korc M (2001) Soluble type II transforming growth factor-beta (TGF-beta) receptor inhibits TGF-beta signaling in COLO-357 pancreatic cancer cells in vitro and attenuates tumor formation. Clin Cancer Res 7(9): 2931–2940

Rowland-Goldsmith MA, Maruyama H, Matsuda K, Idezawa T, Ralli M, Ralli S, Korc M (2002) Soluble type II transforming growth factor-beta receptor attenuates expression of metastasis-associated genes and suppresses pancreatic cancer cell metastasis. Mol Cancer Ther 1(3): 161–167

Rubin LL, de Sauvage FJ (2006) Targeting the Hedgehog pathway in cancer. Nat Rev Drug Discov 5(12):1026–1033. doi:nrd2086 [pii] 10.1038/nrd2086

Ruffell B, DeNardo DG, Affara NI, Coussens LM (2010) Lymphocytes in cancer development: polarization towards pro-tumor immunity. Cytokine Growth Factor Rev 21(1):3–10. doi:S1359-6101(09)00109-9 [pii] 10.1016/j.cytogfr.2009.11.002

Ruiz i Altaba A, Palma V, Dahmane N (2002a) Hedgehog-Gli signalling and the growth of the brain. Nat Rev Neurosci 3(1):24–33. doi:10.1038/nrn704 nrn704 [pii]

Ruiz i Altaba A, Sanchez P, Dahmane N (2002b) Gli and hedgehog in cancer: tumours, embryos and stem cells. Nat Rev Cancer 2(5):361–372. doi:10.1038/nrc796

Samowitz WS, Powers MD, Spirio LN, Nollet F, van Roy F, Slattery ML (1999) Beta-catenin mutations are more frequent in small colorectal adenomas than in larger adenomas and invasive carcinomas. Cancer Res 59(7):1442–1444

Scales SJ, de Sauvage FJ (2009) Mechanisms of Hedgehog pathway activation in cancer and implications for therapy. Trends Pharmacol Sci 30(6):303–312. doi:S0165-6147(09)00069-8 [pii] 10.1016/j.tips.2009.03.007

Schlingensiepen KH, Jaschinski F, Lang SA, Moser C, Geissler EK, Schlitt HJ, Kielmanowicz M, Schneider A (2011) Transforming growth factor-beta 2 gene silencing with trabedersen (AP 12009) in pancreatic cancer. Cancer Sci 102(6):1193–1200. doi:10.1111/j.1349-7006.2011.01917.x

Schmidt L, Duh FM, Chen F, Kishida T, Glenn G, Choyke P, Scherer SW, Zhuang Z, Lubensky I, Dean M, Allikmets R, Chidambaram A, Bergerheim UR, Feltis JT, Casadevall C, Zamarron A, Bernues M, Richard S, Lips CJ, Walther MM, Tsui LC, Geil L, Orcutt ML, Stackhouse T, Lipan J, Slife L, Brauch H, Decker J, Niehans G, Hughson MD, Moch H, Storkel S, Lerman MI, Linehan WM, Zbar B (1997) Germline and somatic mutations in the tyrosine kinase domain of the MET proto-oncogene in papillary renal carcinomas. Nat Genet 16(1):68–73. doi:10.1038/ng0597-68

Shawber C, Boulter J, Lindsell CE, Weinmaster G (1996) Jagged2: a serrate-like gene expressed during rat embryogenesis. Dev Biol 180(1):370–376. doi:S0012-1606(96)90310-3 [pii] 10.1006/dbio.1996.0310

Shutter JR, Scully S, Fan W, Richards WG, Kitajewski J, Deblandre GA, Kintner CR, Stark KL (2000) Dll4, a novel Notch ligand expressed in arterial endothelium. Genes Dev 14(11): 1313–1318

Sierra JR, Tsao MS (2011) c-MET as a potential therapeutic target and biomarker in cancer. Ther Adv Med Oncol 3(1 Suppl):S21–S35. doi:10.1177/1758834011422557 10.1177_1758834011422557 [pii]

Singh S, Srivastava SK, Bhardwaj A, Owen LB, Singh AP (2010) CXCL12-CXCR4 signalling axis confers gemcitabine resistance to pancreatic cancer cells: a novel target for therapy. Br J Cancer 103(11):1671–1679. doi: http://www.nature.com/bjc/journal/v103/n11/suppinfo/6605968s1.html

Soman NR, Correa P, Ruiz BA, Wogan GN (1991) The TPR-MET oncogenic rearrangement is present and expressed in human gastric carcinoma and precursor lesions. Proc Natl Acad Sci U S A 88(11):4892–4896

Sonnenberg E, Meyer D, Weidner KM, Birchmeier C (1993) Scatter factor/hepatocyte growth factor and its receptor, the c-met tyrosine kinase, can mediate a signal exchange between mesenchyme and epithelia during mouse development. J Cell Biol 123(1):223–235

Suzuki K, Hayashi N, Yamada Y, Yoshihara H, Miyamoto Y, Ito Y, Ito T, Katayama K, Sasaki Y, Ito A et al (1994) Expression of the c-met protooncogene in human hepatocellular carcinoma. Hepatology 20(5):1231–1236. doi:S0270913994003459 [pii]

Takeuchi H, Bilchik A, Saha S, Turner R, Wiese D, Tanaka M, Kuo C, Wang HJ, Hoon DS (2003) c-MET expression level in primary colon cancer: a predictor of tumor invasion and lymph node metastases. Clin Cancer Res 9(4):1480–1488

Tascilar M, Skinner HG, Rosty C, Sohn T, Wilentz RE, Offerhaus GJ, Adsay V, Abrams RA, Cameron JL, Kern SE, Yeo CJ, Hruban RH, Goggins M (2001) The SMAD4 protein and prognosis of pancreatic ductal adenocarcinoma. Clin Cancer Res 7(12):4115–4121

Terabe M, Ambrosino E, Takaku S, O'Konek JJ, Venzon D, Lonning S, McPherson JM, Berzofsky JA (2009) Synergistic enhancement of CD8+ T cell-mediated tumor vaccine efficacy by an anti-transforming growth factor-beta monoclonal antibody. Clin Cancer Res 15(21):6560–6569. doi:1078-0432.CCR-09-1066 [pii] 10.1158/1078-0432.CCR-09-1066

Thayer SP, di Magliano MP, Heiser PW, Nielsen CM, Roberts DJ, Lauwers GY, Qi YP, Gysin S, Fernandez-del Castillo C, Yajnik V, Antoniu B, McMahon M, Warshaw AL, Hebrok M (2003) Hedgehog is an early and late mediator of pancreatic cancer tumorigenesis. Nature 425(6960):851–856. doi:10.1038/nature02009 nature02009 [pii]

Trusolino L, Bertotti A, Comoglio PM (2010) MET signalling: principles and functions in development, organ regeneration and cancer. Nat Rev Mol Cell Biol 11(12):834–848. doi:nrm3012 [pii] 10.1038/nrm3012

Wang Z, Zhang Y, Li Y, Banerjee S, Liao J, Sarkar FH (2006) Down-regulation of Notch-1 contributes to cell growth inhibition and apoptosis in pancreatic cancer cells. Mol Cancer Ther 5(3):483–493. doi:5/3/483 [pii] 10.1158/1535-7163.MCT-05-0299

Wang L, Heidt DG, Lee CJ, Yang H, Logsdon CD, Zhang L, Fearon ER, Ljungman M, Simeone DM (2009a) Oncogenic function of ATDC in pancreatic cancer through Wnt pathway activation and beta-catenin stabilization. Cancer Cell 15(3):207–219. doi:10.1016/j.ccr.2009.01.018

Wang P, Fan J, Chen Z, Meng ZQ, Luo JM, Lin JH, Zhou ZH, Chen H, Wang K, Xu ZD, Liu LM (2009b) Low-level expression of Smad7 correlates with lymph node metastasis and poor prognosis in patients with pancreatic cancer. Ann Surg Oncol 16(4):826–835. doi:10.1245/s10434-008-0284-5

Wang Z, Li Y, Kong D, Banerjee S, Ahmad A, Azmi AS, Ali S, Abbruzzese JL, Gallick GE, Sarkar FH (2009c) Acquisition of epithelial-mesenchymal transition phenotype of gemcitabine-resistant pancreatic cancer cells is linked with activation of the notch signaling pathway. Cancer Res 69(6):2400–2407. doi:10.1158/0008-5472.can-08-4312

Weidner KM, Di Cesare S, Sachs M, Brinkmann V, Behrens J, Birchmeier W (1996) Interaction between Gab1 and the c-Met receptor tyrosine kinase is responsible for epithelial morphogenesis. Nature 384(6605):173–176. doi:10.1038/384173a0

Wells JM, Esni F, Boivin GP, Aronow BJ, Stuart W, Combs C, Sklenka A, Leach SD, Lowy AM (2007) Wnt/beta-catenin signaling is required for development of the exocrine pancreas. BMC Dev Biol 7:4. doi:10.1186/1471-213X-7-4

Wilentz RE, Iacobuzio-Donahue CA, Argani P, McCarthy DM, Parsons JL, Yeo CJ, Kern SE, Hruban RH (2000) Loss of expression of Dpc4 in pancreatic intraepithelial neoplasia: evidence that DPC4 inactivation occurs late in neoplastic progression. Cancer Res 60(7):2002–2006

Willert K, Jones KA (2006) Wnt signaling: is the party in the nucleus? Genes Dev 20(11):1394–1404. doi:20/11/1394 [pii] 10.1101/gad.1424006

Yokoi K, Fidler IJ (2004) Hypoxia increases resistance of human pancreatic cancer cells to apoptosis induced by gemcitabine. Clin Cancer Res 10(7):2299–2306. doi:10.1158/1078-0432.ccr-03-0488

Zarnegar R, Michalopoulos GK (1995) The many faces of hepatocyte growth factor: from hepatopoiesis to hematopoiesis. J Cell Biol 129(5):1177–1180. doi:10.1083/jcb.129.5.1177

Zhou S, Fujimuro M, Hsieh JJ, Chen L, Miyamoto A, Weinmaster G, Hayward SD (2000) SKIP, a CBF1-associated protein, interacts with the ankyrin repeat domain of NotchIC To facilitate NotchIC function. Mol Cell Biol 20(7):2400–2410

Mouse Models of Pancreatic Ductal Adenocarcinoma

Filip Bednar and Marina Pasca di Magliano

Abstract In the past several years, numerous genetically engineered mice have been used to model pancreatic cancer. These models differed based on the approach used (some use transgenes while others used homologous recombination), as well as on the oncogene or combination of oncogenes. The expression of an oncogenic form of Kras in the mouse pancreas at physiological levels has led to models that not only develop pancreatic ductal adenocarcinoma (PDA), but that mimic the progression of the human disease, including pre-carcinogenic stages, such as Pancreatic Intraepithelial Neoplasia (PanIN), and activation of specific signaling pathways. Thanks to genetically engineered mouse models we have started to discern the contribution of different signaling pathways to initiation and progression, and in some cases maintenance, of pancreatic cancers. We have also started dissecting the importance of the interactions between the tumor cells and their surrounding microenvironment. Notwithstanding the sophistication of the current models, further modification of the approaches used could be implemented, for example to develop mice with clonal tumors, such as seen in human patients. Moreover, applying genomic approaches to the study of the mouse models might shed light on their ability to recapitulate specific subsets of human tumors.

F. Bednar
Department of Surgery, University of Michigan, Ann Arbor, MI, USA

M.P. di Magliano, Ph.D. (✉)
Department of Surgery, University of Michigan, Ann Arbor, MI, USA

Department of Cell and Developmental Biology, University of Michigan,
1500 East Medical Center Drive, Ann Arbor, MI 48109-5936, USA

Program in Cellular and Molecular Biology, University of Michigan, Ann Arbor, MI, USA

Comprehensive Cancer Center, University of Michigan, Ann Arbor, MI, USA
e-mail: marinapa@umich.edu; marinapa@med.umich.edu

D.M. Simeone and A. Maitra (eds.), *Molecular Genetics of Pancreatic Cancer*,
DOI 10.1007/978-1-4614-6549-2_7,

Introduction

Pancreatic ductal adenocarcinoma (PDA), the most common form of pancreatic cancer, is one of the most lethal human malignancies. The American Cancer Society estimates that 43,920 people will be diagnosed with the disease in 2012 and 37,390 people will die from their cancer (www.cancer.org). Pancreatic cancer is rapidly lethal. Curative surgery and adjuvant chemoradiotherapy is an option only for approximately 10–20 % of newly diagnosed patients. Even for these patients, the 5-year survival rate is only 20–30 %. Most patients, however, have locally advanced or widely metastatic disease at the time of diagnosis, and typically survive only 6–12 months (Desai et al. 2009; Ferrone et al. 2008; Hsu et al. 2010; Katz et al. 2009). Moreover, the prognosis for pancreatic cancer patients has not significantly improved over the past 40 years. The possibility to develop new targeted therapies for this disease relies on understanding its biology, and therefore requires basic and translational research. This chapter summarizes a subset of the ongoing research in pancreatic cancer, namely, the development and analysis of genetically engineered mouse models of this disease.

Several experimental models (Mansour et al. 1988; Thomas and Capecchi 1987), each with its own inherent advantages and disadvantages, can be used to explore fundamental cancer biology questions. Here we limit our discussion to genetically engineered mouse models, but we will compare their use with other widely used alternatives. The possibility to generate mouse models of cancer followed the discovery of oncogenes as well as the development of techniques to modify the mouse genome. Two main groups of genetically modified mice, each with its own subcategories, have been widely used in carcinogenesis studies: transgenic mice (Gordon et al. 1980; Palmiter and Brinster 1986), and models based on gene targeting in mouse ES cells (Doetschman et al. 1987). Below we discuss the use of both of these approaches to generate genetically defined mouse models of PDA that are directly based on our understanding of the human disease.

Pancreatic Cancer Progression

Human pancreatic adenocarcinoma is widely believed to be preceded by precursor lesions. The most common precursor lesions resemble pancreatic ducts and are known as Pancreatic Intraepithelial Neoplasia (PanINs). PanINs are classified as 1A, 1B, 2, and 3 based on histological characteristics (Hruban et al. 2004). Namely, PanIN1As are lined by a columnar epithelium with abundant accumulation of intracellular mucins; the cells forming PanIN1B are similar, but they form papillary structures within the duct lumen; PanIN2s show loss of cellular polarity and form a pseudostratified epithelium. PanIN3s are also known as carcinoma in situ. Other, less common, precursor lesions for pancreatic cancer are mucinous cystic neoplasms (MCNs) and intraductal papillary mucinous neoplasms (IPMNs) [for comprehensive reviews, see (Adsay 2008; Hezel et al. 2006)]. Since MCNs and IPMNs can form clinically detectable lesions, they can often be surgically removed prior to the development of pancreatic adenocarcinoma; in addition, they are believed to infrequently progress to adenocarcinoma; in contrast, PanINs can currently not be detected and are rarely found in the absence of invasive tumors (Adsay 2008; Tanaka et al. 2006).

The PanIN progression describes changes in the pancreatic epithelium, the compartment that will ultimately form the tumor cells. At each step of the process, however, the changes in the epithelium are accompanied by changes in the surrounding tissue. The most prominent is the accumulation of an abundant fibroinflammatory stroma, also known as a desmoplastic reaction, of mesenchymal origin, surrounding the epithelial lesions (Korc 2007). The stroma forms a complex microenvironment with each component having an important function. While all tumor types are formed by tumor and non-tumor cells, pancreatic cancer is unique in the abundance of the stroma, which can form the vast majority of the tumor volume. Therefore, an ideal mouse model of pancreatic cancer should follow the stepwise changes in the epithelial cells (such as the PanIN progression), from normal to cancer, as well as elicit the formation of extensive stroma, in order to develop tumors that are similar to the human counterpart. The following sections describe the development of such models, followed by describing their potential use for basic and translational research.

Developing Mouse Models of Pancreatic Cancer

RIP-Tag Insulinoma and Ela-Tag Mouse Models

Before gene targeting in ES cells was developed, the pioneers of mouse modeling used transgenic mice to express oncogenes in the pancreas. One of the most widely used models of cancer is the RIP-Tag mouse, which uses the rat insulin promoter/enhancer to express the SV40 large T antigen (Tag) in the insulin producing cells of the pancreas (Hanahan 1985). The RIP-Tag mouse develops islet cell (β cell) tumors that resemble human insulinomas. This model develops step-wise carcinogenesis, with a predictable time-frame and complete penetrance. Given that the model is based on a single transgene, it is easy to cross into other genetically engineered animals. It is therefore no surprise that some of the most fundamental aspects of cancer biology, later referred to as the hallmarks of cancer, have been carefully characterized in this model, including the need for sustained angiogenesis for frank tumor formation and the presence of unlimited replicative potential (Hanahan and Weinberg 2000, 2011).

A transgenic mouse model system targeting the SV40 Tag to the acinar compartment using an elastase promoter also demonstrated progressive stepwise neoplasia (Ornitz et al. 1987). The tumors were composed of disorganized nests of acinar-like cells, which had lost some of their differentiation characteristics during the neoplastic transformation. Although this model did not recapitulate the histological characteristics of human PDA, it proved that, together with the RIP-Tag2 model, specific pancreatic compartments could be targeted for oncogenesis using the proper enhancer/promoter elements. These findings were crucial for the development of the later generations of pancreatic cancer mouse models. Transgenic mice containing oncogenes including myc, Ras, and SV40 Tag therefore provided the initial systems that allowed investigators to study the overall process of tumorigenesis in vivo and paved the way for future model development.

TGFα Overexpression Models

Transforming growth factor α (TGFα) is a known ligand for the epidermal growth factor receptor (EGFR) (Liebmann 2011), which is highly expressed in human pancreatic cancer (Wagner et al. 1998). TGFα was overexpressed in transgenic mice under the control of the metallothionein or the rat elastase promoter (Sandgren et al. 1990). The metallothionein promoter allowed inducible, Zn^{2+} regulated, expression of TGFα, which led to hyperplasia and fibrosis in multiple tissues, including the pancreas. In Elastase-TGFα mice, TGFα is expressed specifically in the pancreatic exocrine cells; however, the expression is constitutive. These mice developed pancreatic fibrosis and acinar-ductal metaplasia (ADM), the replacement of acinar cells with duct-like structures (Sandgren et al. 1990). Older Elastase-TGFα mice developed cystic papillary neoplasms and acinar cell carcinoma (Hruban et al. 2006; Wagner et al. 1998). The contemporary inactivation of one of both alleles of the tumor suppressor p53 accelerated the onset of pancreatic tumor formation, but did not change the tumor type (Wagner et al. 2001). Even though it does not fully recapitulate the progression of human pancreatic adenocarcinoma, the Elastase-TGFα mouse has been a useful model to improve our understanding of this disease. Intriguingly, Elastase-TGFα;$p53^{+/-}$ (or $p53^{-/-}$) mice shared several alterations common to the human tumors such as increased levels of Ras signaling, as well as loss of p16/Ink4a, Smad4, and occasionally Rb1 (Wagner et al. 2001). Thus, the mutational profile of the model recapitulated the genetic changes common in primary human pancreatic cancers, but the histology of this model did not recapitulate the PanIN/PDA progression (Hruban et al. 2006). Despite these reservations, the TGFα models represent some of the earliest informative examples of animal models of pancreatic neoplasia.

PyMT Model

An alternative method to obtain tissue-specific expression of a transgene is based on the tissue-specific expression of the viral receptor TVA; the mice can then be infected with the avian leukosis virus (ALV) engineered to express a gene of interest (Federspiel et al. 1994). The virus will only infect the TVA-expressing cells, thus ensuring tissue-specific delivery of the transgene. The main advantages of this system are, first, the possibility to induce expression of the gene of interest in the adult organism, and, second, to generate a single transgenic model that can then be infected with several viruses, or combination of viruses, expressing different genes of interest. The replication-competent avian sarcoma-leukosis virus (RCAS-TVA) system was used to generate transgenic elastase-*tv-a* mice where TVA is expressed in the pancreatic acinar cells. These mice were then infected the mice with an ALV expressing the polyoma virus middle T antigen (PyMT) or the *c-myc* oncogene (Lewis et al. 2003). Expression of PyMT led to cystic papillary neoplasms with some ductal differentiation and to acinar cell carcinomas. The acinar cell carcinomas were more common in TVA transgenic mice that were bred onto the Ink4a$^{-/-}$ background. Expression of c-Myc had an altogether different effect; it led to the formation of insulin-positive pancreatic neuroendocrine tumors (Lewis et al. 2003). Therefore, this model did not recapitulate the formation of PDA, but it highlighted the role for specific oncogenes, inducing distinct transcriptional networks, in the initiation of different tumor types from the same cell of origin.

Oncogenic Kras Models

A turning point in the approach to pancreatic cancer modeling came with an enhanced focus on Kras, which had previously been found to be a frequently mutated gene in human pancreatic adenocarcinoma (Almoguera et al. 1988). A decade of subsequent work identified additional genetic alterations that are commonly found in this disease and this work eventually was summarized in a progression model of pancreatic neoplasia (Hruban et al. 2000) that, with small additions, is still valid today. Recent genome sequencing efforts have confirmed that Kras mutations appear early during the progression to pancreatic adenocarcinoma (Kanda et al. 2012), and that mutations of the INK4A, p53, DPC4/SMAD4, and BRCA2 tumor suppressor loci occur at later stages and precede the development of invasive PDA [reviewed in (Maitra and Hruban 2008)]. The wealth of genetic information that accumulated during the past two decades was used to generate mouse models of pancreatic cancer that seek to reproduce the genetics of the human disease, and the oncogenic form of KRAS took center stage in a series of mouse models.

The normal form of Kras, one of the three members of the Ras family of GTPases, is activated in response to extracellular signals that are transduced through a receptor tyrosine kinase; Kras transmits the signal to its downstream effectors and at the same time hydrolyzes GTP to GDP, thus inactivating itself [reviewed in (Pylayeva-Gupta et al. 2011)], in one of the many mechanisms that ensure tight regulation of cellular signals. In human pancreatic adenocarcinoma, the most common mutations of Kras are the G12D and G12V single amino acid substitutions (Moskaluk et al. 1997). Both mutant forms lead to a protein that is unable to hydrolyze GTP to GDP, and is therefore “locked” in a constitutively active state. Given its frequent presence in early lesions, oncogenic Kras had been hypothesized to be the initiating driver mutation for PDA (Hruban et al. 2000). A series of transgenic mice were generated to test this hypothesis, and with the goal to develop a model of pancreatic cancer faithful to the human disease.

In one of the first models, the G12D allele of Kras was expressed under the control of the elastase promoter, thus targeting it to acinar cells in the pancreas (Grippo et al. 2003). ELA-Kras^{G12D} mice developed frequent ADM, and cystic papillary neoplasms, but no PanINs or ductal adenocarcinomas. These results suggested that mutant Kras had been targeted to the wrong cellular compartment, it was expressed at inappropriate levels, or that Kras was insufficient to drive PanIN and PDA formation in the absence of additional genetic alterations. Alternatively, the elastase promoter may be inactivated during PanIN formation leading to the inactivation of oncogenic Kras expression and potential reversion of the PanINs back to the acinar fate. A second transgenic model utilized the Kras^{G12V} allele under the control of the cytokeratin-19 (CK-19) promoter, which leads to expression in the pancreatic ductal epithelium, as well as in the intestinal villi, colonocytes, and gastric isthmus cells (Brembeck et al. 2001). CK19-Kras^{G12V} mice developed periductal lymphocytic infiltrates and occasional ductal hyperplasia (Brembeck et al. 2003), but no precursor lesions or tumors. Yet an alternative approach to mutant Kras expression of the pancreas was the generation of Mist1-Kras^{G12D} mice by inserting the oncogenic Kras allele into the Mist1 locus (Tuveson et al. 2006). Mist1, a basic helix-loop-helix transcription factor, is expressed in the acinar cells of the pancreas, the serous cells of the salivary glands, the chief cells of the stomach, and secretory cells within the male reproductive system and in other organs (Pin et al. 2000). Mist1 plays a key role in the maintenance of the pancreatic acinar cells and Mist1$^{-/-}$ mice progressively lose their acinar compartment (Pin et al. 2001). When Kras^{G12D} was inserted into the Mist1 locus, the pancreas developed acinar metaplasia and cystic papillary neoplasms; the mice also develop hepatocellular carcinoma. Metastatic acinar cell carcinomas were the most common invasive neoplasms in this model (Tuveson et al. 2006). Mist1-Kras^{G12D} mice bred with p53$^{+/-}$ mice developed the same tumor spectrum, with more rapid onset of tumor formation and shorter survival compared to either of the parental strains (Tuveson et al. 2006). This model did not recapitulate the progression of human pancreatic cancer, but it did introduce the use of targeted constructs, rather than transgenes, to obtain oncogene expression.

Oncogenic Kras Expression from Its Endogenous Locus: Modeling PanIN Formation

The first model to closely mimic the human disease, later to be known as the KC mouse (Aguirre et al. 2003; Hingorani et al. 2003), was based on the use of the Cre/LoxP system to control transgene expression. Cre recombinase can specifically recognize the LoxP sequence; when two LoxP sequences are inserted in series, Cre will eliminate the sequence between the two sites and join the ends of the DNA leaving one LoxP site behind (Sauer and Henderson 1988; Sternberg and Hamilton 1981). The KC mouse was generated by crossing two genetically modified strains: the first one expresses the Cre recombinase in a pancreas-specific manner [either Ptf1a/p48-Cre (Kawaguchi et al. 2002) or Pdx1-Cre were used in the initial description of this model, with similar results]; the second allele is a "knock-in" allele where the oncogenic variant $Kras^{G12D}$ was inserted in the Kras endogenous locus, and preceded by a STOP cassette flanked by LoxP sites (Jackson et al. 2001). Thus the STOP cassette could be removed in tissue-specific manner, allowing pancreas-specific expression of oncogenic Kras from its endogenous locus. Both Pdx1-Cre and Ptf1a-Cre are expressed during the earliest stages of pancreas development (Gu et al. 2003; Heiser et al. 2006; Kawaguchi et al. 2002) and in all of the pancreatic cell lineages; therefore, oncogenic Kras is expressed in the whole pancreas throughout development. Since recombination is irreversible and inherited by all the daughter cells, the pancreatic epithelium in this model keeps expressing mutant Kras (provided that the Kras endogenous locus is active) even after the expression of Pdx1 and Ptf1a is restricted, respectively, to the endocrine compartment and to the acinar cells of the pancreas in the adult animal. Notwithstanding the pancreas-wide expression of oncogenic Kras, KC mice are born at the expected Mendelian ratio and with a normal pancreas; however, shortly after weaning, they develop PanIN1A lesions that over time progress to higher grade PanINs lesions (Fig. 1a). These lesions share the histologic features of human PanINs and mimic their progression (Hruban et al. 2006). A subset of KC mice develops frank PDA between 6 and 12 months of age. These findings demonstrated that oncogenic Kras drives the initiation of pancreatic adenocarcinoma, when expressed at the appropriate level in the appropriate cell type(s). Thus, the KC mouse was the first model to closely mimic the progression of the human disease, and has since served as the basis for many studies on pancreatic cancer biology, some of which we highlight later in this chapter.

Modeling Invasive and Metastatic Pancreatic Adenocarcinoma: The KC; Ink4a$^{-/-}$ and KPC Models

Although KC mice develop PDA, they do so with low penetrance and long latency, indicating the need for additional genetic or epigenetic events to occur. Pancreatic cancer in human is characterized by the loss of multiple tumor suppressor genes (Hezel et al. 2006), which are likely restricting the formation of PanINs and PDA even when Kras mutations are present. The first model with high penetrance of PDA formation to be described combined the KC mouse with the inactivation of the Ink4a locus. The Ink4a locus codes for two tumor suppressor genes, $p16^{Ink4a}$ and $p19^{Arf}$ in mice [reviewed in (Sherr 2004)]. The Ink4a locus is almost invariably silenced by promoter methylation or by inactivating mutations in the vast majority of human pancreatic adenocarcinomas and higher grade PanINs (Moskaluk et al. 1997;

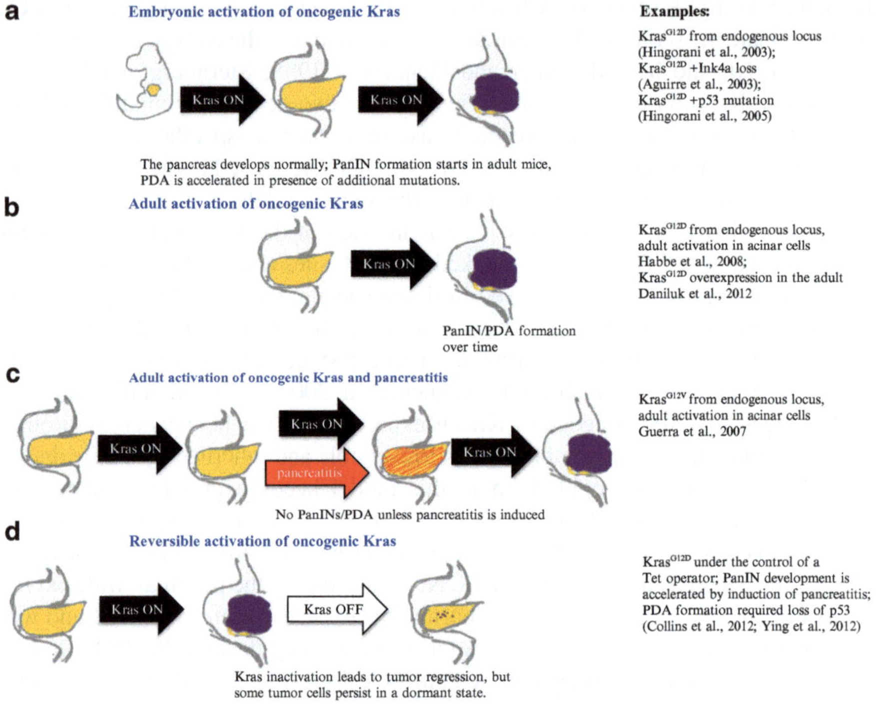

Fig. 1 Different strategies to generate Kras-based models of pancreatic cancer. (**a**) Embryonic activation of Kras has, surprisingly, no effect on pancreas development and leads to carcinogenesis in adult mice. (**b**) Adult activation leads to carcinogenesis spontaneously in some models. (**c**) Other studies have shown carcinogenesis following adult activation only upon induction of pancreatitis. (**d**) Reversible expression of Kras has revealed a role for this oncogene in tumor maintenance

Schutte et al. 1997; Wilentz et al. 1998). Additionally patients with the Familial Atypical Multiple Mole Melanoma (FAMMM) Syndrome, caused by germline mutations in p16, are predisposed to the development of pancreatic cancer (Borg et al. 2000; Goldstein et al. 1995). When KC mice are crossed with a conditional "knock-out" allele of Ink4a, deleting exons 2 and 3 and thereby inactivating both p16 and p19, the resulting animals develop PanINs that rapidly progress to poorly differentiated and highly metastatic PDA (Aguirre et al. 2003). In these tumors the p53 pathway, another key tumor suppressor often lost in pancreatic cancer, was found to be still active: in fact, tumor cells isolated from these mice were able to upregulate p53 expression in response to ionizing radiation. Thus the Ink4a products serve as a key constraint in the progression of the PanINs to invasive PDA.

The relative importance of the two genes encoded by the Ink4a locus was subsequently addressed using genetically engineered alleles that allowed inactivation of p16 and p19 individually in the context of the KC model (Bardeesy et al. 2006a). This work demonstrated that both p16 and p19 each play key roles in the suppression of PDA formation in mice. Knockouts of different tumor suppressors alone or in combination, such as p16 with and without p53 or p16 with and without p19, also revealed the relative contribution of these tumor suppressor genes to PDA formation. Deletion of both p16 and p19 led to tumor formation with similar latency as the loss of both copies of p53 (Bardeesy et al. 2006a). Addition of p16 deletion to p53 biallelic loss did not lead to faster tumor appearance. Deletion of p16 alone led to a higher latency in tumor formation than in the combined p16/p19 knockout mice. In addition to the timing of tumor development, another interesting observation was the correlation of different tumor suppressor disruptions with the resultant PDA histology. Mice that contained p16 or p16/p19 disruptions alone formed a subset of carcinomas that were highly anaplastic or sarcomatoid in nature (Bardeesy et al. 2006a). In contrast, tumors that developed in mice with a p53 deletion with or without Ink4a locus disruption tended to have the classical ductal morphology. Genomic studies analyzing tumors from the different backgrounds demonstrated recurrent amplifications of loci containing the Kras and c-myc oncogenes regardless of the genetic background (Bardeesy et al. 2006a). These observations correlated with the genomic changes observed in human PDA and validated the genetically engineered mouse models as relevant to the pathology of the human disease.

The involvement of p53 as a key tumor suppressor in PDA has been postulated for a long time and the loss or mutation of p53 occurs late during PanIN progression to frank PDA [reviewed in (Hezel et al. 2006; Hruban et al. 2000)]. Genetically engineered mouse models of PDA have been invaluable in demonstrating the synergy of p53 deletion or mutation with the presence of an oncogenic Kras allele. KC mice were crossed with mice bearing the $Trp53^{R172H}$ allele to generate animals expressing both mutant Kras and mutant p53 in the pancreas (Hingorani et al. 2005), a model currently referred to as the KPC model (Hingorani et al. 2005; Olive and Tuveson 2006). Of note, the design of the p53 mutant allele results in inactivation of one copy of p53 ubiquitously and expression of the mutant form upon Cre recombination (thus in a pancreas-specific manner). Expression of oncogenic Kras in the presence of mutated p53 led to rapid onset of pancreatic neoplasia and the development of PDA with high penetrance and relatively short latency (5–6 months of age). KPC mice have highly metastatic disease to multiple organs including the lymph nodes, liver, lungs, diaphragm, and adrenal glands (Hingorani et al. 2005), with liver and lung being the prevalent metastatic sites, consistent with the human disease. Extra-pancreatic malignancies, including hepatocellular carcinomas, non-small-cell lung carcinomas, lymphomas, and teratocarcinoma, were noted in a subset of animals and are likely the result of the loss of heterozygosity of p53. The pancreatic cancers in the KPC mice ranged from well-differentiated ductal adenocarcinomas to poorly differentiated and anaplastic/sarcomatoid tumors. Whether additional genetic alterations, acquired over the course of the malignant progression, differ between the different tumor typologies observed in these animals remains to be determined. In fact, genetic and chromosomal analysis of the tumors and derived cell lines demonstrated high levels of genetic instability, including large scale genomic rearrangements (Hingorani et al. 2005). Of note, recent expression studies have identified subsets of pancreatic cancer in humans, with the prevalent forms described as "classical" and "quasi-mesenchymal": both forms are recapitulated in KPC mice (Collisson et al. 2011).

A different model based on endogenous expression of mutant Kras is the Ela-tTA/tetO-Cre; LSL-Kras$^{G12V\ lacZ}$ mouse, which allows inducible expression of Kras^{G12V} in the pancreatic epithelial cells (Guerra et al. 2007). Embryonic activation of oncogenic Kras leads to PanIN formation in adult mice; in the p53$^{+/-}$ background the animals develop invasive pancreatic adenocarcinoma. Pathologic analysis of the tumors showed ductal adenocarcinomas of varying differentiation and metastases were observed in the liver, lungs, and regional lymph nodes. Interestingly, activation of oncogenic Kras in adult animals did not lead to PanIN formation, possibly indicating that adult pancreatic cells are less susceptible to transformation. This model is further discussed later within this chapter.

Understanding of the Role of Tumor Suppressors: DPC4/Smad4

The transforming growth factor β (TGFβ) plays many roles in the process of tumorigenesis and metastasis [reviewed in (Massague 2008)]. Signals from the TGFβ receptor family are transduced into the cell nucleus through the Smad transcription factor family. Smad4 serves as the common nuclear binding partner for the all of the Smad transcription complexes (Massague 2008). Nonbiased loss-of-heterozygosity (LOH) analysis of primary human PDA revealed recurrent deletions of chromosome 18q21.1 involving DPC4 (deleted in pancreatic cancer 4)/Smad4, which indicated its possible role as a tumor suppressor in PDA (Hahn et al. 1996). Subsequent histological analysis of human PanIN lesions and PDA showed that DPC4 was lost in highly dysplastic PanIN-3s and frank PDA, while its expression was intact in lower grade PanINs (Wilentz et al. 2000). These data implicated the loss of Smad4 as a later event in pancreatic neoplastic progression. Importantly, the DPC4/Smad4 status of the primary pancreatic cancer also correlated with the type of recurrence seen in patients, with DPC4 loss marking cancers that tended to widely metastasize (Iacobuzio-Donahue et al. 2009). The contribution of Smad4 loss to PDA progression has also been modeled in mice. When both of the Smad4 alleles are deleted in the KC model, all of the mice die by 24 weeks of age from pancreatic cancer (Bardeesy et al. 2006b), thus greatly anticipating both the age of onset and the penetrance of cancer development. The pancreatic cancers in this model frequently arose in the setting of intraductal papillary mucinous neoplasm (IPMN)-like lesions, which also exist in humans and are considered to be precursor lesions of PDA. When the Smad4 allele loss was combined with a loss of the Ink4a locus, the resulting tumors often retained ductal epithelial morphology rather than the frequently seen sarcomatoid histology seen in the Ink4a$^{-/-}$ models. The disruption of TGFβ signaling therefore prevents the pancreatic cancers from progressing to the fully anaplastic state in this model. Consistent with this finding, Tgfbr2 knockout mice crossed onto the KC background had a very rapid onset of tumors comparable with the Ink4a loss model (within 7–10 weeks), but histological analysis revealed the tumors remained well differentiated (Ijichi et al. 2006). Overall, studies in mice are consistent with observations in human patient samples, and indicate that DPC4/Smad4 serves as late-stage checkpoint in pancreatic neoplastic progression.

Oncogenic Kras Expression in the Adult Pancreas

Even the most sophisticated mouse models are not identical to the human disease, and not only because of the obvious observation that mice are not human beings. The KC mouse is one of the cancer models that most faithfully resemble the human disease it seeks to model, but profound differences remain. One of the most obvious is that KC mice express oncogenic Kras in every cell (or the vast majority) of the pancreas, starting during the early embryonic development. This is a very different scenario than the human disease, which is prevalent in older adults, and is believed to arise clonally from a single cell that has acquired an initiating mutation. The first approach to expressing oncogenic Kras in the adult pancreas was achieved with a triple transgenic system, as described earlier in the chapter: Elastase-tTa drives expression of the tetracycline transactivator in the pancreas; when tTa is active (in absence of tetracycline or doxycycline), it activates TetO-Cre; thus, the Cre recombinase is expressed in the pancreas, in an inducible manner (Guerra et al. 2007). When the mice are exposed to doxycycline in their water or chow, the Tet-off transactivator does not bind the tetracycline promoter and the Cre recombinase is silent. When Cre is expressed, it excises the STOP cassette that precedes a $Kras^{G12V}$ mutant inserted in the Kras endogenous locus. Therefore, in this model, mutant Kras can be expressed during embryogenesis, or in the adult pancreas, at will (Guerra et al. 2003, 2007). Intriguingly, different outcomes follow embryonic versus adult activation of oncogenic Kras. When oncogenic Kras was activated during development, the mice developed PanINs and occasional progression to PDA. The added presence of a loss-of-function allele of p53 leads to highly penetrant and rapid PDA formation as previously noted. When Kras was activated on postnatal day 10, a subset of the mice still developed PanINs and eventually progressed to frank PDA. However, when Kras was activated in adult mice (postnatal day 60), it had no effect: the mice never developed any PanIN lesions or PDA (Guerra et al. 2007). Thus, it appears that the embryonic pancreas is permissive for Kras-induced transformation while the adult organ is relatively refractory to tumor development. This finding highlights the need for additional genetic or environmental factors to synergize with the oncogenic Kras to induce PanIN formation.

Chronic pancreatitis is one of the known risk factors for PDA development in humans, and thus constituted a potential candidate as an environmental factor that could synergize with oncogenic Kras to induce tumor formation. Indeed, when the oncogenic Kras allele was activated in adult mice after the induction of chronic pancreatitis, all of the animals developed high-grade PanIN lesions and 30 % of them progressed to invasive PDA (Fig. 1c). These findings highlight the existence of contributing environmental factors to PDA pathogenesis and help explain the epidemiological role of chronic pancreatitis in the onset of pancreatic cancer. Moreover, this study showed that adult acinar or centroacinar cells, the only cell types to express the elastase promoter in the adult pancreas, can constitute the cell of origin for PDA.

Intriguingly, different results were obtained when the LSL-KrasG12D mouse was crossed with elastase-CreER or Mist1-CreER, two acinar-specific drivers that express an inactive form of Cre that can be activated at will by administering tamoxifen or its derivatives to the mice. When Cre recombination is induced in the adult acinar cells, these double-transgenic mice develop PanINs with high penetrance (Habbe et al. 2008), with no need to induce pancreatitis, in contrast to the absolute need for pancreatitis in the KrasG12V model (Fig. 1b). Possibly, the KrasG12D is a more potent oncogenic effect than KrasG12V, although further studies will be needed to test this possibility. Yet a different model used a transgenic approach to express KrasG12D in the adult pancreas using a CAG promoter, thus allowing for higher expression levels that those of the endogenous locus (Daniluk et al. 2012). This model highlighted the issue of dosage of Kras, or Kras activation rather than expression. In fact, the levels of Kras activity increase during pancreatic carcinogenesis in humans, and expression of mutant Kras from the endogenous locus, present in every cell of the pancreas since the earliest stages of embryogenesis, results only in PanIN formation over several months of age, indicating the need for additional genetic or epigenetic events, or potentially the need to reach a threshold of Kras activity.

A second important question that the different models have addressed is that of the cell of origin for pancreatic cancer. The fact that activation of Cre in adult acinar cells can lead to PanIN formation seems to point at an acinar, or potentially centroacinar, cell of origin for pancreatic adenocarcinoma. De-differentiation of acinar cells into a duct-like cell type, a process known as ADM—often as a result of pancreatic injury—has been suggested to be required for Kras-driven transformation (Morris et al. 2010b). However, a controversy on the cell of origin of pancreatic cancer remains. When the ductal-specific driver CK19-CreER was used to activate the LSL-KrasG12D allele rare PaNIN lesions were observed albeit with low frequency and with a long latency (Ray et al. 2011), indicating susceptibility of ductal cells to malignant transformation. Pancreatic adenocarcinoma formation from other cell types, such as insulin positive beta-cells, has been observed in the context of pancreatic injury, but it appears not to occur in physiological conditions (Gidekel Friedlander et al. 2009). Thus mouse models have provided some indication as to the cell of origin of pancreatic cancer, and no doubt lineage tracing studies will further expand on the subject in the future.

Reversible Expression of Oncogenic Kras: Role in Pancreatic Cancer Progression and Maintenance

In the previous paragraphs, we have described the studies that led to the identification of Kras as the fundamental initiating driver for pancreatic adenocarcinoma. However, the role of this oncogene in disease progression, and, in cancer maintenance, had until recently not been addressed. In fact, the question of which oncogenes are essential for tumor maintenance is highly relevant to identifying potential therapeutic targets. Indeed, in the Kras-based model described above, activation of oncogenic Kras is irreversible. Since drugs targeting oncogenic Kras are not available, this question could only be addressed through genetic inactivation of the oncogene.

Recently, our group has described a new mouse model, named iKras*, that allows tissue-specific, inducible and reversible expression of oncogenic Kras in the pancreas (Collins et al. 2012). The iKras* mouse is based on three transgenes that allow pancreas-specific Cre recombination using Ptf1a-Cre (Kawaguchi et al. 2002), Cre-inducible expression of rtTa from the Rosa 26 locus (Soriano 1999) and doxycycline-dependent regulation of TetO-Kras^{G12D} (Fisher et al. 2001). This model, therefore, does not use the endogenous Kras locus to express the oncogenic variant. Nevertheless, the resulting iKras* mouse develops PanIN lesions with high frequency if Kras is activated during embryonic development. If Kras is activated in the adult mouse, PanIN formation is sporadic and occurs with long latency, but can be efficiently anticipated by inducing acute pancreatitis, an approach that was first described in the KC model (Morris et al. 2010a). Inactivation of Kras* in low-grade, and newly formed PanINs leads to complete regression of the lesions by, at least in part, redifferentiation of PanIN duct-like cells into acinar cells and simultaneous remodeling of the stroma. When Kras* is inactivated at later stages, when the PanINs have acquired characteristic of higher grade lesions, and have existed in the tissue for several weeks, however, the epithelial cells undergo apoptosis, having become dependent on sustained oncogenic activity for survival. iKras* mice develop invasive adenocarcinoma at a very low frequency and with long latency; however, when crossed to a loss-of-function allele of p53, they develop adenocarcinoma rapidly and with high frequency. Inactivation of oncogenic Kras in these invasive tumors leads to sustained regression. Therefore, Kras is not only essential for pancreatic cancer initiation, but also important for tumor maintenance (Fig. 1d). Moreover, analysis of this model showed that Kras regulates several aspects of the biology of pancreatic cancer, including the maintenance of an inflammatory microenvironment and the presence of an active desmoplastic stroma.

Recently, the role of Kras in pancreatic cancer maintenance has been explained by the oncogene's ability to regulate the tumor's anabolic metabolism and to increase glucose uptake (Ying et al. 2012). Thus, Kras induces a metabolic switch in tumor cells which could potentially be exploited by targeting metabolic genes therapeutically. Further studies will be needed to test the requirement for Kras in presence of other tumor suppressor gene mutations, such as point mutations of p53 and/or loss of p16/p19; moreover, it will be essential to determine what is the fate of the tumor cells upon Kras* inactivation. In fact, both studies suggest that some of the tumor cells do not undergo apoptosis, and potentially remain dormant into the tissue (Collins et al. 2012; Ying et al. 2012). In the future these models might prove useful to model Kras inhibition in pancreatic cancer, and to elucidate potential mechanisms of escape from Kras dependency that might limit the application of Kras inhibitors, were they to be developed, in human patients.

Mouse Models as Discovery Tools

Mouse Models and Developmental Pathways in PDA

Hedgehog Signaling

Hedgehog (Hh) signaling regulates multiple developmental processes, including the patterning of the gastrointestinal tract [reviewed in (van den Brink 2007)]. During pancreas development, Hh signals have to be specifically excluded from the region of the foregut that will give rise to the pancreas. Mice that overexpress Sonic Hedgehog (Shh), one of the Hh ligands, under the control of the Ipf1/Pdx1 promoter fail to develop a pancreas (Apelqvist et al. 1997). In some of the mice where limited pancreatic tissue can be detected, cells of the endocrine and exocrine compartment are present, but do not form cohesive islets or acini and are instead interspersed with duct-like structures that strongly resemble early PanINs (Thayer et al. 2003). Therefore, timing and strength of Hh signaling plays a key role in pancreas development.

Genetically engineered mouse models have played a key role in delineating the role of Hh signaling in PDA biology. Hh ligands and pathway components were found to be upregulated in PanIN 2–3 lesions, PDA, as well as in pancreatic cancer cell lines (Thayer et al. 2003), providing the first indication of ligand-driven activation of the Hedgehog pathway in cancer. Secreted Hh ligands could potentially act in an autocrine or paracrine fashion to activate the pathway. To address whether Hh ligand acted in a cell-autonomous fashion on the tumor epithelium, a conditional Smoothened knockout mouse model was crossed into the p48-Cre, LSL-Kras^{G12D}, p53$^{fl/+}$ murine PDA model. Pancreatic epithelium-specific knockout of Smoothened did not affect overall pancreatic development and did not inhibit PanIN and PDA development (Nolan-Stevaux et al. 2009). Surprisingly, the authors noted overexpression of Gli transcription factor and expression of Hh pathway target genes in the epithelial cells, notwithstanding the inactivation of Smo, thus demonstrating the lack of ligand-driven pathway activation in the epithelial cells. They further demonstrated that Gli expression and activity was driven by noncanonical Kras and TGFβ signaling and was required for the maintenance of the neoplastic state (Nolan-Stevaux et al. 2009). Thus it appears that Gli expression and activity in the tumor cells is Hedgehog-ligand independent.

Concurrent experiments with pancreatic cancer cell lines and other Hh-overexpressing tumors revealed that tumor epithelial cells were often insensitive to Hh ligands (Yauch et al. 2008). In contrast, ligand-dependent pathway activation was noted in the surrounding stroma (Yauch et al. 2008). Hh signaling inhibition abrogated the growth pancreatic cancer xenografts in immune-compromised mice, suggesting that the activated Hh pathway in the tumor stroma functionally supports tumor growth (Yauch et al. 2008). When the Hedgehog activity reporter allele Ptch1-lacZ was introduced into the Pdx1-Cre, LSL-Kras^{G12D}, Ink4a$^{fl/fl}$ mouse model of PDA, Hh pathway activation was again only seen in the tumor stroma (Tian et al. 2009).

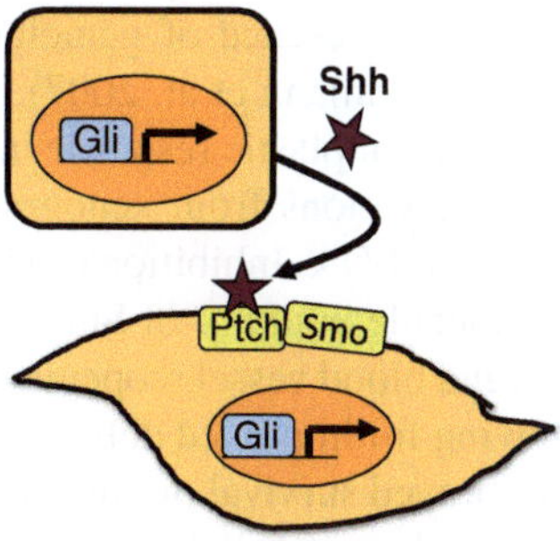

Fig. 2 Paracrine activation of Hedgehog signaling in pancreatic cancer. The tumor cells secrete Hedgehog ligands, prevalently Sonic Hedgehog (Shh); the surrounding fibroblasts express the receptor Patched (Ptch). Binding of Shh to Ptch releases Ptch repression of Smoothened (Smo); a signaling cascade ensues that leads to activation of target genes

Analysis of microdissected tissues from human PDAs demonstrated relative Gli1 upregulation in the tumor stroma as compared to the tumor epithelium (Tian et al. 2009). Taken together, these experiments support a model where the activation of an oncogenic version of Kras together with additional noncanonical pathways leads to the Gli factor activation and Hh ligand expression in the tumor epithelium. The secreted Hh ligands then act in a paracrine fashion on the infiltrating tumor stroma, which in turn supports tumor epithelial cell survival and growth.

Noncanonical activation of the Gli factors in the tumor epithelium also has profound consequences on pancreatic tumorigenesis. As previously noted, disruption of Gli1 in pancreatic cancer cell lines inhibited in vitro colony formation and increased pancreatic cancer cell apoptosis (Nolan-Stevaux et al. 2009). When the CLEG2 allele, an N-terminally deleted dominant active version of Gli2, was introduced into the pancreatic epithelium in the presence of oncogenic $Kras^{G12D}$, the pancreata rapidly developed the full spectrum of PanINs, multiple cystic lesions, and eventually undifferentiated carcinomas (Pasca di Magliano et al. 2006). Additionally Gli1 expression from the Rosa26 locus on the p48-Cre, LSL-$Kras^{G12D}$ background led to rapid onset of PanIN and classical PDA formation (Rajurkar et al. 2012). Conversely, inhibition of Gli signaling by a dominant negative Gli3 construct in the pancreatic epithelium abrogated PanIN and PDA formation (Rajurkar et al. 2012). Therefore, Gli activity plays a key role in pancreatic tumorigenesis by acting in both the stromal and epithelial compartments of the tumor (Fig. 2).

The notion that Hedgehog signaling acts in a paracrine fashion in pancreatic cancer led to studies exploring the possibility of inhibiting the Hh pathway to target the stroma as a new therapeutic modality. Human pancreatic cancer cell xenografts in nude mice were treated with cyclopamine with or without gemcitabine, a standard antimetabolite used in pancreatic cancer therapy, and subsequent tumor growth and metastatic spread were assessed. Inhibition of Hh signaling by cyclopamine led to decreased tumor growth in the presence of gemcitabine and almost complete inhibition of metastatic spread (Feldmann et al. 2007). Disruption of Hh signaling also led to the inhibition of the desmoplastic response in orthotopic models of PDA (Bailey et al. 2008). These observations from xenografts models were replicated and extended in the KPC model of PDA. Inhibition of Hh signaling in de novo KPC tumors by IPI-926, a small molecule inhibitor of Smo, led to disruption of the stromal component of the tumors and blood vessel reopening and significantly improved perfusion of the tumors, allowing for improved delivery of gemcitabine (Olive et al. 2009). These effects led to improved survival of mice treated with the IPI-926 compound and gemcitabine compared to gemcitabine alone. These data suggested that the Hh pathway could serve as a therapeutic target leading to tumor stroma remodeling and improved delivery of cytotoxic agents.

Wnt/β-Catenin Signaling

Wnt signaling has been implicated in the biology of multiple malignancies but its role in PDA initiation and progression remains only incompletely understood. Parallel efforts to delineate the role of Wnt/β-catenin signaling in pancreatic development and PDA pathogenesis in genetically engineered mouse models have begun to offer some clues. Conditionally knocking out β-catenin by deleting exons 3–6 in a pancreatic progenitor-specific manner with a Pdx1-Cre recombinase construct inhibited pancreatic development resulting in severe pancreatic hypoplasia primarily due to the non-proliferation of the acinar compartment (Wells et al. 2007). β-catenin was found to regulate the expression of the Ptf1a/p48 transcription factor, which is key to the specification of the exocrine pancreatic lineage (Wells et al. 2007). Additional work utilized three distinct pancreas-specific Cre drivers and the dominant active stabilized version of β-catenin lacking exon 3 to address the effects of different Wnt pathway activation timing on pancreatic development. When a Pdx1-Cre^{early} construct, expressed and active by E10.5, was used to drive β-catenin activation, the resulting mice suffered from severe pancreatic agenesis and early death by 7 days of age (Heiser et al. 2006). In contrast when the Pdx1-Cre^{late} construct, active by E11.5–12.5 and more mosaic in function, was used, the mice developed a grossly normal pancreas and were viable. As the animals aged, however, they developed significant pancreatic hyperplasia driven by the expansion of the acinar compartment. When the Ptf1a/p48-Cre construct was used to stabilize β-catenin, the mice also developed pancreatic hyperplasia but now demonstrated multiple ductal lesions not seen in the Pdx1-Cre mouse strains. They also developed tumors histologically and molecularly similar to benign human lesions termed solid pseudopapillary tumors (Heiser et al. 2008). Together, this data indicates that the timing and dosage of Wnt pathway activation and the identity of target cells losing or gaining β-catenin expression directly influence the final outcome of pancreatic development. Therefore, Wnt signaling has to be tightly regulated throughout the process of pancreatic growth to yield a properly developed functional gland in the adult animal.

Several lines of evidence implicate Wnt signaling in PDA biology, but again the role of this pathway during disease progression is incompletely understood. The pathway is active in PanINs and PDA that develop in the Pdx1-Cre, LSL-Kras^{G12D} and Pdx1-Cre, LSL-Kras, p53$^{fl/+}$ mouse models (Pasca di Magliano et al. 2007). In addition multiple Wnt ligands are expressed in human pancreatic cancer cell lines. The Wnt activity levels in pancreatic cancer cell lines are, however, much lower than those observed in colon cancer, where the pathway is dysregulated due to mutations in key components. Inhibition of Wnt signaling in pancreatic cancer cells led to decreased proliferation and increased apoptosis (Pasca di Magliano et al. 2007). The mechanism of pathway activation in pancreatic cancer might be variable. On one hand, pathway activation may be due to ligand overexpression (Nawroth et al. 2007; Pasca di Magliano et al. 2007). On the other hand, the key component, β-catenin, can be stabilized by disruption of the β-catenin degradation complex through the formation of a complex between Dishevelled-2 (Dvl-2) and ataxia-telangiectasia group D complementing gene (ATDC/TRIM29) (Wang et al. 2009). Two principal studies utilizing GEMMs have yielded further insight into Wnt regulation of PDA biology. When the dominant active form of β-catenin was expressed in the p48-Cre, LSL-Kras^{G12D} model of PDA, PanINs and PDA did not form (Heiser et al. 2008). In contrast, the mice developed lesions resembling human intraductal tubular neoplasms. Further dissection of the underlying mechanisms involved in tumor development demonstrated that in the process of ADM and early PanIN formation, oncogenic Kras inhibits β-catenin expression, which is normally necessary for proper redifferentiation of progenitor-like cells back to acinar cells during pancreatitis recovery (Morris et al. 2010a). When the dominant active form of β-catenin is present during early pancreatitis, ADM and PanIN formation are inhibited and the active Wnt pathway in essence overrides the oncogenic Kras signal (Morris et al. 2010a). However, some uncertainty still remains as to the interplay of the Wnt signaling pathway and Kras during formation of PanINs. In fact, stabilized β-catenin might not be appropriate to model the activation of Wnt signaling in pancreatic cancer as it greatly exceeds the levels of activation observed in this disease.

Notch Signaling

The Notch signaling pathway plays a key role in the fate specification of the pancreatic exocrine compartment (Afelik et al. 2012; Apelqvist et al. 1999; Jensen et al. 2000; Murtaugh et al. 2003; Nakhai et al. 2008; Sumazaki et al. 2004) Notch pathway components were noted to be upregulated in human PDA when compared to the normal pancreas (Miyamoto et al. 2003). Genetic Notch pathway manipulation in the mouse models of PDA has begun to yield insights into its function in tumor initiation and progression. As in human pancreatic cancers, Notch pathway components were overexpressed in the Pdx1-Cre, LSL-Kras, Trp53$^{fl/+}$ model of PDA (Plentz et al. 2009). When these mice were treated with a gamma secretase inhibitor preventing proper proteolytic processing of the Notch receptors after ligand binding, PanIN progression, and PDA onset were significantly inhibited (Plentz et al. 2009).

Genetic disruption of the Notch pathway by conditional deletion of the Notch1 and Notch2 receptors in the presence of an oncogenic Kras allele has yielded results suggesting different roles for their involvement in PDA initiation and progression (Hanlon et al. 2010; Mazur et al. 2010). When Notch1 was conditionally deleted in the pancreas in the Pdx1-Cre, LSL-Kras model of PDA, mild acceleration of PanIN progression was seen, suggesting that Notch1 may in fact function as a tumor suppressor in this mouse model (Hanlon et al. 2010). In an independent study that analyzed the conditional deletion of Notch1 in the Ptf1a/p48-Cre, LSL-Kras model demonstrated a trend towards worse survival in the Notch1 knockout mice (Mazur et al. 2010). In contrast, when Notch2 was deleted in the mouse pancreas, survival significantly improved. The animals that did eventually develop PDA in the setting of Notch2 deletion more often demonstrated undifferentiated "sarcomatoid" pathology and myc upregulation (Mazur et al. 2010). This data suggests that distinct Notch receptors may play nonredundant and possibly opposing roles in the PDA pathogenesis.

Notch pathway activation in the setting of an oncogenic Kras allele has also been modeled by expressing the Notch1 intracellular domain (N1IC) from the Rosa26 locus in the Pdx1-CreERT, LSL-Kras and Ela-CreERT, LSL-Kras models (De La O et al. 2008). When the Cre recombinase and N1IC expression was activated during embryonic development (E10.5) in the Pdx1-CreERT model, rapid acceleration of PanIN formation was seen in the mouse cohorts compared to the control Pdx1-CreERT, LSL-Kras mice. The N1IC domain by itself was not capable of inducing PanIN/PDA formation. However, when N1IC expression was induced in the adult acinar compartment in the presence of the oncogenic Kras G12D allele (Ela-CreERT model), rapid and efficient ADM was seen within 2 weeks of Cre activation by tamoxifen (De La O et al. 2008). The genetic gain-of-function experiments suggest that Notch signaling may play a key role in the reprogramming process involved in the generation of the final duct-like fate seen in the neoplastic tissue similar to its role in normal pancreatic development.

Mouse Models in Preclinical Applications

Despite marked advances in our understanding of cancer biology we have not been able to translate these to effective therapeutics in the vast majority of patients. The field of pancreatic cancer in particular is littered with many examples of failed Phase III clinical trials that were originally based on promising preclinical data, albeit much of which was generated using commercially available, highly passaged cell lines. Genetically engineered mouse models of PDA have the potential to fill the void between basic bench research and clinical therapeutics and serve as high fidelity preclinical models.

The contribution of the tumor microenvironment to tumor growth has been recently highlighted as an emerging hallmark of cancer (Hanahan and Weinberg 2011). The role of the microenvironment and its regulation by Hh signaling in PDA has already been noted earlier in this chapter (Bailey et al. 2008; Yauch et al. 2008). A study by Olive et al. directly tested the clinical applicability of this concept using the Pdx1-Cre, $Kras^{LSL-G12D/+}$, $Trp53^{R172H/+}$ (KPC) model of PDA (Olive et al. 2009). The authors utilized the small molecule Smo inhibitor IPI-926 (Infinity Pharmaceuticals) to disrupt the Hh signaling activation in the stroma. In the process they observed significantly increased perfusion of the treated tumors with concomitant increase in blood vessel number. This led to a markedly improved delivery of the standard PDA chemotherapeutic gemcitabine along with a rise in tumor cell apoptosis. Most importantly, the tumor perfusion effect translated into increased survival of the mice treated with the IPI-926/gemcitabine regimen versus either drug alone or the untreated controls, albeit with modest survival differences (about 2 weeks survival advantage, from 11 to 25 days) . This study combined knowledge of basic PDA biology previously described in mouse and human tumor transplantation models with the concept of using mice with autochthonous tumors as a preclinical model to validate a possible new treatment strategy, which was subsequently translated into human clinical trials. Despite this proof of concept experiment it is important to note that the subsequent human clinical trial of IPI-926 in pancreatic cancer had to be halted due to more rapid tumor growth seen in the IPI-926-treated cohort, highlighting our still incomplete understanding of how to best preclinically model human PDA.

Subsequent efforts at targeting the tumor microenvironment in GEMMs of PDA focused on the role of the tumor extracellular matrix in PDA function. Two concurrent studies demonstrated that both human and murine PDAs in the KPC model exhibit highly elevated levels of the extracellular matrix component hyaluronan (Jacobetz et al. 2013; Provenzano et al. 2012). Hyaluronan is a high molecular weight polymer of *N*-acetyl glucosamine and glucuronic acid groups, is highly anionic, and sequesters small molecule solutes and water. The authors hypothesized that its presence led to the high increase in the interstitial tissue fluid pressure (IFP) seen in human and murine PDA (Provenzano et al. 2012). Both groups then used pegylated hyaluronidase to digest the stromal hyaluronan. This intervention decreased the IFP and led to the reopening and reperfusion of the already-existing tumor vasculature. Along with this, increased structural permeability of the vessels was noted (Jacobetz et al. 2013). Similar to the tumors treated with the Hh antagonist IPI-926, the tumors were better perfused with chemotherapeutics with subsequent inhibition of tumor growth. These interventions again led to increased survival of the mouse cohorts treated with the combination hyaluronidase/gemcitabine therapy compared to either drug alone or the controls in both studies (survival benefit of hyaluronidase/gemcitabine 91.5 days in average; gemcitabine alone 55.5 days in average). These preclinical findings from GEMMs have now led to a multi-institutional trial testing the combination therapy in human patients with advanced unresectable PDA.

These results highlight the usefulness of GEMMs in studying the biology of PDA and subsequently extending these observations to develop new potentially useful therapeutic approaches to this deadly disease. It is important to point out that despite having fairly high-fidelity models of PDA in mice, there still probably exist many differences between the murine models and human biology which may affect the outcomes of new therapies first tested in GEMMs, followed by testing in human clinical trials. Since no model is perfect at fully recapitulating all aspects of a disease, it will most likely be important to combine several approaches including GEMMs and orthotopic primary human PDA xenograft models to fully study a therapeutic strategy before translation into human clinical trials.

Conclusions

GEMMs have yielded impressive new insights into PDA biology over the past decade. It is without question that they serve as one of the best tools for basic scientific discovery in many fields, including tumor biology. The challenge that still exists, however, is to take new knowledge gained from the mouse models and best utilize it to improve disease outcomes in patients. We have to ultimately define the best systems that will allow us to translate basic findings into more relevant and effective therapeutic approaches to pancreatic cancer. It is still unclear to what extent the GEMMs can contribute in the drug discovery arena. It will be important to further modify the existing models of PDA such as the KC and KPC models to even better mimic the processes underlying the human disease. For example, the KC and KPC models rely on embryonic activation of the multitude of oncogenic drivers. This is clearly not the case in the vast majority of human patients. Models using inducible Cre drivers are the first step to further model the disease in a more biologically relevant manner. Another key difference between most models, as currently designed, and the human patients is that in the former the oncogenes are expressed tissue-wide, while in the patients the mutations presumably appear in a single cell. Using CreER or other similar approaches, it will be possible to restrict the activation of the oncogenes to a small number of cells in mice, thus obtaining a single tumor in a field of otherwise normal cells. However, the penetrance of tumor formation is likely to be lower, and studies of that nature will need resources and space to maintain very large mouse colonies. A different aspect that will have to be investigated in detail is how closely mouse tumors recapitulate the genetic and epigenetic alterations of their human counterparts. Likely, future efforts aimed at sequencing mouse tumors will give us an answer to this question. Nevertheless, it will be important to develop models that utilize the primary human tissue as a complementary approach to genetically engineered mouse models in preclinical and early stage clinical studies.

One of the key components that will be the most difficult to recapitulate is the contribution of the immune system to the process of tumorigenesis and immunosurveillance. GEMMs benefit from having an intact endogenous immune system that fairly closely correlates with the responses seen in human tumors. All models currently using human tissue lack any contribution from the human immune compartment. Incorporation of a humanized immune system in mice combined with human tumor tissue orthotopic xenotransplantation may provide a way to study human tumors in a more biologically appropriate system that would complement observations made in GEMMs. It is possible that a combination of these approaches may lead to the development of new effective treatments for this deadly disease.

References

Adsay NV (2008) Cystic neoplasia of the pancreas: pathology and biology. J Gastrointest Surg 12:401–404

Afelik S, Qu X, Hasrouni E, Bukys MA, Deering T, Nieuwoudt S, Rogers W, Macdonald RJ, Jensen J (2012) Notch-mediated patterning and cell fate allocation of pancreatic progenitor cells. Development 139:1744–1753

Aguirre AJ, Bardeesy N, Sinha M, Lopez L, Tuveson DA, Horner J, Redston MS, DePinho RA (2003) Activated Kras and Ink4a/Arf deficiency cooperate to produce metastatic pancreatic ductal adenocarcinoma. Genes Dev 17:3112–3126

Almoguera C, Shibata D, Forrester K, Martin J, Arnheim N, Perucho M (1988) Most human carcinomas of the exocrine pancreas contain mutant c-K-ras genes. Cell 53:549–554

Apelqvist A, Ahlgren U, Edlund H (1997) Sonic hedgehog directs specialised mesoderm differentiation in the intestine and pancreas. Curr Biol 7:801–804

Apelqvist A, Li H, Sommer L, Beatus P, Anderson DJ, Honjo T, Hrabe de Angelis M, Lendahl U, Edlund H (1999) Notch signalling controls pancreatic cell differentiation. Nature 400:877–881

Bailey JM, Swanson BJ, Hamada T, Eggers JP, Singh PK, Caffery T, Ouellette MM, Hollingsworth MA (2008) Sonic hedgehog promotes desmoplasia in pancreatic cancer. Clin Cancer Res 14:5995–6004

Bardeesy N, Aguirre AJ, Chu GC, Cheng KH, Lopez LV, Hezel AF, Feng B, Brennan C, Weissleder R, Mahmood U et al (2006a) Both p16(Ink4a) and the p19(Arf)-p53 pathway constrain progression of pancreatic adenocarcinoma in the mouse. Proc Natl Acad Sci USA 103:5947–5952

Bardeesy N, Cheng KH, Berger JH, Chu GC, Pahler J, Olson P, Hezel AF, Horner J, Lauwers GY, Hanahan D, DePinho RA (2006b) Smad4 is dispensable for normal pancreas development yet critical in progression and tumor biology of pancreas cancer. Genes Dev 20:3130–3146

Borg A, Sandberg T, Nilsson K, Johannsson O, Klinker M, Masback A, Westerdahl J, Olsson H, Ingvar C (2000) High frequency of multiple melanomas and breast and pancreas carcinomas in CDKN2A mutation-positive melanoma families. J Natl Cancer Inst 92:1260–1266

Brembeck FH, Moffett J, Wang TC, Rustgi AK (2001) The keratin 19 promoter is potent for cell-specific targeting of genes in transgenic mice. Gastroenterology 120:1720–1728

Brembeck FH, Schreiber FS, Deramaudt TB, Craig L, Rhoades B, Swain G, Grippo P, Stoffers DA, Silberg DG, Rustgi AK (2003) The mutant K-ras oncogene causes pancreatic periductal lymphocytic infiltration and gastric mucous neck cell hyperplasia in transgenic mice. Cancer Res 63:2005–2009

Collins MA, Bednar F, Zhang Y, Brisset JC, Galban S, Galban CJ, Rakshit S, Flannagan KS, Adsay NV, Pasca di Magliano M (2012) Oncogenic Kras is required for both the initiation and maintenance of pancreatic cancer in mice. J Clin Invest 122:639–653

Collisson EA, Sadanandam A, Olson P, Gibb WJ, Truitt M, Gu S, Cooc J, Weinkle J, Kim GE, Jakkula L et al (2011) Subtypes of pancreatic ductal adenocarcinoma and their differing responses to therapy. Nat Med 17:500–503

Daniluk J, Liu Y, Deng D, Chu J, Huang H, Gaiser S, Cruz-Monserrate Z, Wang H, Ji B, Logsdon CD (2012) An NF-kappaB pathway-mediated positive feedback loop amplifies Ras activity to pathological levels in mice. J Clin Invest 122:1519–1528

De La O J, Emerson LL, Goodman JL, Froebe SC, Illum BE, Curtis AB, Murtaugh LC (2008) Notch and Kras reprogram pancreatic acinar cells to ductal intraepithelial neoplasia. Proc Natl Acad Sci USA 105:18907–18912

Desai S, Ben-Josef E, Griffith KA, Simeone D, Greenson JK, Francis IR, Hampton J, Colletti L, Chang AE, Lawrence TS, Zalupski MM (2009) Gemcitabine-based combination chemotherapy followed by radiation with capecitabine as adjuvant therapy for resected pancreas cancer. Int J Radiat Oncol Biol Phys 75:1450–1455

Doetschman T, Gregg RG, Maeda N, Hooper ML, Melton DW, Thompson S, Smithies O (1987) Targetted correction of a mutant HPRT gene in mouse embryonic stem cells. Nature 330:576–578

Federspiel MJ, Bates P, Young JA, Varmus HE, Hughes SH (1994) A system for tissue-specific gene targeting: transgenic mice susceptible to subgroup A avian leukosis virus-based retroviral vectors. Proc Natl Acad Sci USA 91:11241–11245

Feldmann G, Dhara S, Fendrich V, Bedja D, Beaty R, Mullendore M, Karikari C, Alvarez H, Iacobuzio-Donahue C, Jimeno A et al (2007) Blockade of hedgehog signaling inhibits pancreatic cancer invasion and metastases: a new paradigm for combination therapy in solid cancers. Cancer Res 67:2187–2196

Ferrone CR, Brennan MF, Gonen M, Coit DG, Fong Y, Chung S, Tang L, Klimstra D, Allen PJ (2008) Pancreatic adenocarcinoma: the actual 5-year survivors. J Gastrointest Surg 12:701–706

Fisher GH, Wellen SL, Klimstra D, Lenczowski JM, Tichelaar JW, Lizak MJ, Whitsett JA, Koretsky A, Varmus HE (2001) Induction and apoptotic regression of lung adenocarcinomas by regulation of a K-Ras transgene in the presence and absence of tumor suppressor genes. Genes Dev 15:3249–3262

Gidekel Friedlander SY, Chu GC, Snyder EL, Girnius N, Dibelius G, Crowley D, Vasile E, DePinho RA, Jacks T (2009) Context-dependent transformation of adult pancreatic cells by oncogenic K-Ras. Cancer Cell 16:379–389

Goldstein AM, Fraser MC, Struewing JP, Hussussian CJ, Ranade K, Zametkin DP, Fontaine LS, Organic SM, Dracopoli NC, Clark WH Jr et al (1995) Increased risk of pancreatic cancer in melanoma-prone kindreds with p16INK4 mutations. N Engl J Med 333:970–974

Gordon JW, Scangos GA, Plotkin DJ, Barbosa JA, Ruddle FH (1980) Genetic transformation of mouse embryos by microinjection of purified DNA. Proc Natl Acad Sci USA 77:7380–7384

Grippo PJ, Nowlin PS, Demeure MJ, Longnecker DS, Sandgren EP (2003) Preinvasive pancreatic neoplasia of ductal phenotype induced by acinar cell targeting of mutant Kras in transgenic mice. Cancer Res 63:2016–2019

Gu G, Brown JR, Melton DA (2003) Direct lineage tracing reveals the ontogeny of pancreatic cell fates during mouse embryogenesis. Mech Dev 120:35–43

Guerra C, Mijimolle N, Dhawahir A, Dubus P, Barradas M, Serrano M, Campuzano V, Barbacid M (2003) Tumor induction by an endogenous K-ras oncogene is highly dependent on cellular context. Cancer Cell 4:111–120

Guerra C, Schuhmacher AJ, Canamero M, Grippo PJ, Verdaguer L, Perez-Gallego L, Dubus P, Sandgren EP, Barbacid M (2007) Chronic pancreatitis is essential for induction of pancreatic ductal adenocarcinoma by K-Ras oncogenes in adult mice. Cancer Cell 11:291–302

Habbe N, Shi G, Meguid RA, Fendrich V, Esni F, Chen H, Feldmann G, Stoffers DA, Konieczny SF, Leach SD, Maitra A (2008) Spontaneous induction of murine pancreatic intraepithelial neoplasia (mPanIN) by acinar cell targeting of oncogenic Kras in adult mice. Proc Natl Acad Sci USA 105:18913–18918

Hahn SA, Schutte M, Hoque AT, Moskaluk CA, da Costa LT, Rozenblum E, Weinstein CL, Fischer A, Yeo CJ, Hruban RH, Kern SE (1996) DPC4, a candidate tumor suppressor gene at human chromosome 18q21.1. Science 271:350–353

Hanahan D (1985) Heritable formation of pancreatic beta-cell tumours in transgenic mice expressing recombinant insulin/simian virus 40 oncogenes. Nature 315:115–122

Hanahan D, Weinberg RA (2000) The hallmarks of cancer. Cell 100:57–70

Hanahan D, Weinberg RA (2011) Hallmarks of cancer: the next generation. Cell 144:646–674

Hanlon L, Avila JL, Demarest RM, Troutman S, Allen M, Ratti F, Rustgi AK, Stanger BZ, Radtke F, Adsay V et al (2010) Notch1 functions as a tumor suppressor in a model of K-ras-induced pancreatic ductal adenocarcinoma. Cancer Res 70:4280–4286

Heiser PW, Lau J, Taketo MM, Herrera PL, Hebrok M (2006) Stabilization of beta-catenin impacts pancreas growth. Development 133:2023–2032

Heiser PW, Cano DA, Landsman L, Kim GE, Kench JG, Klimstra DS, Taketo MM, Biankin AV, Hebrok M (2008) Stabilization of beta-catenin induces pancreas tumor formation. Gastroenterology 135:1288–1300

Hezel AF, Kimmelman AC, Stanger BZ, Bardeesy N, Depinho RA (2006) Genetics and biology of pancreatic ductal adenocarcinoma. Genes Dev 20:1218–1249

Hingorani SR, Petricoin EF, Maitra A, Rajapakse V, King C, Jacobetz MA, Ross S, Conrads TP, Veenstra TD, Hitt BA et al (2003) Preinvasive and invasive ductal pancreatic cancer and its early detection in the mouse. Cancer Cell 4:437–450

Hingorani SR, Wang L, Multani AS, Combs C, Deramaudt TB, Hruban RH, Rustgi AK, Chang S, Tuveson DA (2005) Trp53R172H and KrasG12D cooperate to promote chromosomal instability and widely metastatic pancreatic ductal adenocarcinoma in mice. Cancer Cell 7:469–483

Hruban RH, Goggins M, Parsons J, Kern SE (2000) Progression model for pancreatic cancer. Clin Cancer Res 6:2969–2972

Hruban RH, Takaori K, Klimstra DS, Adsay NV, Albores-Saavedra J, Biankin AV, Biankin SA, Compton C, Fukushima N, Furukawa T et al (2004) An illustrated consensus on the classification of pancreatic intraepithelial neoplasia and intraductal papillary mucinous neoplasms. Am J Surg Pathol 28:977–987

Hruban RH, Adsay NV, Albores-Saavedra J, Anver MR, Biankin AV, Boivin GP, Furth EE, Furukawa T, Klein A, Klimstra DS et al (2006) Pathology of genetically engineered mouse models of pancreatic exocrine cancer: consensus report and recommendations. Cancer Res 66:95–106

Hsu CC, Herman JM, Corsini MM, Winter JM, Callister MD, Haddock MG, Cameron JL, Pawlik TM, Schulick RD, Wolfgang CL et al (2010) Adjuvant chemoradiation for pancreatic adenocarcinoma: the Johns Hopkins Hospital-Mayo Clinic collaborative study. Ann Surg Oncol 17:981–990

Iacobuzio-Donahue CA, Fu B, Yachida S, Luo M, Abe H, Henderson CM, Vilardell F, Wang Z, Keller JW, Banerjee P et al (2009) DPC4 gene status of the primary carcinoma correlates with patterns of failure in patients with pancreatic cancer. J Clin Oncol 27:1806–1813

Ijichi H, Chytil A, Gorska AE, Aakre ME, Fujitani Y, Fujitani S, Wright CV, Moses HL (2006) Aggressive pancreatic ductal adenocarcinoma in mice caused by pancreas-specific blockade of transforming growth factor-beta signaling in cooperation with active Kras expression. Genes Dev 20:3147–3160

Jackson EL, Willis N, Mercer K, Bronson RT, Crowley D, Montoya R, Jacks T, Tuveson DA (2001) Analysis of lung tumor initiation and progression using conditional expression of oncogenic K-ras. Genes Dev 15:3243–3248

Jacobetz MA, Chan DS, Neesse A, Bapiro TE, Cook N, Frese KK, Feig C, Nakagawa T, Caldwell ME, Zecchini HI et al (2013) Hyaluronan impairs vascular function and drug delivery in a mouse model of pancreatic cancer. Gut 62(1):112–120

Jensen J, Pedersen EE, Galante P, Hald J, Heller RS, Ishibashi M, Kageyama R, Guillemot F, Serup P, Madsen OD (2000) Control of endodermal endocrine development by Hes-1. Nat Genet 24:36–44

Kanda M, Matthaei H, Wu J, Hong SM, Yu J, Borges M, Hruban RH, Maitra A, Kinzler K, Vogelstein B, Goggins M (2012) Presence of somatic mutations in most early-stage pancreatic intraepithelial neoplasia. Gastroenterology 142(4):730–733

Katz MH, Wang H, Fleming JB, Sun CC, Hwang RF, Wolff RA, Varadhachary G, Abbruzzese JL, Crane CH, Krishnan S et al (2009) Long-term survival after multidisciplinary management of resected pancreatic adenocarcinoma. Ann Surg Oncol 16:836–847

Kawaguchi Y, Cooper B, Gannon M, Ray M, MacDonald RJ, Wright CV (2002) The role of the transcriptional regulator Ptf1a in converting intestinal to pancreatic progenitors. Nat Genet 32:128–134

Korc M (2007) Pancreatic cancer-associated stroma production. Am J Surg 194:S84–S86

Lewis BC, Klimstra DS, Varmus HE (2003) The c-myc and PyMT oncogenes induce different tumor types in a somatic mouse model for pancreatic cancer. Genes Dev 17:3127–3138

Liebmann C (2011) EGF receptor activation by GPCRs: an universal pathway reveals different versions. Mol Cell Endocrinol 331:222–231

Maitra A, Hruban RH (2008) Pancreatic cancer. Annu Rev Pathol 3:157–188

Mansour SL, Thomas KR, Capecchi MR (1988) Disruption of the proto-oncogene int-2 in mouse embryo-derived stem cells: a general strategy for targeting mutations to non-selectable genes. Nature 336:348–352

Massague J (2008) TGFbeta in cancer. Cell 134:215–230

Mazur PK, Einwachter H, Lee M, Sipos B, Nakhai H, Rad R, Zimber-Strobl U, Strobl LJ, Radtke F, Kloppel G et al (2010) Notch2 is required for progression of pancreatic intraepithelial neoplasia and development of pancreatic ductal adenocarcinoma. Proc Natl Acad Sci USA 107: 13438–13443

Miyamoto Y, Maitra A, Ghosh B, Zechner U, Argani P, Iacobuzio-Donahue CA, Sriuranpong V, Iso T, Meszoely IM, Wolfe MS et al (2003) Notch mediates TGF alpha-induced changes in epithelial differentiation during pancreatic tumorigenesis. Cancer Cell 3:565–576

Morris JP 4th, Cano DA, Sekine S, Wang SC, Hebrok M (2010a) Beta-catenin blocks Kras-dependent reprogramming of acini into pancreatic cancer precursor lesions in mice. J Clin Invest 120:508–520

Morris JP 4th, Wang SC, Hebrok M (2010b) KRAS, Hedgehog, Wnt and the twisted developmental biology of pancreatic ductal adenocarcinoma. Nat Rev Cancer 10:683–695

Moskaluk CA, Hruban RH, Kern SE (1997) p16 and K-ras gene mutations in the intraductal precursors of human pancreatic adenocarcinoma. Cancer Res 57:2140–2143

Murtaugh LC, Stanger BZ, Kwan KM, Melton DA (2003) Notch signaling controls multiple steps of pancreatic differentiation. Proc Natl Acad Sci USA 100:14920–14925

Nakhai H, Siveke JT, Klein B, Mendoza-Torres L, Mazur PK, Algul H, Radtke F, Strobl L, Zimber-Strobl U, Schmid RM (2008) Conditional ablation of Notch signaling in pancreatic development. Development 135:2757–2765

Nawroth R, van Zante A, Cervantes S, McManus M, Hebrok M, Rosen SD (2007) Extracellular sulfatases, elements of the wnt signaling pathway, positively regulate growth and tumorigenicity of human pancreatic cancer cells. PLoS One 2:e392

Nolan-Stevaux O, Lau J, Truitt ML, Chu GC, Hebrok M, Fernandez-Zapico ME, Hanahan D (2009) GLI1 is regulated through smoothened-independent mechanisms in neoplastic pancreatic ducts and mediates PDAC cell survival and transformation. Genes Dev 23:24–36

Olive KP, Tuveson DA (2006) The use of targeted mouse models for preclinical testing of novel cancer therapeutics. Clin Cancer Res 12:5277–5287

Olive KP, Jacobetz MA, Davidson CJ, Gopinathan A, McIntyre D, Honess D, Madhu B, Goldgraben MA, Caldwell ME, Allard D et al (2009) Inhibition of Hedgehog signaling enhances delivery of chemotherapy in a mouse model of pancreatic cancer. Science 324:1457–1461

Ornitz DM, Hammer RE, Messing A, Palmiter RD, Brinster RL (1987) Pancreatic neoplasia induced by SV40 T-antigen expression in acinar cells of transgenic mice. Science 238:188–193

Palmiter RD, Brinster RL (1986) Germ-line transformation of mice. Annu Rev Genet 20:465–499

Pasca di Magliano M, Sekine S, Ermilov A, Ferris J, Dlugosz AA, Hebrok M (2006) Hedgehog/Ras interactions regulate early stages of pancreatic cancer. Genes Dev 20:3161–3173

Pasca di Magliano M, Biankin AV, Heiser PW, Cano DA, Gutierrez PJ, Deramaudt T, Segara D, Dawson AC, Kench JG, Henshall SM et al (2007) Common activation of canonical Wnt signaling in pancreatic adenocarcinoma. PLoS One 2:e1155

Pin CL, Bonvissuto AC, Konieczny SF (2000) Mist1 expression is a common link among serous exocrine cells exhibiting regulated exocytosis. Anat Rec 259:157–167

Pin CL, Rukstalis JM, Johnson C, Konieczny SF (2001) The bHLH transcription factor Mist1 is required to maintain exocrine pancreas cell organization and acinar cell identity. J Cell Biol 155:519–530

Plentz R, Park JS, Rhim AD, Abravanel D, Hezel AF, Sharma SV, Gurumurthy S, Deshpande V, Kenific C, Settleman J et al (2009) Inhibition of gamma-secretase activity inhibits tumor progression in a mouse model of pancreatic ductal adenocarcinoma. Gastroenterology 136(1741–1749):e1746

Provenzano PP, Cuevas C, Chang AE, Goel VK, Von Hoff DD, Hingorani SR (2012) Enzymatic targeting of the stroma ablates physical barriers to treatment of pancreatic ductal adenocarcinoma. Cancer Cell 21:418–429

Pylayeva-Gupta Y, Grabocka E, Bar-Sagi D (2011) RAS oncogenes: weaving a tumorigenic web. Nat Rev Cancer 11:761–774

Rajurkar M, De Jesus-Monge WE, Driscoll DR, Appleman VA, Huang H, Cotton JL, Klimstra DS, Zhu LJ, Simin K, Xu L et al (2012) The activity of Gli transcription factors is essential for Kras-induced pancreatic tumorigenesis. Proc Natl Acad Sci USA 109:E1038–E1047

Ray KC, Bell KM, Yan J, Gu G, Chung CH, Washington MK, Means AL (2011) Epithelial tissues have varying degrees of susceptibility to Kras(G12D)-initiated tumorigenesis in a mouse model. PLoS One 6:e16786

Sandgren EP, Luetteke NC, Palmiter RD, Brinster RL, Lee DC (1990) Overexpression of TGF alpha in transgenic mice: induction of epithelial hyperplasia, pancreatic metaplasia, and carcinoma of the breast. Cell 61:1121–1135

Sauer B, Henderson N (1988) Site-specific DNA recombination in mammalian cells by the Cre recombinase of bacteriophage P1. Proc Natl Acad Sci USA 85:5166–5170

Schutte M, Hruban RH, Geradts J, Maynard R, Hilgers W, Rabindran SK, Moskaluk CA, Hahn SA, Schwarte-Waldhoff I, Schmiegel W et al (1997) Abrogation of the Rb/p16 tumor-suppressive pathway in virtually all pancreatic carcinomas. Cancer Res 57:3126–3130

Sherr CJ (2004) Principles of tumor suppression. Cell 116:235–246

Soriano P (1999) Generalized lacZ expression with the ROSA26 Cre reporter strain. Nat Genet 21:70–71

Sternberg N, Hamilton D (1981) Bacteriophage P1 site-specific recombination. I. Recombination between loxP sites. J Mol Biol 150:467–486

Sumazaki R, Shiojiri N, Isoyama S, Masu M, Keino-Masu K, Osawa M, Nakauchi H, Kageyama R, Matsui A (2004) Conversion of biliary system to pancreatic tissue in Hes1-deficient mice. Nat Genet 36:83–87

Tanaka M, Chari S, Adsay V, Fernandez-del Castillo C, Falconi M, Shimizu M, Yamaguchi K, Yamao K, Matsuno S (2006) International consensus guidelines for management of intraductal papillary mucinous neoplasms and mucinous cystic neoplasms of the pancreas. Pancreatology 6:17–32

Thayer SP, di Magliano MP, Heiser PW, Nielsen CM, Roberts DJ, Lauwers GY, Qi YP, Gysin S, Fernandez-del Castillo C, Yajnik V et al (2003) Hedgehog is an early and late mediator of pancreatic cancer tumorigenesis. Nature 425:851–856

Thomas KR, Capecchi MR (1987) Site-directed mutagenesis by gene targeting in mouse embryo-derived stem cells. Cell 51:503–512

Tian H, Callahan CA, DuPree KJ, Darbonne WC, Ahn CP, Scales SJ, de Sauvage FJ (2009) Hedgehog signaling is restricted to the stromal compartment during pancreatic carcinogenesis. Proc Natl Acad Sci USA 106:4254–4259

Tuveson DA, Zhu L, Gopinathan A, Willis NA, Kachatrian L, Grochow R, Pin CL, Mitin NY, Taparowsky EJ, Gimotty PA et al (2006) Mist1-KrasG12D knock-in mice develop mixed differentiation metastatic exocrine pancreatic carcinoma and hepatocellular carcinoma. Cancer Res 66:242–247

van den Brink GR (2007) Hedgehog signaling in development and homeostasis of the gastrointestinal tract. Physiol Rev 87:1343–1375

Wagner M, Luhrs H, Kloppel G, Adler G, Schmid RM (1998) Malignant transformation of duct-like cells originating from acini in transforming growth factor transgenic mice. Gastroenterology 115:1254–1262

Wagner M, Greten FR, Weber CK, Koschnick S, Mattfeldt T, Deppert W, Kern H, Adler G, Schmid RM (2001) A murine tumor progression model for pancreatic cancer recapitulating the genetic alterations of the human disease. Genes Dev 15:286–293

Wang L, Heidt DG, Lee CJ, Yang H, Logsdon CD, Zhang L, Fearon ER, Ljungman M, Simeone DM (2009) Oncogenic function of ATDC in pancreatic cancer through Wnt pathway activation and beta-catenin stabilization. Cancer Cell 15:207–219

Wells JM, Esni F, Boivin GP, Aronow BJ, Stuart W, Combs C, Sklenka A, Leach SD, Lowy AM (2007) Wnt/beta-catenin signaling is required for development of the exocrine pancreas. BMC Dev Biol 7:4

Wilentz RE, Geradts J, Maynard R, Offerhaus GJ, Kang M, Goggins M, Yeo CJ, Kern SE, Hruban RH (1998) Inactivation of the p16 (INK4A) tumor-suppressor gene in pancreatic duct lesions: loss of intranuclear expression. Cancer Res 58:4740–4744

Wilentz RE, Iacobuzio-Donahue CA, Argani P, McCarthy DM, Parsons JL, Yeo CJ, Kern SE, Hruban RH (2000) Loss of expression of Dpc4 in pancreatic intraepithelial neoplasia: evidence that DPC4 inactivation occurs late in neoplastic progression. Cancer Res 60:2002–2006

Yauch RL, Gould SE, Scales SJ, Tang T, Tian H, Ahn CP, Marshall D, Fu L, Januario T, Kallop D et al (2008) A paracrine requirement for hedgehog signalling in cancer. Nature 455:406–410

Ying H, Kimmelman AC, Lyssiotis CA, Hua S, Chu GC, Fletcher-Sananikone E, Locasale JW, Son J, Zhang H, Coloff JL et al (2012) Oncogenic Kras maintains pancreatic tumors through regulation of anabolic glucose metabolism. Cell 149:656–670

The Genetics of Pancreatic Cancer Progression

Christine A. Iacobuzio-Donahue

Abstract Pancreatic cancer is caused by mutations in specific cancer genes. While these cancer genes are often categorized into those that are inherited versus somatically acquired, they may also be categorized into those that occur during carcinogenesis versus those that accumulate during clonal progression. This newfound approach to understanding pancreatic cancer genetics now opens the door to understanding those events that play a role specifically in progression to metastatic disease.

Introduction

Clinicopathologic Features

Pancreatic ductal adenocarcinoma (PDAC, pancreatic cancer) is the eighth leading cause of cancer-related deaths in the world corresponding to a >94 % mortality rate (Jemal et al. 2011; Bosetti et al. 2012; Malvezzi et al. 2011). Most patients present with advanced stage disease at the time of diagnosis leaving relatively few patients as candidates for potentially curative resection (Hidalgo 2010). Unfortunately, even in patients who undergo pancreatic resection, both local and systemic recurrences are common with a median post-resection survival of less than 24 months (Katz et al. 2009; Winter et al. 2006, 2012). Perhaps not surprisingly most patients diagnosed with pancreatic cancer will die of locally advanced (Stage III) or metastatic disease (Stage IV) (Stathis and Moore 2010) (Fig. 1), indicating the urgent need to understand pancreatic cancer progression so as to improve upon these statistics.

C.A. Iacobuzio-Donahue, M.D., Ph.D. (✉)
Department of Pathology, Johns Hopkins Medical Institutions, Baltimore, MD 21231, USA
e-mail: ciacobu@jhmi.edu

D.M. Simeone and A. Maitra (eds.), *Molecular Genetics of Pancreatic Cancer*,
DOI 10.1007/978-1-4614-6549-2_8,

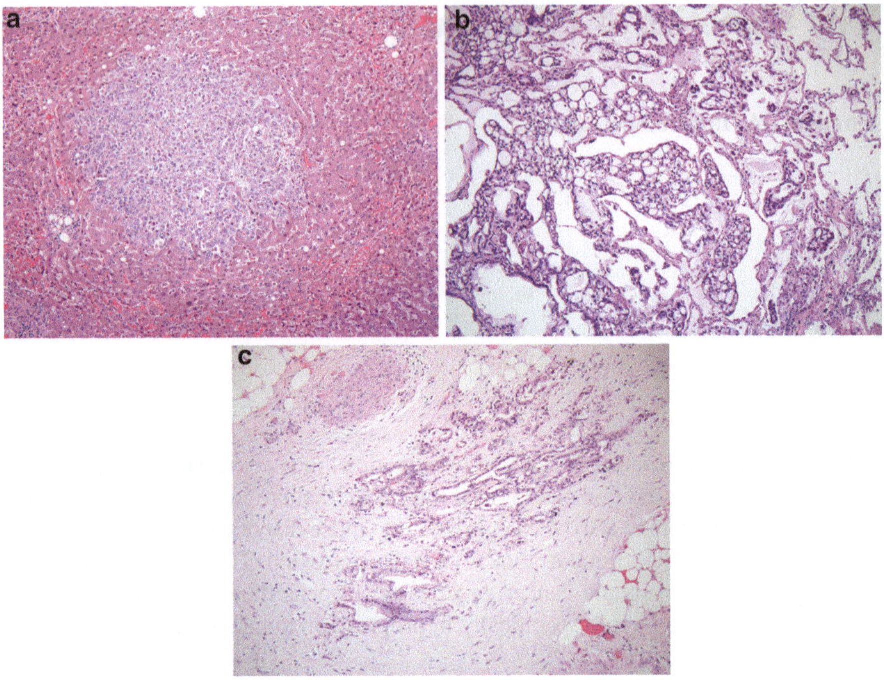

Fig. 1 Histologic images of the most common sites of metastatic disease in patients with pancreatic cancer. (**a**) Liver metastasis. (**b**) Lung metastasis. (**c**) Peritoneal metastasis. All images ×100 magnification

Pancreatic Cancer as a Model System for Metastasis

To date, the vast majority of pancreatic cancer research has been devoted to understanding the biology of the neoplastic cell in association with carcinogenesis and/or invasion within the primary site, including methods for therapeutic targeting [reviewed in (Hidalgo 2010; Stathis and Moore 2010; Maitra and Hruban 2008; Vincent et al. 2011)]. Such studies are undoubtedly important and have revealed a wealth of information for this disease. By contrast, studies of pancreatic cancer progression and of the metastatic phenotype specifically are relatively rare, leading investigators to depend on cell culture systems or mouse models of metastatic progression (Hingorani et al. 2005; Aguirre et al. 2003; Ijichi et al. 2006; Feldmann et al. 2007; Morton et al. 2010; Little et al. 2012) without relevant human tissues for correlation. Thus, metastasis research as an independent field with the pancreatic cancer community is in its infancy, making it fertile ground for novel discoveries.

In addition to the enormous medical need to understand progressive disease, pancreatic cancer itself is an ideal model tumor type to understand the dynamics of cancer progression and metastasis in general. First, the genetics of pancreatic cancer are among the most well described for any solid tumor type, and indicate that pancreatic cancers arise due to the progressive accumulation of activating mutations in oncogenes and inactivating mutations in tumor suppressor genes (Maitra and Hruban 2008; Jones et al. 2008a, 2009, 2012). Second, the precursor lesions that give rise to invasive ductal adenocarcinomas have also been well described (Hruban et al. 2001; Matthaei et al. 2011), and when studied in the context of pancreatic cancer genetics it is clear that the increasing cytologic atypia of these precursors is well associated with accumulation of alterations in specific genes (Hruban et al. 2000; Maitra et al. 2003). Third, metastasis is a common feature of pancreatic cancer with up to 50 % of patients presenting with metastatic disease at the time of initial diagnosis (Stathis and Moore 2010). Moreover, even patients who undergo surgery commonly develop metastatic disease (Sohn et al. 2000; Iacobuzio-Donahue et al. 2009). Finally, because metastatic pancreatic cancer is not a surgical disease, it is not uncommon for patients to have their primary carcinoma in situ for comparison to coexistent metastases, particular in the setting of performing an autopsy for research purposes as we have reported (Embuscado et al. 2005).

Genetics of Pancreatic Carcinogenesis

One cannot discuss the genetics of pancreatic cancer progression without an understanding of the genetic alterations that occur during pancreatic carcinogenesis. This is discussed in detail in Chap. 2, and therefore, only a brief overview is presented here.

High Frequency Genetic Alterations

Much of the genetic basis of pancreatic cancer has largely been elucidated using a candidate gene approach (Redston et al. 1994; Caldas et al. 1994; Hahn et al. 1996). Traditionally this approach has relied on conventional dideoxy sequencing and has identified the four genes most commonly associated with pancreatic cancer as *KRAS, CDKN2A* (*p16*)*, TP53*, and *SMAD4* (*DPC4*). *KRAS* is a member of the RAS family of guanosine triphosphate (GTP)-binding proteins that, when bound to GTP, mediate a wide range of cellular functions including proliferation, cell survival, and cytoskeletal remodeling. Activating mutations in *KRAS* impair its GTPase activity, resulting in a constitutively active oncoprotein independent of extracellular or intracellular signals (Schubbert et al. 2007). Mutations of *KRAS* are not only the most common genetic alteration in pancreatic cancer (>99 %) but also detectable as early as PanIN-1A lesions (Kanda et al. 2012).

By contrast, *CDKN2A, TP53 and SMAD4* are tumor suppressor genes that are inactivated in approximately 90 %, 80 %, and 55 % of pancreatic cancers, respectively (Jones et al. 2008a; Redston et al. 1994; Caldas et al. 1994; Hahn et al. 1996; Schutte et al. 1997). *CDKN2A* inactivation occurs by one of three mechanisms such as homozygous deletion, intragenic mutation with loss of the second allele, or least commonly epigenetic silencing of gene expression by promoter methylation

(Caldas et al. 1994; Schutte et al. 1997). *CDKN2A* is a member of the cyclin-dependent kinase (CDK) inhibitor family; in its absence cells proceed unchecked through the G1-S checkpoint mediated by CDKs such as CDK4 and CDK6 (Kim and Sharpless 2006). *TP53* and *SMAD4* inactivation occur through mutation and loss of the second allele, or by homozygous deletion (Jones et al. 2008a). In normal cells, the p53 protein is a critical regulator of many cellular functions, including the G1-S cell cycle checkpoint, maintenance of G2-M arrest, and the induction of apoptosis following cellular stress (Riley et al. 2008). In the presence of DNA damage loss of p53 function allows cells to survive and divide, leading to an accumulation of additional genetic abnormalities (Vogelstein et al. 2000). Smad4 protein is a central mediator of the transforming growth factor-β (TGF-β) canonical signaling pathway that functions in cellular growth and differentiation (Massagué 2008), and loss of *SMAD4* thus inhibits Smad-dependent TGF-β signaling, allowing an escape from TGF-β induced growth inhibition (Padua and Massagué 2009).

Exomic Sequencing

Sequencing of the cancer exome has provided greater insight into the mutational spectrum of human cancer beyond the success of candidate gene approaches. This strategy has been applied to the study of pancreatic cancers with the goal of identifying the complete spectrum of somatic mutations beyond the most commonly altered genes described above, and to identify the molecular pathways important for this tumor type (Jones et al. 2008a, 2009). It is important to note that initial exome sequencing efforts of pancreatic cancer were performed by high throughput dideoxy sequencing (Jones et al. 2008a) and the data thus reflects the sensitivity of this approach. With the use of high resolution next generation methods, even greater numbers of mutations per carcinoma can be expected (Meyerson et al. 2010).

Overall, the pancreatic cancer exome determined by dideoxy sequencing is notable for an average of 63 alterations per cancer genome, the majority of which correspond to single base changes, or point mutations. These base changes most often caused missense mutations, many of which were silent. Homozygous deletions and amplifications were also found. The numbers of homozygous deletions per pancreatic cancer genome was variable, but were more numerous than amplifications. Review of the genetic alterations in each cancer analyzed, coupled with estimations of passenger mutation rates, resulted in a list of 91 candidate cancer genes, or CAN genes, in pancreatic cancer. This list included all genes previously known to play a significant role in pancreatic cancer through common mutation or copy number change (*KRAS, CDKN2A, TP53*, and *SMAD4*). In addition, because of the unbiased approach afforded by exomic sequencing, numerous other genes of potential biological interest were discovered such as *ARID1A* and *MLL3* (Jones et al. 2012; Balakrishnan et al. 2007). Low frequency mutational targets such as *MKK4, TGFBR2*, and *STK11* were also identified consistent with prior reports (Su et al. 1998; Goggins et al. 1998; Su et al. 1999).

Genetics of Pancreatic Cancer Progression

Locally advanced or metastatic pancreatic cancer is not a surgical disease (Stathis and Moore 2010; de Jong et al. 2010a, b), and it is for this reason that tissues of advanced stage human pancreatic cancer have not been available for study. This need has been addressed by our labs' use of a rapid autopsy protocol in which patients with end stage pancreatic cancer consent premortem to an autopsy for the purpose of collecting high quality cancer tissues for research (Iacobuzio-Donahue et al. 2009; Embuscado et al. 2005). From each patient the entire primary carcinoma and up to 20 different metastases from different organ sites are collected in a variety of methods, allowing study of the metastatic process in humans in unprecedented detail, including the genetics of advanced stage disease.

Patterns of Metastatic Failure

In an initial survey of the dynamics of pancreatic cancer metastasis, 76 patients with pancreatic cancer who underwent a rapid autopsy were studied (Iacobuzio-Donahue et al. 2009). These patients represented the full spectrum of those encountered in clinical practice and included those initially diagnosed with Stage I/II resectable disease, Stage III locally advanced and Stage IV metastatic disease. Patients were also treated with a variety of chemotherapy regimens in keeping with standard of care based on the stage of diagnosis. Surprisingly, careful review of each patient's terminal stage disease revealed two distinct patterns of spread (Fig. 2). In the first pattern (*w*idely *m*etastatic *p*ancreatic *c*ancer, WMPC) patients died with widespread

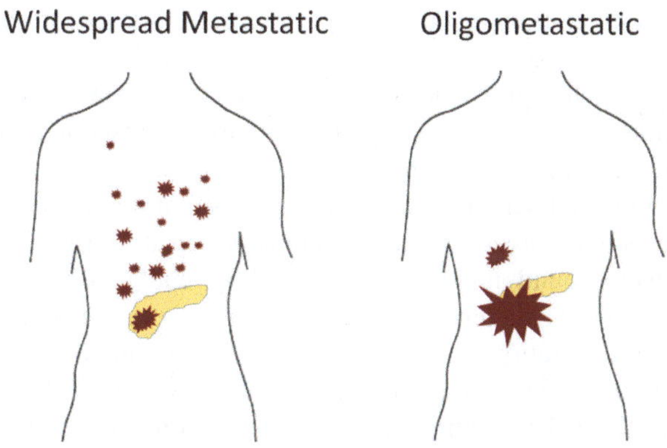

Fig. 2 Patterns of metastatic failure. At autopsy, approximately two-thirds of patients have widely metastatic pancreatic cancer that is defined as >10 distant metastases but often numbers in the tens to >100 of deposits. By contrast, patients with oligometastatic pancreatic cancer have few metastases (no more than 10), and the cause of death is most often due to the large primary carcinoma that invades into adjacent vital structures

metastatic disease to multiple organ sites. This pattern was seen in approximately two thirds of patients, with the number of metastases generally in the tens to hundreds of deposits. Moreover, when cause of death could be determined, patients with WMPC most often died of complications of organ failure due to replacement by tumor. In the second pattern (*o*ligo*m*etastatic *p*ancreatic *c*ancer, OMPC), patients predominantly died with localized disease in association with few metastases, typically fewer than <10. These patients more often died of destruction of local vital structures such as mesenteric vessels leading to ischemia, or diaphragmatic infiltration leading to respiratory failure; metastatic disease was not the primary cause of death in most OMPC patients. Of interest, in a subset of patients with OMPC only a bulky primary carcinoma and no metastases were found at autopsy, further supporting the notion that the locally destructive primary carcinoma was the cause of death in these patients. Irrespective of the extent of metastatic burden in WMPC and OMPC, the most common sites of metastatic failure were the liver, peritoneum and lung (Yachida and Iacobuzio-Donahue 2009). Collectively, these observations indicate that there is a range of metastatic efficiencies in pancreatic cancer and not all patients die of aggressive metastasis. It is also important to note that these patterns of metastatic failure were unrelated to clinicopathologic features at diagnosis, a finding also shown by Hishinuma et al. (2006), indicating that the current staging modalities do not fully capture the metastatic phenotype of pancreatic cancer.

Genetic Correlates of Metastatic Efficiency

While deregulation of a variety of cellular programs or pathways have been suggested to play a role in metastasis in general, including pancreatic cancer metastasis (Padua and Massagué 2009; Polyak and Weinberg 2009; Subarsky and Hill 2003), very few genetic alterations have been specifically implicated in the formation of metastatic disease (Mudali et al. 2006). Thus, given the wealth of information gleaned from candidate approaches and exomic sequencing, a logical question is the extent to which the genes identified by these surveys correspond to the distinct patterns of spread seen at autopsy, including organ specific metastasis.

Data from autopsied patients provide clues to the nature of genetic alterations in promoting metastatic spread (Iacobuzio-Donahue et al. 2009). For example, no relationships were found for *KRAS* or *CDKN2A* and pattern of failure, likely because these genes are altered in the vast majority of pancreatic cancers, and because oncogenic mutations in *KRAS* and inactivating mutations in *CDKN2A* have similar downstream effects, respectively (Hruban et al. 2000; Kanda et al. 2012; Wilentz et al. 1998). By contrast, significant relationships were noted for both *TP53* and *SMAD4* based on univariate analyses in that the frequency of *TP53* or *SMAD4* inactivation was significantly higher in WMPC than in OMPC. Moreover, carcinomas for which *TP53* and *SMAD4* mutations were coexistent had the greatest metastatic burden at autopsy (>100 deposits), suggesting a degree of synergy when both genes are inactivated in the same carcinoma. When considering that both *TP53* and *SMAD4* inactivation occurs in PanIN3 lesions, metastatic efficiency may be established, in part, even before the development of the invasive carcinoma. However, it is likely that additional modifiers of metastatic efficiency also exist (Nguyen and Massagué 2007) for which *TP53* and *SMAD4* are but two examples.

The mechanisms by which *TP53* and *SMAD4* inactivation promote metastasis may include both dependent and independent factors. In normal cells the p53 protein is a critical regulator of numerous cellular functions, including regulation of the G1-S cell cycle checkpoint, maintenance of G2-M arrest, and the induction of apoptosis due to cellular stress (Vogelstein et al. 2000). Loss of p53 function allows cells to survive and divide despite the presence of damaged DNA, thus allowing the accumulation of additional genetic abnormalities and hence genetic instability (Goh et al. 2011). How *TP53* inactivation specifically promotes metastasis has yet to be elucidated. However, studies in mouse models of pancreatic cancer based on conditional inactivation of *Trp53* by deletion versus missense mutation indicates that only PDACs with mutant *Trp53* exhibited invasive activity in vitro (Morton et al. 2010). A similar finding was reported by Neilsen et al. (2011) including the observation that mutant p53 proteins utilize p63 to facilitate invasion by secretion of pro-invasive factors into the tumor microenvironment.

By contrast, the mechanisms by which *SMAD4* loss promotes metastasis are better characterized. In normal cells SMAD4 protein mediates canonical TGF-β signals from specific cell surface receptors to the nucleus, thereby controlling cellular growth and differentiation (Massagué 2008). Thus, loss of *SMAD4* inhibits canonical TGF-β signaling, allowing an escape from TGF-β induced growth inhibition and apoptosis (Massagué 2008; Padua and Massagué 2009; Siegel and Massagué 2003). Available TGF-β ligand may then function as a tumor promoting factor on the cancer cells, stimulate formation of an immunosuppressive microenvironment, and promote angiogenesis and epithelial–mesenchymal transition (Siegel and Massagué 2003; Jonson et al. 2001; Pertovaara et al. 1994).

Genetic Alterations in Pancreatic Cancer Metastasis

As a more unbiased approach towards understanding the genetic features of pancreatic cancer metastasis, the exomes of seven metastases have been studied in detail (Yachida et al. 2010). These seven metastases were a subset of the 24 pancreatic cancers studied by Jones et al. (2008a), providing a unique opportunity to compare the genetics of distant metastases from treated patients who died of their disease to surgically resected and treatment naïve tumors. Overall, distant metastases have similar numbers of genetic alterations than surgically resected pancreatic cancers, and the types of alterations (missense mutations, nonsense mutations, deletions, amplifications, etc.) are also similar in frequency among resectable and late stage disease (Fig. 3). This suggests that disease progression is not associated with an enhanced genetic mutation rate or change in mutation spectrum.

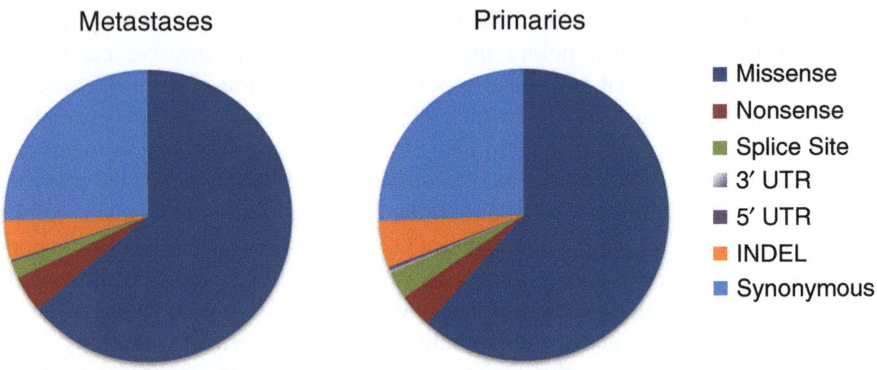

Fig. 3 Mutational spectra in metastases versus primary carcinomas. Shown are the proportions of each type of intragenic mutation in each subset of samples analyzed by whole exome sequencing (Jones et al. 2008a)

Review of the specific genes targeted by somatic alteration in distant metastases versus resectable carcinomas has also been informative. Yachida et al. (2010) found that each metastasis exome contained numerous mutations, yet virtually none of these alterations were shared among two or more patients. Because somatic alterations may occur at the chromosomal level, Campbell et al. (2010) also analyzed these metastases by massively parallel paired-end sequencing to identify the scope of rearrangements in each metastasis as compared to primary carcinomas. Similar to that of whole exome sequencing, specific genetic events that promote metastasis were not found, the majority of gene rearrangements identified occurred early during tumor evolution, and beyond known driver genes no rearrangements were found in common among two or more cancers.

Comparative Lesion Sequencing

Use of a single cancer sample with matched normal has traditionally been used for identification of cancer genes; however, this approach is insufficient for genetic studies of metastatic disease as it does not account for the clonally heterogeneous nature of the primary neoplasm (Fidler and Hart 1982). Thus, studies of metastasis ideally rely upon use of two or more distinct samples derived from a given patients' cancer to perform comparative lesion sequencing, a simple yet powerful method to evaluate the clonal relatedness of different carcinoma samples within a single individual. These samples may be any number of synchronous metastases, the primary carcinoma and a subsequent metastatic recurrence, or even samples taken from geographically distinct regions of a single primary carcinoma (Gerlinger et al. 2012; Navin et al. 2011).

In essence, genetic alterations present in one cancer sample are analyzed in additional geographically or temporally distinct samples from that same patient, a method not dissimilar from other phylogenetic approaches in the biosciences (Murchison et al. 2012; Garcia-Porta et al. 2012; Krumbholz et al. 2009). Using this approach, genetic alterations found by any method can be classified into two categories. The first category corresponds to alterations present in all samples analyzed for a patient, and these are called founder events. Moreover, because they were present in both the primary carcinoma and the matched metastases, a logical assumption is that most if not all of these alterations accumulated within the precursor that ultimately gave rise to that pancreatic cancer and are thus present in the majority, if not all, of the cells of the tumor (Fig. 4). Thus, founder alterations are genetic lineage markers of the original parental clone of cells that formed that carcinoma. Consistent with this notion, founder alterations are represented by known driver mutations important for pancreatic cancer formation (*KRAS, CDKN2A, TP53* and *SMAD4*), as well as many CAN genes also identified by whole exome sequencing (Jones et al. 2008a). Founder alterations more commonly include tumor suppressor genes as well that are inactivated by mutation and allelic loss.

By contrast, progressor alterations are genetic lineage markers of specific subclones that arise during clonal evolution of the primary carcinoma (Fig. 4). This is because progressor alterations are found in a subset of samples analyzed, yet founder alterations are found in all samples for that patient that include those with progressor alterations. For example, Yachida et al. (2010) noted that while subclones containing founder alterations were present through the primary carcinoma, a subset of samples that had both founder and progressor alterations were present in a geographically restricted area of the primary carcinoma. Thus progressor alterations occurred after founders and logically represent subclonal evolution beyond the parental clone. The finding of shared progressor alterations in the primary and metastases in the same patient is also strong evidence that metastases arise from a preexisting primary carcinoma.

Timeline of Metastasis Formation

Beyond pancreatic carcinogenesis, the dynamics of pancreatic cancer progression are increasingly complex. In this context mathematical modeling based on genetic data has provided an invaluable tool to understand the metastatic process.

A major issue in pancreatic cancer management is if the poor prognosis of patients with this disease is because they are diagnosed too late in the natural history of the disease, or if pancreatic cancer is rapidly metastatic shortly after it forms. To address this question, a computational model was created that relied on data generated from exomic sequencing to estimate three critical times in the genetic evolution of pancreatic cancer for these seven patients (Fig. 4). The first time interval (T1) corresponded for the time taken from the initiating mutation in a normal ductal epithelial cell (i.e., *KRAS*) until the development of the founder cell that contained all somatic mutations present in the parental clone that eventually became the infiltrating carcinoma. The second time interval (T2) corresponded to the subsequent

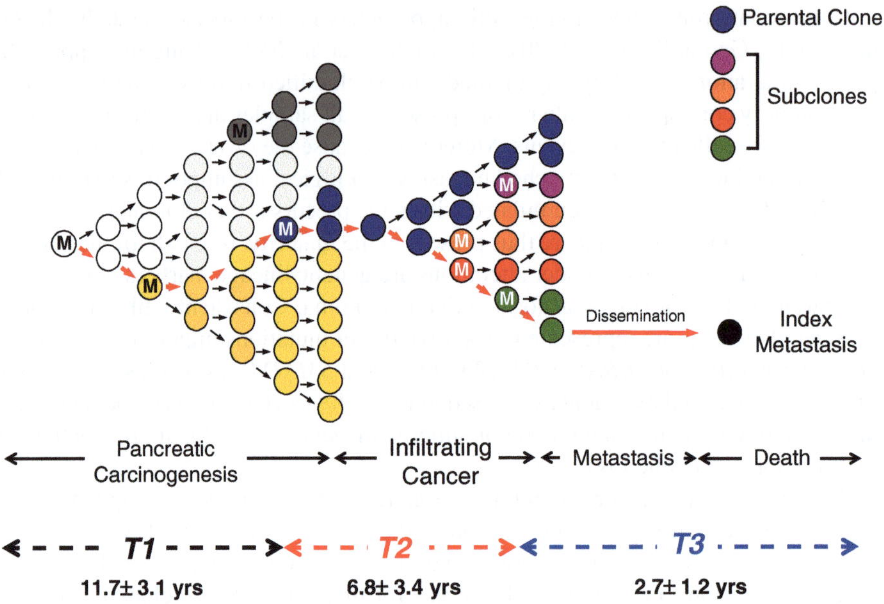

Fig. 4 Clonal evolution of pancreatic cancer. Carcinogenesis, and time T1, begins with an initiating alteration (M) in a normal cell that provides a selective advantage. Over time, waves of clonal expansion occur in association with the acquisition of additional mutations, corresponding to the progression model of pancreatic intraepithelial neoplasia (PanIN). This clonal expansion will generate the founder cell within a PanIN lesion (*blue clone*) that will eventually become the parental clone and hence initiate the infiltrating carcinoma. The birth of this cell corresponds to the beginning of time T2. Following additional waves of clonal expansion from the parental clone, subclones are generated within the infiltrating carcinoma. The birth of the cell within the primary carcinoma that will become the metastatic subclone (*green clone*) corresponds to the start of time T3. The estimated average time for each interval is indicated at the bottom of the illustration and corresponds to a total of 21.2 years from tumor initiation until the patient's death from metastatic disease. *Red arrows* indicate the lineage of the index metastasis from its origin in a normal cell

time taken for the development of the founder cell within the primary carcinoma that contained all mutations present in the metastatic subclone that eventually seeded the index metastasis in that patient (i.e., the metastasis that was exome sequenced), and the third time interval (T3) corresponds for the subsequent time until the patients' death. Based on this model, the conservative estimate of 11.7 years, 6.8 years, and 2.7 years per interval, respectively, was arrived at corresponding to an average of ~21 years from the initiating mutation until the patients' death. Unfortunately, most patients with pancreatic cancer are diagnosed well towards the end of this time span (Hidalgo 2010; Stathis and Moore 2010), indicating that the overall poor prognosis is likely due to diagnosis occurring far too late in the natural history of the disease. Nonetheless, pancreatic cancer is quite similar to other tumor types that have a long latency from initiation to patient death that is on the order of decades, not months to years (Jones et al. 2008b), indicating a prolonged window of opportunity for early detection while still in the curative stage.

Summary

Focused studies of pancreatic cancer progression are only useful if it can be used to improve the survival of patients with pancreatic cancer (McDermott et al. 2011). As we are now in the era of whole genome analyses (Meyerson et al. 2010), such studies are expected to be fruitful towards development of screening modalities to identify patients before they develop metastatic disease (Vincent et al. 2011), and therapeutic developments targeting the metastatic phenotype in those who present with metastases.

References

Aguirre AJ, Bardeesy N, Sinha M, Lopez L, Tuveson DA, Horner J et al (2003) Activated Kras and Ink4a/Arf deficiency cooperate to produce metastatic pancreatic ductal adenocarcinoma. Genes Dev 17(24):3112–3126

Balakrishnan A, Bleeker FE, Lamba S, Rodolfo M, Daniotti M, Scarpa A et al (2007) Novel somatic and germline mutations in cancer candidate genes in glioblastoma, melanoma, and pancreatic carcinoma. Cancer Res 67(8):3545–3550

Bosetti C, Bertuccio P, Negri E, La Vecchia C, Zeegers MP, Boffetta P (2012) Pancreatic cancer: overview of descriptive epidemiology. Mol Carcinog 51(1):3–13

Caldas C, Hahn SA, da Costa LT, Redston MS, Schutte M, Seymour AB et al (1994) Frequent somatic mutations and homozygous deletions of the p16 (MTS1) gene in pancreatic adenocarcinoma. Nat Genet 8(1):27–32

Campbell PJ, Yachida S, Mudie LJ, Stephens PJ, Pleasance ED, Stebbings LA et al (2010) The patterns and dynamics of genomic instability in metastatic pancreatic cancer. Nature 467(7319):1109–1113

de Jong MC, Farnell MB, Sclabas G, Cunningham SC, Cameron JL, Geschwind J-F et al (2010a) Liver-directed therapy for hepatic metastases in patients undergoing pancreaticoduodenectomy: a dual-center analysis. Ann Surg 252(1):142–148

de Jong MC, Tsai S, Cameron JL, Wolfgang CL, Hirose K, van Vledder MG et al (2010b) Safety and efficacy of curative intent surgery for peri-ampullary liver metastasis. J Surg Oncol 102(3): 256–263

Embuscado EE, Laheru D, Ricci F, Yun KJ, de Boom Witzel S, Seigel A et al (2005) Immortalizing the complexity of cancer metastasis: genetic features of lethal metastatic pancreatic cancer obtained from rapid autopsy. Cancer Biol Ther 4(5):548–554

Feldmann G, Dhara S, Fendrich V, Bedja D, Beaty R, Mullendore M et al (2007) Blockade of hedgehog signaling inhibits pancreatic cancer invasion and metastases: a new paradigm for combination therapy in solid cancers. Cancer Res 67(5):2187–2196

Fidler IJ, Hart IR (1982) Biological diversity in metastatic neoplasms: origins and implications. Science 217(4564):998–1003

Garcia-Porta J, Litvinchuk SN, Crochet PA, Romano A, Geniez PH, Lo-Valvo M et al (2012) Molecular phylogenetics and historical biogeography of the west-palearctic common toads (Bufo bufo species complex). Mol Phylogenet Evol 63(1):113–130

Gerlinger M, Rowan AJ, Horswell S, Larkin J, Endesfelder D, Gronroos E et al (2012) Intratumor heterogeneity and branched evolution revealed by multiregion sequencing. N Engl J Med 366(10):883–892

Goggins M, Shekher M, Turnacioglu K, Yeo CJ, Hruban RH, Kern SE (1998) Genetic alterations of the transforming growth factor beta receptor genes in pancreatic and biliary adenocarcinomas. Cancer Res 58(23):5329–5332

Goh AM, Coffill CR, Lane DP (2011) The role of mutant p53 in human cancer. J Pathol 223(2):116–126
Hahn SA, Schutte M, Hoque AT, Moskaluk CA, da Costa LT, Rozenblum E et al (1996) DPC4, a candidate tumor suppressor gene at human chromosome 18q21.1. Science 271(5247):350–353
Hidalgo M (2010) Pancreatic cancer. N Engl J Med 362(17):1605–1617
Hingorani SR, Wang L, Multani AS, Combs C, Deramaudt TB, Hruban RH et al (2005) Trp53R172H and KrasG12D cooperate to promote chromosomal instability and widely metastatic pancreatic ductal adenocarcinoma in mice. Cancer Cell 7(5):469–483
Hishinuma S, Ogata Y, Tomikawa M, Ozawa I, Hirabayashi K, Igarashi S (2006) Patterns of recurrence after curative resection of pancreatic cancer, based on autopsy findings. J Gastrointest Surg 10(4):511–518
Hruban RH, Goggins M, Parsons J, Kern SE (2000) Progression model for pancreatic cancer. Clin Cancer Res 6(8):2969–2972
Hruban RH, Adsay NV, Albores-Saavedra J, Compton C, Garrett ES, Goodman SN et al (2001) Pancreatic intraepithelial neoplasia: a new nomenclature and classification system for pancreatic duct lesions. Am J Surg Pathol 25(5):579–586
Iacobuzio-Donahue CA, Fu B, Yachida S, Luo M, Abe H, Henderson CM et al (2009) DPC4 gene status of the primary carcinoma correlates with patterns of failure in patients with pancreatic cancer. J Clin Oncol 27(11):1806–1813
Ijichi H, Chytil A, Gorska AE, Aakre ME, Fujitani Y, Fujitani S et al (2006) Aggressive pancreatic ductal adenocarcinoma in mice caused by pancreas-specific blockade of transforming growth factor-beta signaling in cooperation with active Kras expression. Genes Dev 20(22): 3147–3160
Jemal A, Bray F, Center MM, Ferlay J, Ward E, Forman D (2011) Global cancer statistics. CA Cancer J Clin 61(2):69–90
Jones S, Zhang X, Parsons DW, Lin JC-H, Leary RJ, Angenendt P et al (2008a) Core signaling pathways in human pancreatic cancers revealed by global genomic analyses. Science 321(5897):1801–1806
Jones S, Chen W-D, Parmigiani G, Diehl F, Beerenwinkel N, Antal T et al (2008b) Comparative lesion sequencing provides insights into tumor evolution. Proc Natl Acad Sci USA 105(11):4283–4288
Jones S, Hruban RH, Kamiyama M, Borges M, Zhang X, Parsons DW et al (2009) Exomic sequencing identifies PALB2 as a pancreatic cancer susceptibility gene. Science 324(5924):217
Jones S, Li M, Parsons DW, Zhang X, Wesseling J, Kristel P et al (2012) Somatic mutations in the chromatin remodeling gene ARID1A occur in several tumor types. Hum Mutat 33(1):100–103
Jonson T, Albrechtsson E, Axelson J, Heidenblad M, Gorunova L, Johansson B et al (2001) Altered expression of TGFB receptors and mitogenic effects of TGFB in pancreatic carcinomas. Int J Oncol 19(1):71–81
Kanda M, Matthaei H, Wu J, Hong SM, Yu J, Borges M et al (2012) Presence of somatic mutations in most early-stage pancreatic intraepithelial neoplasia. Gastroenterology 142(4):730–733.e9
Katz MHG, Wang H, Fleming JB, Sun CC, Hwang RF, Wolff RA et al (2009) Long-term survival after multidisciplinary management of resected pancreatic adenocarcinoma. Ann Surg Oncol 16(4):836–847
Kim WY, Sharpless NE (2006) The regulation of INK4/ARF in cancer and aging. Cell 127(2): 265–275
Krumbholz A, Bininda-Emonds ORP, Wutzler P, Zell R (2009) Phylogenetics, evolution, and medical importance of polyomaviruses. Infect Genet Evol 9(5):784–799
Little EC, Wang C, Watson PM, Watson DK, Cole DJ, Camp ER (2012) Novel immunocompetent murine models representing advanced local and metastatic pancreatic cancer. J Surg Res 176(2):359–366
Maitra A, Hruban RH (2008) Pancreatic cancer. Annu Rev Pathol 3:157–188
Maitra A, Adsay NV, Argani P, Iacobuzio-Donahue C, De Marzo A, Cameron JL et al (2003) Multicomponent analysis of the pancreatic adenocarcinoma progression model using a pancreatic intraepithelial neoplasia tissue microarray. Mod Pathol 16(9):902–912

Malvezzi M, Arfé A, Bertuccio P, Levi F, La Vecchia C, Negri E (2011) European cancer mortality predictions for the year 2011. Ann Oncol 22(4):947–956

Massagué J (2008) TGFbeta in cancer. Cell 134(2):215–230

Matthaei H, Schulick RD, Hruban RH, Maitra A (2011) Cystic precursors to invasive pancreatic cancer. Nat Rev Gastroenterol Hepatol 8(3):141–150

McDermott U, Downing JR, Stratton MR (2011) Genomics and the continuum of cancer care. N Engl J Med 364(4):340–350

Meyerson M, Gabriel S, Getz G (2010) Advances in understanding cancer genomes through second-generation sequencing. Nat Rev Genet 11(10):685–696

Morton JP, Timpson P, Karim SA, Ridgway RA, Athineos D, Doyle B et al (2010) Mutant p53 drives metastasis and overcomes growth arrest/senescence in pancreatic cancer. Proc Natl Acad Sci USA 107(1):246–251

Mudali SV, Fu B, Lakkur SS, Luo M, Embuscado EE, Iacobuzio-Donahue CA (2006) Patterns of EphA2 protein expression in primary and metastatic pancreatic carcinoma and correlation with genetic status. Clin Exp Metastasis 23(7–8):357–365

Murchison EP, Schulz-Trieglaff OB, Ning Z, Alexandrov LB, Bauer MJ, Fu B et al (2012) Genome sequencing and analysis of the tasmanian devil and its transmissible cancer. Cell 148(4):780–791

Navin N, Kendall J, Troge J, Andrews P, Rodgers L, McIndoo J et al (2011) Tumour evolution inferred by single-cell sequencing. Nature 472(7341):90–94

Neilsen PM, Noll JE, Suetani RJ, Schulz RB, Al-Ejeh F, Evdokiou A et al (2011) Mutant p53 uses p63 as a molecular chaperone to alter gene expression and induce a pro-invasive secretome. Oncotarget 2(12):1203–1217

Nguyen DX, Massagué J (2007) Genetic determinants of cancer metastasis. Nat Rev Genet 8(5):341–352

Padua D, Massagué J (2009) Roles of TGFbeta in metastasis. Cell Res 19(1):89–102

Pertovaara L, Kaipainen A, Mustonen T, Orpana A, Ferrara N, Saksela O et al (1994) Vascular endothelial growth factor is induced in response to transforming growth factor-beta in fibroblastic and epithelial cells. J Biol Chem 269(9):6271–6274

Polyak K, Weinberg RA (2009) Transitions between epithelial and mesenchymal states: acquisition of malignant and stem cell traits. Nat Rev Cancer 9(4):265–273

Redston MS, Caldas C, Seymour AB, Hruban RH, da Costa L, Yeo CJ et al (1994) p53 mutations in pancreatic carcinoma and evidence of common involvement of homocopolymer tracts in DNA microdeletions. Cancer Res 54(11):3025–3033

Riley T, Sontag E, Chen P, Levine A (2008) Transcriptional control of human p53-regulated genes. Nat Rev Mol Cell Biol 9(5):402–412

Schubbert S, Shannon K, Bollag G (2007) Hyperactive Ras in developmental disorders and cancer. Nat Rev Cancer 7(4):295–308

Schutte M, Hruban RH, Geradts J, Maynard R, Hilgers W, Rabindran SK et al (1997) Abrogation of the Rb/p16 tumor-suppressive pathway in virtually all pancreatic carcinomas. Cancer Res 57(15):3126–3130

Siegel PM, Massagué J (2003) Cytostatic and apoptotic actions of TGF-beta in homeostasis and cancer. Nat Rev Cancer 3(11):807–821

Sohn TA, Yeo CJ, Cameron JL, Koniaris L, Kaushal S, Abrams RA et al (2000) Resected adenocarcinoma of the pancreas-616 patients: results, outcomes, and prognostic indicators. J Gastrointest Surg 4(6):567–579

Stathis A, Moore MJ (2010) Advanced pancreatic carcinoma: current treatment and future challenges. Nat Rev Clin Oncol 7(3):163–172

Su GH, Hilgers W, Shekher MC, Tang DJ, Yeo CJ, Hruban RH et al (1998) Alterations in pancreatic, biliary, and breast carcinomas support MKK4 as a genetically targeted tumor suppressor gene. Cancer Res 58(11):2339–2342

Su GH, Hruban RH, Bansal RK, Bova GS, Tang DJ, Shekher MC et al (1999) Germline and somatic mutations of the STK11/LKB1 Peutz–Jeghers gene in pancreatic and biliary cancers. Am J Pathol 154(6):1835–1840

Subarsky P, Hill RP (2003) The hypoxic tumour microenvironment and metastatic progression. Clin Exp Metastasis 20(3):237–250

Vincent A, Herman J, Schulick R, Hruban RH, Goggins M (2011) Pancreatic cancer. Lancet 378(9791):607–620

Vogelstein B, Lane D, Levine AJ (2000) Surfing the p53 network. Nature 408(6810):307–310

Wilentz RE, Geradts J, Maynard R, Offerhaus GJ, Kang M, Goggins M et al (1998) Inactivation of the p16 (INK4A) tumor-suppressor gene in pancreatic duct lesions: loss of intranuclear expression. Cancer Res 58(20):4740–4744

Winter JM, Cameron JL, Campbell KA, Arnold MA, Chang DC, Coleman J et al (2006) 1423 Pancreaticoduodenectomies for pancreatic cancer: a single-institution experience. J Gastrointest Surg 10(9):1199–1210, discussion 1210–1

Winter JM, Brennan MF, Tang LH, D'Angelica MI, Dematteo RP, Fong Y et al (2012) Survival after resection of pancreatic adenocarcinoma: results from a single institution over three decades. Ann Surg Oncol 19(1):169–175

Yachida S, Iacobuzio-Donahue CA (2009) The pathology and genetics of metastatic pancreatic cancer. Arch Pathol Lab Med 133(3):413–422

Yachida S, Jones S, Bozic I, Antal T, Leary R, Fu B et al (2010) Distant metastasis occurs late during the genetic evolution of pancreatic cancer. Nature 467(7319):1114–1117

Epigenetic Alterations in Pancreatic Cancer

Michael Ayars and Michael Goggins

Abstract Pancreatic cancer remains one of the deadliest malignancies. In addition to genetic alterations a wide variety of epigenetic aberrations have been identified in pancreatic neoplasms some of which are thought to play an important role in neoplastic development and maintenance. Newer technologies are helping to better characterize cancer epigenomes. Efforts are underway to identify epigenetic alterations that would make optimal diagnostic markers and therapeutic targets. In this chapter, we discuss recent findings in the field of pancreatic cancer epigenetics and the implications they hold for future research.

Introduction/Background

Pancreatic cancer has the lowest survival rate of any solid cancer and is the 4th most common cause of cancer death in the USA. In 2012, it is estimated that 43,920 Americans will be diagnosed and 37,930 will die of pancreatic cancer (Siegel et al. 2012). From date of diagnosis, the 1 year survival rate is 50 % and 5 year survival is 6 %. Much of the lethality of pancreatic cancer is owed to its late

M. Ayars
Department of Pathology, The Sol Goldman Pancreatic Cancer Research Center, Johns Hopkins Medical Institutions, 1550 Orleans Street, Baltimore, MD 21231, USA

M. Goggins, M.D. (✉)
Department of Pathology, The Sol Goldman Pancreatic Cancer Research Center, Johns Hopkins Medical Institutions, 1550 Orleans Street, Baltimore, MD 21231, USA

Departments of Pathology, Medicine, and Oncology, The Sol Goldman Pancreatic Cancer Research Center, Johns Hopkins Medical Institutions, 1550 Orleans Street, Baltimore, MD 21231, USA
e-mail: mgoggins@jhmi.edu

D.M. Simeone and A. Maitra (eds.), *Molecular Genetics of Pancreatic Cancer*,
DOI 10.1007/978-1-4614-6549-2_9, © Springer Science+Business Media New York 2013

diagnosis. It is estimated that 85 % of patients only present in advanced stages of the disease at which it is unresectable. Further, the 5-year survival of patients who undergo surgical resection with curative intent is only 23 % (Siegel et al. 2011). Other forms of treatment have little success in halting or slowing the progression of metastatic disease. A key step in curtailing the mortality of pancreatic cancer will be identifying useful diagnostic and prognostic markers to detect it at earlier stages as well as new targets for treatment.

Cancer has long been perceived as a genetic disease, but the past decade has seen a dramatic shift in our understanding of the role played by epigenetics (Baylin and Jones 2011). Epigenetic changes are defined as any heritable changes to gene expression that are not accompanied by changes in DNA sequence (Jones and Baylin 2007). In normal cells, epigenetic mechanisms are employed in development to silence and activate expression of specific genes at specific times. In cancer cells, epigenetic abnormalities contribute to the overexpression of oncogenes and suppression of tumor suppressor genes. They are conserved and frequently observed in both adenocarcinoma and precursor lesions.

Since cancer is a disease of pathways, it is perhaps not surprising that for pancreatic cancers, epigenetic and genetic abnormalities are mostly nonoverlapping with respect to the genes that are targeted. There are many examples of genes that are rarely mutated, but frequently silenced epigenetically and vice versa (Chan et al. 2008; Schuebel et al. 2007). Discoveries such as these emphasize the need for an integrative approach to studying cancer: one that explores both genetic and epigenetic aberrations and how they are coordinated in tumorigenesis Table 1.

To this end, the use of next-generation sequencing techniques has dramatically expanded our knowledge of the extent of epigenetic abnormalities in cancer. High throughput sequencing has made it possible to map genome-wide chromatin states (Mikkelsen et al. 2007; Cui et al. 2009) and explore methylation maps with single-base resolution (Lister et al. 2009). These developments offer powerful tools to dissect some of the complex interplay and complementation between genomic and epigenomic factors. In this chapter, we review recent research into the epigenetic hallmarks of pancreatic cancer and their potential role in advancing diagnosis and treatment. Although investigations of the epigenetic abnormalities of pancreatic neoplasms have focused on the epigenetics in pancreatic ductal adenocarcinoma (PDAC) and its precursors, novel epigenetic abnormalities have also been uncovered in neuroendocrine tumors.

Methylation

DNA methylation is the result of covalent addition of a methyl group to the 5′ carbon of a cytosine in a CpG dinucleotide. Across much of the genome length, the occurrence of CpG dinucleotides is much rarer than would be expected from the GC content (Lander et al. 2001). This is attributed to the mutability of methylated cytosine, which results in a loss of germ line CpGs over time (Lunter and Hein 2004).

Table 1 List of selected genes that are genetically altered in pancreatic cancer

Gene symbol	Gene name	Epigenetic alteration	Chromosome site	Known or predicted function	Methylation in pancreatic cancer cell lines, no. (%)	Methylation in primary or xenografted pancreatic cancer, no. (%)	Source, y
PENK	Preproenkephalin	Hypermethylation	8q23-q24	Neuropeptide precursor	11/11 (100)	43/47 (91)	Ueki et al. (2000) Fukushima et al. (2003)
UCHL1	Ubiquitin carboxyl-terminal esterase L1 (ubiquitin thiolesterase)	Hypermethylation	4p14	Ubiquitin hydroxylase	22/22 (100)	42/42 (100)	Sato et al. (2003a)
MDF-1	MAD (yeast Mitosis Arrest DeFicient) related	Hypermethylation	11q13	Glycogen metabolism	45/47 (96)	Not determined	Omura et al. (2008)
NPTX2	Neuronal pentraxin II	Hypermethylation	7q21.3-q22.1	Neuronal transport	21/22 (95)	20/20 (100)	Sato et al. (2003a)
SPARC/ON	Secreted protein, acidic, cysteine-rich (osteonectin)	Hypermethylation	5q31.3-q32	Cell-cycle progression inhibition, cell–matrix interaction	16/17 (94)	21/24 (88)	Sato et al. (2003b)
RPRM	Reprimo, TP53-dependent G2 arrest mediator candidate	Hypermethylation	2q23.3	P53-induced G2/M cell-cycle arrest	20/22 (91)	16/20 (80)	Saito et al. (2006)
BNIP3	BCL2/adenovirus E1B 19 kDa interacting protein 3	Hypermethylation	10q26.3	Hypoxia-induced cell death	9/10 (90)	8/10 (80)	Okami et al. (2004)
miR9-1	MicroRNA 9-1	Hypermethylation	1q22	miRNA translation control	42/47 (89)	Not determined	Omura et al. (2008)
SERPINB5	Serpin peptidase inhibitor, clade B, member 5 (maspin)	Hypomethylation	18q21.3	Regulation of cell motility and cell death	20/23 (87)	32/34 (94)	Sato et al. (2003a, b)

(continued)

Table 1 (continued)

Gene symbol	Gene name	Epigenetic alteration	Chromosome site	Known or predicted function	Methylation in pancreatic cancer cell lines, no. (%)	Methylation in primary or xenografted pancreatic cancer, no. (%)	Source, y
							Fitzgerald et al. (2003)
							Ohike et al. (2003)
CCND2	Cyclin D2	Hypermethylation	12p13	Cell-cycle control	19/22 (86)	71/109 (65)	Matsubayashi et al. (2003)
ZNF415	Zinc finger protein 415	Hypermethylation	19q13.42		40/47 (86)	Not determined	Omura et al. (2008)
CLDN4	Claudin-4	Hypomethylation	7q11.23	Cell adhesion/invasion	17/20 (85)	33/37 (89)	Sato et al. (2003a, b)
SFN	Stratifin (14-3-3 Ï*f*)	Hypomethylation	1p35	P53-induced G2/M cell-cycle arrest	17/20 (85)	36/37 (97)	Sato et al. (2003a, b)
							Iacobuzio-Donahue et al. (2003)
LCN2	Lipocalin-2	Hypomethylation	9q34	Epithelial differentiation	17/20 (85)	34/37 (92)	Sato et al. (2003a, b)
TFPI2	Tissue factor pathway inhibitor 2	Hypermethylation	7q22	Serine protease inhibitor	14/17 (82)	102/140 (73)	Sato et al. (2003a, b)
CNTNAP2	Contactin-associated protein-like 2	Hypermethylation	7q35-q36	Higher cortical function	39/47 (82)	Not determined	Omura et al. (2008)
CDKN1C/ p57	Cyclin-dependent kinase inhibitor 1C	Hypermethylation	11p15.5	Cyclin-dependent kinase inhibitor	7/9 (78)	Not determined	Sato et al. (2005)
SIP1	Survival of motor neuron protein-interacting protein 1	Hypermethylation	14q13-q21	Assembly of spliceosomal snRNP	11/15 (73)	34/35 (97)	Li et al. (2010a, b)
ELOVL4	Elongation of very-long-chain fatty acids (FEN1/Elo2, SUR4/Elo3, yeast)â€"like 4	Hypermethylation	6q14	Fatty acid synthesis	32/47 (68)	Not determined	Omura et al. (2008)

TFF2	Trefoil factor 2	Hypomethylation	21q22.3	Secretory polypeptide/ epithelial repair	13/20 (65)	31/37 (84)	Sato et al. (2003a, b)
FOXE1	Forkhead box E1 (thyroid transcription factor 2)	Hypermethylation	9q22	Thyroid transcription factor	14/22 (64)	15/20 (75)	Sato et al. (2003a, b)
S100P	S100 calcium-binding protein P	Hypomethylation	4p16	Cell-cycle progression and differentiation	13/23 (57)	30/34 (88)	Sato et al. (2003a, b)
RARB	Retinoic acid receptor, Î^{2}	Hypermethylation	3p24	Cell-growth control	5/9 (56)	4/36 (11)	Ueki et al. (2000)
S100A4	S100 calcium-binding protein A4	Hypomethylation	1q21	Motility, invasion, tubulin polymerization	10/20 (50)	28/37 (76)	Rosty et al. (2002) Sato et al. (2003a, b)
CDKN2A/ p16	Cyclin-dependent kinase inhibitor 2A	Hypermethylation	9P21	Cyclin-dependent kinase inhibitor	3/9 (33)	5/36 (14)	Schutte et al. (1997) Ueki et al. (2000)
MSLN	Mesothelin	Hypomethylation	16p13.3	Cell surface antigen/cell adhesion	8/20 (40)	34/37 (29)	Sato et al. (2003a, b)
SOCS1	Suppressor of cytokine signaling 1	Hypermethylation	16p13.13	Inhibitor of JAK/STAT pathway	6/19 (32)	13/60 (22)	Fukushima et al. (2003)
PSCA	Prostate stem cell antigen	Hypomethylation	8q24.2	Cell surface antigen/cell differentiation	6/20 (30)	20/37 (54)	Sato et al. (2003a, b)
CADM1/ TSLC1	Cell adhesion molecule 1	Hypermethylation	11q23.2	Cell–cell, cell–matrix interaction	4/17 (24)	25/91 (27)	Jansen et al. (2002)
MLH1	MutL homolog 1	Hypermethylation		DNA repair.	2/36 (6)	0/9 (0)	Ueki et al. (2000)
NDRG1	N-myc downstream regulated	Hypermethylation	8q24.3	Hormone responses, cell growth, differentiation.	0/6[a]	Not determined	Angst et al. (2010)
CDH1	Epithelial cadherin	Hypermethylation	16q22.1	Cell adhesion/invasion	1/36 (3)	2/9 (22)	Ueki et al. (2000)

Modified from Hong et al. (2011)

Abbreviation: MAPK, mitogen-activated protein kinase

[a]NDRG1 was upregulated by treatment with 5-AzA and TSA, but none of the cell lines examined by bisulfite sequencing showed promoter hypermethylation

Notably, remaining CpG sites are much denser in gene-rich regions of the genome. These high densities cluster into CpG islands, regions 0.5–4 kb in length that are heavily saturated in CpGs and found in the promoters of ~70 % of mammalian genes (Bestor et al. 1988; Yen et al. 1992). In stark contrast to the ~80 % methylation observed in genomic CpG sites, promoter CpG islands are almost uniformly unmethylated (Takai and Jones 2002), though there are exceptions in tissue-specific patterns and some developmental processes (Angst et al. 2010) such as X inactivation and genomic imprinting.

Hypermethylation of promoter CpG sites is commonly associated with gene silencing. This is achieved through alterations in DNA conformation that render local sequence inaccessible to the transcription complex. Methylation also recruits methylcytosine-binding proteins (MBDs) that can recruit histone deacetylases (HDACs) to remodel local chromatin, imposing gene silencing by other means (Jones et al. 1998).

The enzymes that facilitate the process of DNA methylation are the DNA methyltransferases (DNMTs), which in mammals have three prominent members: DNMT1, DNMT3a, and DNMT3b (Bestor et al. 1988; Yen et al. 1992; Okano et al. 1998). In normal cells, DNMT1, the most abundant methyltransferase, is responsible for methylating the hemimethylated daughter strand during replication, preserving the parental methylation pattern (Robert et al. 2003). DNMT3a and DNMT3b on the other hand are characterized by their activity on de novo strands in early development (Dodge et al. 2002; Okano et al. 1999). The importance of DNA methylation in normal development has been highlighted by transgenic mice. Homozygous deletions of DNMT1 and DNMT3b are embryonic lethal, while mice with homozygous deletion of DNMT3a become runted and die at 4 weeks of age.

Overexpression of DNMTs is thought to be a key causative factor in the aberrant methylation of cancer cells. DNMT1 is overexpressed in ~80 % of pancreatic cancer cells (Li et al. 2010a) and its degree of overexpression has been correlated to disease progression (Wang et al. 2009a). The cause of DNMT1 overexpression is not well understood, but has previously been attributed to aberrant signaling via mutant *KRAS*. *KRAS* mutations are observed in approximately 95 % of PDACs (Brune et al. 2006; Jones et al. 2008) and these mutations are usually the earliest mutations identifiable in precursor lesions (DiGiuseppe et al. 1994; Kanda et al. 2012). *KRAS* wild-type pancreatic cancers also have DNMT1 overexpression (Li et al. 2010a). More recently, GLI1, a key transcription factor of the hedgehog signaling pathway, has been shown to bind to the promoter region of *DNMT1* and to induce expression of DNMT1 (He et al. 2011).

DNMT upregulation is likely to promote aberrant methylation marks. Once made, methylation marks are maintained with little turnover, so it is unlikely that a large increase in DNMT expression is a prerequisite for aberrant hypermethylation. Dysregulation of DNMT activity in cancer cells also extends beyond expression changes of the enzyme. O'Hagan et al. recently demonstrated that induced oxidative stress results in the relocalization of DNMT1, DNMT3B, and members of the polycomb repressive complex 4 from GC-poor areas to GC-rich areas of damaged chromatin (O'Hagan et al. 2011). They proposed that this mechanism might explain why cancer cells feature both global hypomethylation and hypermethylation of CpG islands. Continuously inducing this relocalization through constant oxidative damage would be expected to result in just such a genome-wide pattern.

In cancer cells, the consequences of aberrant methylation patterns are varied and far-reaching. Given that promoter methylation is often associated with gene silencing, it is unsurprising that cancer cells have long been characterized with global hypomethylation, implying a systemic loss of regulation. Vitamin B12 and folate deficiency are associated with decreased levels of the methyl-group donor *S*-adenosylmethionine that can result in widespread hypomethylation, but methyl-group deficiency is not the primary mechanism of global hypomethylation in cancers. Additionally, demethylation causes a reduction in thymidylate synthesis from uracil which leads to genomic instability by triggering double-stranded breaks and translocations of mobile DNA elements (Chen et al. 1998). In mouse models methyl group deficiency has been directly linked to tumorigenesis (Gaudet et al. 2003).

The cancer methylome resembles therefore a landscape of hypomethylation with hypermethylated CpG islands often located in the promoter regions of tumor suppressor genes. In an intriguing reflection of the driver/passenger model for genetic mutations, Carvalho et al. found evidence that the hypermethylation of a small proportion of CpG islands is critical to the viability of cancer cells (De Carvalho et al. 2012). Depletion of DNMTs by genetic or chemical methods resulted in demethylation of most CpG islands, with a small subset consistently preserved by functional selection. These results emphasize the importance of aberrant methylation to not just malignant transformation, but cancer cell survival.The complexity of known methylation mechanisms has been expanded by the characterization of the ten-eleven translocation (TET) enzyme family. TET enzymes facilitate three sequential oxidation reactions that begin with converting 5-methylcytosine to 5-hydroxymethylcytosine and ultimately result in DNA demethylation. Through this process, they act as potent tumor suppressors and are downregulated in a variety of cancers (Jones and Baylin 2007; Yang and Liu 2013). Further, 5hmC levels are consistently and dramatically reduced in these cancers, highlighting it as a potentially very valuable biomarker for cancer development.

In pancreatic cancer, an extensive number of genes are regulated by aberrant methylation patterns. In one study, Omura et al. used a methylated CpG island amplification (MCA) array to identify 606 differentially methylated genes in comparing the Panc-1 pancreatic cancer cell line with the HPDE normal cell line (Omura et al. 2008). In a similar study, Vincent et al. identified CpG islands frequently, differentially methylated in pancreatic cancer samples compared to normal tissue using a CpG island microarray (Vincent et al. 2011). These loci were correlated with previous expression data to identify genes in which methylation status is predictive of epigenetic silencing or induction.

Investigating genes previously shown to be overexpressed in pancreatic cancer, Sato et al. found that many of their promoters were hypomethylated in comparison to normal tissue (Rosty et al. 2003). These genes included *TFF2*, *CLDN4*, *LCN2*, *MSLN* and *PSCA*. A later study identified *SERPINB5* and *S100P* as two additional genes subject to overexpression through hypomethylation by oligonucleotide microarray

(Sato et al. 2004). S100A4 is a protein connected to metastasis and poor differentiation (Rosty et al. 2002). S100A4 has three CpG sites in its first intron; hypomethylation of these sites occurs in the majority of pancreatic cancers and is significantly associated with overexpression.

Conversely, many tumor suppressor genes show silencing through aberrant promoter CpG island hypermethylation in pancreatic cancers. The first of these to be discovered was CDKN2A/p16 (Schutte et al. 1997). CDKNA2A plays an important role in cell cycle regulation by inhibiting the cyclin-dependent kinase 4-cyclin D2 complex and is silenced in over 95 % of pancreatic cancers. Genetic inactivation of CDKN2A/p16 has previously been characterized in pancreatic and other cancer types (Schutte et al. 1997), but it is now understood that hypermethylation of *CDKN2A* for much of the inactivation of CDKN2A in pancreatic cancers lacking genetic inactivation (Rosty et al. 2003). Other tumor suppressor genes that undergo genetic inactivation, such as *SMAD4/DPC4*, *TP53*, *and STK/LKB1*, have not been shown to be subject to epigenetic silencing by DNA methylation.

SPARC, which encodes a matricellular glycoprotein involved in tissue remodeling, cell matrix interactions, differentiation, migration, and angiogenesis, is silenced in pancreatic cancer cells by aberrant methylation (Rosty et al. 2003). Concurrently, SPARC expression in fibroblasts adjacent to pancreatic cancer cells is increased (Rosty et al. 2003). SPARC is frequently methylated and silenced in early pancreatic tumors (Gao et al. 2010), while exposure to conditioned media containing secreted Sparc inhibited growth of pancreatic cancer cells (Sato et al. 2003a).

BNIP3 is a proapoptotic gene that is commonly downregulated in several cancers, including pancreatic cancer (Giatromanolaki et al. 2004; Abe et al. 2005; Okami et al. 2004). The BNIP3 promoter coincides with a CpG island that is methylated in most pancreatic cancer cell lines (Abe et al. 2005). In addition to its pro-cell death activity, loss of BNIP3 expression has been connected to gemcitabine resistance (Akada et al. 2005).

Past attempts to use the demethylating agents 5-Aza-Dc and Decitabine therapeutically met with disappointing results for solid tumors. Used at high doses for their ability to induce DNA damage and apoptosis, they were frequently met with toxicity and ambiguous impact (Abele et al. 1987; Issa and Kantarjian 2009). Treating leukemias with the drugs at lower doses has offered much more appealing outcomes, but the mechanism remains unclear. Applying these lower doses to solid tumors, Tsai et al. has demonstrated genomewide promoter DNA demethylation, reexpression of critical tumor suppressor genes, and inhibition of tumor cell growth without immediate toxicity (Tsai et al. 2012).

Perhaps most promisingly of all, the effects appeared specifically targeted to stem-like and tumorigenic subpopulations of cancer cells that are notoriously resistant to existing therapies. Some caution is necessary in assessing the impact of epigenetic therapies, as the effects can be unpredictable. In another study, treatment with 5-Aza-dC actually increased the invasive potential of four out of five pancreatic cancer cell lines through the induction of silenced matrix metalloproteinases (Sato et al. 2003b). The complexity of epigenetic alterations in pancreatic and other cancers and the nonspecific effects of many epigenetic agents indicate the need for additional investigations before the routine use of these agents in the clinical setting.

MLH1, a gene tied to the microsatellite instability of hereditary nonpolyposis colon cancer (HNPCC) and pancreatic medullary carcinomas, undergoes DNA methylation (Ueki et al. 2000; Nakata et al. 2002; Yamamoto et al. 2001). Other cancer-related genes that are silenced in pancreatic cancer include *NDRG1* (Angst et al. 2010), *CDH1* (Ueki et al. 2000), *CCND2* (Matsubayashi et al. 2003), *TFPI2* (Sato et al. 2005), *SOCS-1* (Fukushima et al. 2003), and *TSLC1/IGSF4* (Jansen et al. 2002). Most hypermethylated genes are affected on an individual basis (Easwaran et al. 2010); however, zones of regional, continuous, long-range epigenetic silencing have also been described in several cancer types (Clark 2007).

Pancreatic cancer is characterized by its often-asymptomatic progression from neoplastic precursors to invasive cancer and eventually metastatic disease. Due to the high resistance of advanced disease to conventional therapy, there is considerable interest in identifying biomarkers that can improve detection of precursor lesions. Methylation profiles of these lesions may offer important diagnostic and prognostic tools for clinical screening.

In one study, methylation-specific PCR (MSP) for eight genes aberrantly hypermethylated in PDAC was used to assay the methylation status of the most common precursor lesion, pancreatic intraductal neoplasia (PanINs). Even among the earliest grade lesions (PanIN-1A), aberrant methylation patterns were commonly detected (Sato et al. 2008). Some of the genes investigated showed an increase in methylation frequency by neoplastic grade. A study by Hong et al. used methylation CpG island amplification and Agilent CpG island microarray (MCAM) to generate a methylation profile for intrapapillary mucinous neoplasms (IPMNs). Over a thousand genes were hypermethylated in one or more IPMNs and as with PanINs, methylation increased with neoplastic grade (Hong et al. 2012). Some individual genes, including *BNIP3* and *PTCHD2*, were also found to be aberrantly hypermethylated more frequently or only in high-grade compared to low-grade IPMNs. These results indicate that epigenetic dysregulation is present in the earliest precursors to pancreatic cancer and continues during pancreatic tumorigenesis.

Due to the complex landscape of repressors, insulators, and activators present in the genome, aberrant methylation can have "opposite" effects on relevant genes. Hypermethylation of repressor or insulator sites can actually amplify proximal gene expression in some cases. Imprinting is a normal developmental process in which alleles inherited from one parent are repressed epigenetically. A classic example of imprinting is insulin-like growth factor 2 (IGF2) and H19, which are oppositely imprinted and expressed in a monoallelic fashion from the paternal and maternal chromosomes, respectively. IGF2 is a potent growth-promoting hormone that causes Wilms' tumor development (Md Zin et al. 2011). One study found that hypermethylation of the IGF2 densely methylated region 2 (DMR2) in insulinomas caused a loss of imprinting and overexpression of IGF2 (Dejeux et al. 2009).

For many genes, regulatory methylation is even more complex. Loss of E-cadherin (*CDH1*), a component of adherens junctions between cells, is highly predictive of an undifferentiated and more aggressive cell type (Winter et al. 2008). *CDH1* is rarely inactivated by intragenic mutation or methylation, but it is subject to silencing by the *SIP1* repressor, which is itself suppressed by miR-200. In most pancreatic cancer cells, hypomethylation and overexpression of miR-200 or promoter hypermethylation of *SIP1* cause loss of the repressor and permit normal expression of *E-cadherin* (Li et al. 2010b). Consequently, total loss of *E-cadherin* and diffuse dedifferentiation within a tumor are rarely observed. It has recently been shown, however, that focal loss of *E-cadherin* and pockets of dedifferentiation within a tumor are much more common and prognostically significant (Hong et al. 2011). *E-cadherin* loss is mediated by hypomethylation and overexpression of miR-200 or by histone deacetylation in the CDH1 promoter (Aghdassi et al. 2012). Interestingly, heterogeneity and instability of epigenetic loss of *E-cadherin* has been previously described in various cancer cell lines (Graff et al. 2000) and regulation of *E-cadherin* transcriptional repressors has been shown to have environmental dependence (Klymkowsky and Savagner 2009). Taken together, these findings suggest that local conditions within the tumor may promote epigenetic aberrations in focal subsets of cells.

Telomerase reverse transcriptase (hTERT) extends the length of telomere ends and is often dysregulated in cancer. Researchers have long suspected that *hTERT* is regulated by DNA methylation due to CpG sites in its promoter region, but study has been complicated by contradictory findings on whether hypermethylation induces or silences transcription (Daniel et al. 2012). Renaud et al. showed that hypermethylation of the first exon of *hTERT* prevents binding of the CTCF repressor, which would otherwise silence transcription (Renaud et al. 2007). Hypomethylation of the hTERT promoter site at specific sites allows the CTCF repressor to bind (Zinn et al. 2007); however, total hypermethylation of the promoter region prevents formation of the transcription complex and silences expression (Dessain et al. 2000; Choi et al. 2007; Devereux et al. 1999). *hTERT* therefore requires a specific pattern of hypo- and hypermethylation for transcription to occur. In a study by Alpani et al. researchers found that in pancreatic cancer cells, the hTERT promoter was methylated, resulting in expression of hTERT (Kumari et al. 2009). In normal controls, the promoter was unmethylated and the gene was silenced.

Recent advancements in high resolution mapping techniques for methylation have elucidated the conservation of intragenic methylation patterns. In contrast to the canonical association between promoter methylation and silencing, intragenic methylation appears to promote transcription efficiency. Early studies have consistently observed that the combination of unmethylated promoter regions with methylated gene bodies is conserved in highly expressed genes on a genome-wide scale (Hellman and Chess 2007; Ball et al. 2009; Rauch et al. 2009).

The impact of intragenic methylation patterns is not yet well understood. Previously understood links between DNA methylation and histone modification would predict that densely methylated regions would incur repressive histone marks, but this does not account for the observed expression patterns. Additionally, Hahn et al. recently showed that despite a correlation between the presence of the H3K36me6 mark and intragenic methylation, there was no direct dependence between the two marks (Hahn et al. 2011). One alternative possibility is that they act to suppress intragenic miRNAs that may target the surrounding gene for silencing.

Another postulated role is that intragenic methylation patterns may control alternative splicing and the use of alternative transcription start sites. In a novel approach exploiting the affinity of MBD2 for methylated DNA, Yegnasubramanian et al. enriched and analyzed genomic fragments by tiling microarrays and compared IGM in cancer and normal samples. Both cancer and normal cells had a high enrichment of fragments localized to intron–exon junctions, and these fragments were hypermethylated with greater frequency in cancer cells (Yegnasubramanian et al. 2011). In another study, Maunakea et al. found a similar pattern and observed that in the SHANK3 gene locus, methylation status correlated with intragenic promoter activity (Maunakea et al. 2010). In this way, intragenic methylation may offer an alternate mechanism for cancer to disrupt tumor suppressor genes and induce dysregulation of oncogenes. These exceptions highlight the importance of caution in interpreting methylation patterns; CpG sites can have very different regulatory roles depending on location.

DNA methylation alterations that are specific for pancreatic adenocarcinoma can be used to help identify pancreatic cancer in specimens where cytology is nondiagnostic. In one study Parsi et al. used quantitative methylation-specific PCR (QMSP) to evaluate biliary and pancreatic strictures. Endoscopically obtained brushings of these strictures were as accurate as cytology at differentiating benign from malignant strictures (Parsi et al. 2008). In another study, methylated DNA markers of pancreatic cancer quantified by QMSP detected in pancreatic juice obtained during ERCP were accurately able to identify individuals with pancreatic cancer (Matsubayashi et al. 2006).

Histone Modification

In the nucleus, DNA is wound around histone proteins into nucleosome structures. Repetitive units of nucleosomes in turn form chromatin. Histones are not simple structural elements: their behavior plays an important role in gene expression by dynamically shifting the chromatin between condensed, transcriptionally inactive states (euchromatin), and open, transcriptionally active states (heterochromatin). This behavior is largely controlled by the enzymatic imposition of post-translational modifications or "marks" to the histone cores and tails including acetylation, phosphorylation, methylation, SUMOylation, and biotinylation. Some marks such as lysine tail acetylation induce transcriptional activation by altering the electrostatic charge of the histone protein (Esteller 2007). Histone methylation is traditionally associated with the recruitment of regulatory proteins and therefore has particularly varied effects based on the location and extent of methylation. H3K4 (methylation of histone 3 lysine 4), H3K36, and H3K79 are activating marks while H3K9me2/me3 (di- or trimethylation of histone 3 lysine 9), H4K20me3, and H3K27 me2/me3 are inactivating marks (Kouzarides 2007; Lohse et al. 2011). The combination of marks that dictate the genomic transcriptional landscape have been proposed to form a complex "histone code" (Jenuwein and Allis 2001; Strahl and Allis 2000; Lachner and Jenuwein 2002), a key epigenetic mechanism in normal development as well as tumorigenesis.

Expression state of genes is strongly modulated by local nucleosome architecture, which alters in response to combinations of histone marks (Mikkelsen et al. 2007; Kouzarides 2007). Transcriptionally active gene promoters have nucleosome-depleted regions (NDRs) thought to be produced by the migration of flanking nucleosomes that have a high density of acetylated lysine residues, H3K4me3, and replacement of the H2A residue with the H2A.Z variant (Baylin and Jones 2011; Kelly et al. 2010). Acetylation of histone H4-K16 specifically inhibits the formation of higher-order chromatin structures (Shogren-Knaak et al. 2006).

Mutations in members of the SWI/SNF chromatin remodeling complex have also been identified in pancreatic cancers. Exome-sequencing of PDACs has identified inactivating mutations in the ARID1A tumor suppressor gene (Biankin et al. 2012). ATRX is a protein critical to heterochromatin formation, while DAXX is associated with targeted silencing of genes by hypermethylation. In pancreatic neuroendocrine tumors, mutations in ATRX and DAXX are common and result in alternative lengthening of telomeres (de Wilde et al. 2012).

Expression states are capable of spreading to proximal regions through the action of regulatory proteins recruited by histone marks. The PcG protein heterochromatin-associated protein 1 (HP1) is recruited to methylated H3K9. Once bound, it recruits histone methyltransferases to methylate adjacent H3K9 tails. This creates binding sites for additional copies of HP1, causing the repressive marks to spread and silence nearby genes. Loss of HP1 has been associated with cancer progression (Dialynas et al. 2008).

Until recently, it was thought that histone lysine methylation, like DNA methylation, was an irreversible process used in the stable repression of genes. The discovery and characterization of histone demethylases (Trojer and Reinberg 2006) has overturned this perspective. Early studies suggest that as with other epigenetic enzymes, histone demethylases can exercise both a significant and dualistic role in oncogenesis. Lysine-specific demethylase 1 (LSD1), which reverses H3K4 and H3K9, is overexpressed in a variety of cancer types (Schildhaus et al. 2011; Kauffman et al. 2011). In one study, overexpression of LSD1 was found in breast cancer tissue samples and its pharmacological inhibition reduced cancer cell growth (Lim et al. 2010). Another study found that LSD1 suppressed the metastatic potential and invasion of breast cancer cells in vivo (Wang et al. 2009b). A dichotomous role for such enzymes is perhaps unsurprising given the vast range of targets it regulates and further study will be necessary to effectively incorporate them into meaningful therapy.

Although histone modifications are mediated by different enzyme families, there is a high degree of crosstalk between the histone modification and DNA methylation pathways. Use of chromatin immunoprecipitation has confirmed that methylated DNA is commonly local to deacetylated histones and compact chromatin while unmethylated DNA is common to acetylated histones and open chromatin (Eden et al. 1998). Methylation of H3K4 prevents the binding of DNMT3L, responsible for recruiting DNMT3A and DNMT3B to H3 during developmental DNA methylation (Ooi et al. 2007). In turn, local DNA methylation causes the deacetylation of histone H4 and methylation of H3K9 (Hashimshony et al. 2003). There are also examples of direct interaction between histone modification enzymes and DNA methylation enzymes. The MBDs MECP2 and MBD2 have been shown to recruit HDACs to methylated regions (Jones et al. 1998; Nan et al. 1998). EZH2, a methyltransferase downstream of *Ras*, is overexpressed in pancreatic cancer and downregulates the tumor suppressor genes *E-cadherin* and *RUNX3* through histone H3K27 trimethylation (Fujii et al. 2008). EZH2 also recruits DNMT1, DNMT3A, and DNMT3B to target genes (Vire et al. 2006). These associations provide clues for how aberrant activity in one regulatory arm can have snowballing downstream effects.

Aberrant activity of the enzymes responsible for maintaining histone marks can have sweeping effects on genome-wide expression, with important implications in tumorigenesis. *KRAS2* is the most frequently mutated oncogene in pancreatic cancer (>95 %) and its mutation is one of the earliest events in tumorigenesis (Jones et al. 2008; Kanda et al. 2012). Mutations in KRAS have been implicated as a cause of alterations in histone marks.

Research on chromatin modifications in pancreatic cancer has largely focused on the acetylation state of histone residues maintained by the opposing activities of histone acetyltransferases (HATs) and HDACs. Histone acetylation by HATs neutralizes the positive charge of the histone tail, reducing its binding affinity for DNA and promoting accessibility to transcriptional machinery, a state reversed by HDACs (Yang and Seto 2007). Loss of HAT activity and aberrant increases in HDAC activity have been tied to tumorigenesis in a variety of cancers (Peng and Seto 2011; Ropero and Esteller 2007) presenting appealing targets for therapy. Due to the importance of deacetylation in silencing tumor suppressor genes, HDAC inhibitors have received a lot of attention as potential therapeutic agents. Treatment of pancreatic cancer cell lines has yielded a variety of promising antitumor effects including drastic reductions in cell proliferation, upregulation of p21, and apoptosis (Kumagai et al. 2007; Arnold et al. 2007; Garcia-Morales et al. 2005; Ryu et al. 2006).

Unfortunately, these in vitro results have not been observed in patients. It may be in vitro studies do not take into account nonspecific toxicity that can also occur in normal cells. Although HDAC inhibitors have proven effective in treating hematological malignancies (Byrd et al. 2005; Ellis et al. 2008; Garcia-Manero et al. 2008), success in solid cancers, including pancreatic, has not been observed (Blumenschein et al. 2008). As with DNA methylation inhibitors, effects may vary. In some cases, histone deacetylation may actually promote tumor progression. As a result of findings like these, epigenetic treatment strategies focus on evaluating the combination of HDAC inhibitors with other agents (Garcia-Manero et al. 2008; Pili et al. 2012).

PCG Proteins

One of the key protein families involved in histone modification as a normal or neoplastic process is the polycomb-group (PcG) proteins. In mammals, these proteins are divided into the two functional complexes they form, PRC1 and PRC2, which each play a role in silencing genes. PRC2 proteins catalyze the trimethylation of histone 3 lysine 27 (H3K27me3), an initiating mark in repressive chromatin remodeling. This mark is also thought to recruit PRC1, which in turn monoubiquitinates H2A (Wang et al. 2004), imposing more constitutive silencing. More recently, studies have found that PRC1 can also act independently of PRC2 (Schoeftner et al. 2006; Vincenz and Kerppola 2008). PcG proteins have also been shown to recruit HDACs (Tonini et al. 2004) and "premark" genes for de novo methylation by DNMTs (Vire et al. 2006). The capacity of PcG proteins for silencing both specific genes and large regions of the genome through chromatin remodeling has plain implications in carcinogenesis.

To date, only a few of the PcG proteins have been investigated in pancreatic cancer. In PRC1, BMI1 is a zinc finger protein that interacts with *Myc* to repress *CDKN2A* and dysregulate the cell cycle. In pancreatic cell lines and resected tumors, BMI1 is upregulated and its overexpression correlates with metastases (Song et al. 2010). Additionally, stable RNAi suppression of BMI1 in pancreatic cancer cell lines reduced proliferation, delayed the G1/S transition, and increased sensitivity to apoptotic triggers.

CBX7 is a chromobox family protein that targets PRC1 to specific histone residues and gene promoters. It has a tumor suppressive role in several cancers attributed to inhibition of HDAC activity. CBX7 has been shown to positively regulate *E-cadherin* by preventing HDAC2 inhibition of the *E-cadherin* promoter (Federico et al. 2009). This correlates with the finding in PDAC that CBX7 is depleted in poorly differentiated tumors with loss of E-cadherin expression (Karamitopoulou et al. 2010).

EZH2 is a PRC2 protein responsible for imposing the initiating repressive H3K27 methylation mark on chromatin. This mark has been associated with silencing of a number of tumor suppressor genes including hMLH1, ARHI, and RASSF1A in ovarian cancer (Abbosh et al. 2006). In pancreatic cancer, aberrant EZH2 activity has been linked to loss of p27 (Ougolkov et al. 2008), dysregulating the cell cycle and inducing proliferation. Depletion of EZH2 in pancreatic cancer cells caused reexpression of p27 and inhibited proliferation, but not survival (Ougolkov et al. 2008).

In a broader context, dysregulation of PcG activity may result in a reversal of differentiation milestones that is advantageous to cancer cells. Studies investigating the PcG-associated mark H3K27me3 in several cancer types have revealed silenced genes in which this repressive mark overlaps with activating H3K4me3 (Ohm et al. 2007; Ms et al. 2008). Regions such as these are termed "bivalent domains" and are characteristic of embryonic stem cells. During differentiation, most bivalent domains revert to a univalent state in which one of the two marks is preserved, suggesting that bivalent domains represent a priming state in which a regulatory fate for individual promoters is decided by tissue type (Ku et al. 2008; Zhao et al. 2007). Their existence in cancer cells highlights PcG protein dysregulation as a mechanism by which they may assume a more plastic, stem-cell like phenotype. PcG proteins have only recently become an area of intense study in pancreatic cancer, and already, important mechanisms in the progression of tumors have been elucidated.

miRNAs

MicroRNAs (miRNAs) are small, noncoding RNAs 18–24 nucleotides in length that mediate gene silencing at the translation level through the binding and sequestration or degradation of target mRNA. In the past decade, many miRNAs have been catalogued with broad roles in cellular differentiation, proliferation, and apoptosis (Nakamura 2005). ~1,200 miRNAs have been characterized to date (Kozomara and Griffiths-Jones 2011). Alterations in miRNAs have previously been implicated in the progression of a number of different cancers (Grady et al. 2008; Sassen et al. 2008) including pancreatic cancer (Yu et al. 2012a; Ryu et al. 2011). The timing of alterations during pancreatic PanIN progression has also been described (Yu et al. 2012b).

Although the biogenesis of miRNAs is well understood, the regulation of their expression remains unclear. Intragenic miRNAs are canonically under the control of their overlapping gene's promoter, although there is evidence for exceptions to this rule (Sato et al. 2011; Toyota et al. 2008). Progress has been made in identifying the more elusive promoters for intergenic genes, but many of them have yet to be experimentally confirmed (Chien et al. 2011). Despite these limitations, it is apparent that many miRNAs are regulated by the same epigenetic mechanisms as coding transcripts. Saito et al. demonstrated that treatment of bladder cancer cells with demethylating agents reversed suppression of miR-127, causing the translational inhibition of oncogenic BCL6 (Saito et al. 2006). Other studies that followed expanded the list of epigenetically regulated miRNAs, prompting the use of high-throughput sequencing.

In another study, treatment of two pancreatic cancer cell lines with 5-aza-dC and trichostatin A induced upregulation of 14 different miRNAs (Lee et al. 2009). The five of these that were common to both cell lines were miR-29a, miR-29b, miR-103, miR-107, and miR-320. Methylation-specific PCR confirmed treatment-induced loss of methylation in the 5′ promoter region for miR-107. Retrovirally enforced expression of miR-107 in the same cell lines suppressed cyclin-dependent kinase 6, a putative target of the miRNA, and negatively impacted cell growth. In a similar study, 5-Aza-dc and HDAC inhibitor SAHA were used to treat pancreatic cancer cell lines and pancreatic cancer stem cells (Nalls et al. 2011). The treatment restored expression of miR-34a, a transcriptional target of p53 that putatively targets bcl-2, CDK6, and SIRT1.

Zhang et al. used an miRNA array to compare miRNA expression between pancreatic cancers and adjacent normal tissues (Zhang et al. 2011). miR-132, a miRNA previously associated with pancreatic carcinogenesis, was found to be frequently downregulated in tumor samples. This perturbation was attributed to hypermethylation of the miR-132 promoter preventing the binding of transcription factor Sp1. Further, transfection of miR-132 mimics into cell lines where it was silenced inhibited proliferation, while further depletion of miR-132 had the opposite effect.

Though it is now clear that many miRNAs are regulated by epigenetic mechanisms, it is also apparent that entire epigenetic mechanisms are regulated in turn by miRNAs. In one study, pancreatic cancer cell lines were treated with diflourinated-curcumin (CDF), a synthetic derivative of curcumin (Bao et al. 2012). Reexpression of several suppressed miRNAs was observed including the *let-7* family, miR-26a, miR-101, miR-200b, and miR-200c. Reexpression of *miR-101* resulted in the downregulation of EZH2 and EpCAM, a cell surface adhesion marker tied to invasion. Interestingly, loss of EZH2 causes an upregulation of *let-7*, *miR-200*, and *miR-101* itself; a negative feedback loop that might be exploited therapeutically. Three members of the miR-200 family have been shown to reduce expression of PcG protein BMI1 and upregulated, can reverse EMT in pancreatic cancer cells (Olson et al. 2009). Downregulation of the miR-200 family is a hallmark of metastatic and met-like primary tumors (Olson et al. 2009).

Investigating the expression and effects of individual miRNAs continues to be complicated by the facts that many of them have multiple targets and many miRNAs can target the same gene. A recent review highlights the importance of systems biology approaches to investigating miRNAs (Azmi et al. 2011).

Conclusion and Future Directions

Despite decades of research, pancreatic cancer still carries a devastating mortality rate with little chance of long-term survival. Due to the late presentation of the vast majority of patients, surgical resection is rarely viable and other clinical options remain lackluster in effect. New approaches are needed to better detect and combat pancreatic cancer.

Epigenetics is a rapidly expanding field that with every year is offering new insights into normal and aberrant modulations of gene expression. High throughput sequencing and high resolution mapping techniques have more deeply elucidated the mechanisms of DNA methylation and histone modification than ever before. It is also becoming steadily clear that these regulatory arms are deeply intertwined by crosstalk with each other and with the activities of PcG proteins and miRNAs. The correlation of DNA and histone marks with neoplastic tissue and tumor progression offers an attractive source of biomarkers for diagnosis and patient prognosis. Similarly, the intrinsic reversibility of methylation patterns and histone modifications make them an appealing target for new therapeutic agents.

The recent preliminary publication of data from the Encyclopedia of DNA Elements (ENCODE) has drawn into question many long-standing assumptions about the genetic and epigenetic landscapes. One particularly relevant discovery to epigenetics is that DNA methylation may often occur specifically in regions that are not occupied by transcription factors, suggesting a whole new layer of complexity to this regulatory mechanism (Thurman et al. 2012). It is not yet clear exactly how these discoveries will affect the field of cancer research specifically, but they are expected to have fundamental consequences for future study.

Most of all, it is becoming clear that epigenetic mechanisms of regulation are interdependent with and complementary to genetic ones. A full understanding of the genomic dysregulation necessary to pancreatic and general tumorigenesis will only be possible through an integrative investigation of both fields. Such an understanding will be key to effectively challenging pancreatic cancer in years to come.

Acknowledgements This work was supported by NIH grants (CA62924, R01CA120432, and RC2CA148376), and the Michael Rolfe Foundation.

References

Abbosh PH et al (2006) Dominant-negative histone H3 lysine 27 mutant derepresses silenced tumor suppressor genes and reverses the drug-resistant phenotype in cancer cells. Cancer Res 66(11):5582–5591

Abe T et al (2005) Upregulation of BNIP3 by 5-aza-2′-deoxycytidine sensitizes pancreatic cancer cells to hypoxia-mediated cell death. J Gastroenterol 40(5):504–510

Abele R et al (1987) The EORTC Early Clinical Trials Cooperative Group experience with 5-aza-2'-deoxycytidine (NSC 127716) in patients with colo-rectal, head and neck, renal carcinomas and malignant melanomas. Eur J Cancer Clin Oncol 23(12):1921–1924

Aghdassi A et al (2012) Recruitment of histone deacetylases HDAC1 and HDAC2 by the transcriptional repressor ZEB1 downregulates E-cadherin expression in pancreatic cancer. Gut 61(3):439–448

Akada M et al (2005) Intrinsic chemoresistance to gemcitabine is associated with decreased expression of BNIP3 in pancreatic cancer. Clin Cancer Res 11(8):3094–3101

Angst E et al (2010) Epigenetic regulation affects N-myc downstream-regulated gene 1 expression indirectly in pancreatic cancer cells. Pancreas 39(5):675–679

Arnold NB et al (2007) The histone deacetylase inhibitor suberoylanilide hydroxamic acid induces growth inhibition and enhances gemcitabine-induced cell death in pancreatic cancer. Clin Cancer Res 13(1):18–26

Azmi AS et al (2011) Aberrant epigenetic grooming of miRNAs in pancreatic cancer: a systems biology perspective. Epigenomics 3(6):747–759

Ball MP et al (2009) Targeted and genome-scale strategies reveal gene-body methylation signatures in human cells. Nat Biotechnol 27(4):361–368

Bao B et al (2012) Curcumin analogue CDF inhibits pancreatic tumor growth by switching on suppressor microRNAs and attenuating EZH2 expression. Cancer Res 72(1):335–345

Baylin SB, Jones PA (2011) A decade of exploring the cancer epigenome—biological and translational implications. Nat Rev Cancer 11(10):726–734

Bestor T et al (1988) Cloning and sequencing of a cDNA encoding DNA methyltransferase of mouse cells. The carboxyl-terminal domain of the mammalian enzymes is related to bacterial restriction methyltransferases. J Mol Biol 203(4):971–983

Biankin AV et al (2012) Pancreatic cancer genomes reveal aberrations in axon guidance pathway genes. Nature 491(7424):399–405

Blumenschein GR Jr et al (2008) Phase II trial of the histone deacetylase inhibitor vorinostat (Zolinza, suberoylanilide hydroxamic acid, SAHA) in patients with recurrent and/or metastatic head and neck cancer. Invest New Drugs 26(1):81–87

Brune K et al (2006) Multifocal neoplastic precursor lesions associated with lobular atrophy of the pancreas in patients having a strong family history of pancreatic cancer. Am J Surg Pathol 30(9):1067–1076

Byrd JC et al (2005) A phase 1 and pharmacodynamic study of depsipeptide (FK228) in chronic lymphocytic leukemia and acute myeloid leukemia. Blood 105(3):959–967

Chan TA et al (2008) Convergence of mutation and epigenetic alterations identifies common genes in cancer that predict for poor prognosis. PLoS Med 5(5):e114

Chen RZ et al (1998) DNA hypomethylation leads to elevated mutation rates. Nature 395(6697):89–93

Chien CH et al (2011) Identifying transcriptional start sites of human microRNAs based on high-throughput sequencing data. Nucleic Acids Res 39(21):9345–9356

Choi JH et al (2007) Site-specific methylation of CpG nucleotides in the hTERT promoter region can control the expression of hTERT during malignant progression of colorectal carcinoma. Biochem Biophys Res Commun 361(3):615–620
Clark SJ (2007) Action at a distance: epigenetic silencing of large chromosomal regions in carcinogenesis. Hum Mol Genet 16(1):R88–R95
Cui K et al (2009) Chromatin signatures in multipotent human hematopoietic stem cells indicate the fate of bivalent genes during differentiation. Cell Stem Cell 4(1):80–93
Daniel M, Peek GW, Tollefsbol TO (2012) Regulation of the human catalytic subunit of telomerase (hTERT). Gene 498(2):135–146
De Carvalho DD et al (2012) DNA methylation screening identifies driver epigenetic events of cancer cell survival. Cancer Cell 21(5):655–667
de Wilde RF et al (2012) Loss of ATRX or DAXX expression and concomitant acquisition of the alternative lengthening of telomeres phenotype are late events in a small subset of MEN-1 syndrome pancreatic neuroendocrine tumors. Mod Pathol 25(7):1033–1039
Dejeux E et al (2009) Hypermethylation of the IGF2 differentially methylated region 2 is a specific event in insulinomas leading to loss-of-imprinting and overexpression. Endocr Relat Cancer 16(3):939–952
Dessain SK et al (2000) Methylation of the human telomerase gene CpG island. Cancer Res 60(3):537–541
Devereux TR et al (1999) DNA methylation analysis of the promoter region of the human telomerase reverse transcriptase (hTERT) gene. Cancer Res 59(24):6087–6090
Dialynas GK, Vitalini MW, Wallrath LL (2008) Linking Heterochromatin Protein 1 (HP1) to cancer progression. Mutat Res 647(1–2):13–20
DiGiuseppe JA et al (1994) Detection of K-ras mutations in mucinous pancreatic duct hyperplasia from a patient with a family history of pancreatic carcinoma. Am J Pathol 144(5):889–895
Dodge JE et al (2002) De novo methylation of MMLV provirus in embryonic stem cells: CpG versus non-CpG methylation. Gene 289(1–2):41–48
Easwaran HP et al (2010) Aberrant silencing of cancer-related genes by CpG hypermethylation occurs independently of their spatial organization in the nucleus. Cancer Res 70(20):8015–8024
Eden S et al (1998) DNA methylation models histone acetylation. Nature 394(6696):842
Ellis L et al (2008) Histone deacetylase inhibitor panobinostat induces clinical responses with associated alterations in gene expression profiles in cutaneous T-cell lymphoma. Clin Cancer Res 14(14):4500–4510
Esteller M (2007) Epigenetic gene silencing in cancer: the DNA hypermethylome. Hum Mol Genet 16(1):R50–R59
Federico A et al (2009) Chromobox protein homologue 7 protein, with decreased expression in human carcinomas, positively regulates E-cadherin expression by interacting with the histone deacetylase 2 protein. Cancer Res 69(17):7079–7087
Fitzgerald M, Oshiro M et al (2003) Human pancreatic carcinoma cells activate maspin expression through loss of epigenetic control. Neoplasia 5(5):427–436
Fujii S et al (2008) Enhancer of zeste homologue 2 (EZH2) down-regulates RUNX3 by increasing histone H3 methylation. J Biol Chem 283(25):17324–17332
Fukushima N et al (2003) Aberrant methylation of suppressor of cytokine signalling-1 (SOCS-1) gene in pancreatic ductal neoplasms. Br J Cancer 89(2):338–343
Gao J et al (2010) Methylation of the SPARC gene promoter and its clinical implication in pancreatic cancer. J Exp Clin Cancer Res 29:28
Garcia-Manero G et al (2008) Phase 1 study of the oral isotype specific histone deacetylase inhibitor MGCD0103 in leukemia. Blood 112(4):981–989
Garcia-Morales P et al (2005) Histone deacetylase inhibitors induced caspase-independent apoptosis in human pancreatic adenocarcinoma cell lines. Mol Cancer Ther 4(8):1222–1230
Gaudet F et al (2003) Induction of tumors in mice by genomic hypomethylation. Science 300(5618):489–492

Giatromanolaki A et al (2004) BNIP3 expression is linked with hypoxia-regulated protein expression and with poor prognosis in non-small cell lung cancer. Clin Cancer Res 10(16):5566–5571

Grady WM et al (2008) Epigenetic silencing of the intronic microRNA hsa-miR-342 and its host gene EVL in colorectal cancer. Oncogene 27(27):3880–3888

Graff JR et al (2000) Methylation patterns of the E-cadherin 5′ CpG island are unstable and reflect the dynamic, heterogeneous loss of E-cadherin expression during metastatic progression. J Biol Chem 275(4):2727–2732

Hahn MA et al (2011) Relationship between gene body DNA methylation and intragenic H3K9me3 and H3K36me3 chromatin marks. PLoS One 6(4):e18844

Hashimshony T et al (2003) The role of DNA methylation in setting up chromatin structure during development. Nat Genet 34(2):187–192

He S et al (2011) Expression of DNMT1 and DNMT3a are regulated by GLI1 in human pancreatic cancer. PLoS One 6(11):e27684

Hellman A, Chess A (2007) Gene body-specific methylation on the active X chromosome. Science 315(5815):1141–1143

Hong SM et al (2011) Loss of E-cadherin expression and outcome among patients with resectable pancreatic adenocarcinomas. Mod Pathol 24(9):1237–1247

Hong SM et al (2012) Genome-wide CpG island profiling of intraductal papillary mucinous neoplasms of the pancreas. Clin Cancer Res 18(3):700–712

Iacobuzio-Donahue CA, Maitra A et al (2003) Exploration of global gene expression patterns in pancreatic adenocarcinoma using cDNA microarrays. Am J Pathol 162(4):1151–1162

Issa JP, Kantarjian HM (2009) Targeting DNA methylation. Clin Cancer Res 15(12):3938–3946

Jansen M et al (2002) Aberrant methylation of the 5′ CpG island of TSLC1 is common in pancreatic ductal adenocarcinoma and is first manifest in high-grade PanlNs. Cancer Biol Ther 1(3):293–296

Jenuwein T, Allis CD (2001) Translating the histone code. Science 293(5532):1074–1080

Jones PA, Baylin SB (2007) The epigenomics of cancer. Cell 128(4):683–692

Jones PL et al (1998) Methylated DNA and MeCP2 recruit histone deacetylase to repress transcription. Nat Genet 19(2):187–191

Jones S et al (2008) Core signaling pathways in human pancreatic cancers revealed by global genomic analyses. Science 321(5897):1801–1806

Kanda M et al (2012) Presence of somatic mutations in most early-stage pancreatic intraepithelial neoplasia. Gastroenterology 142(4):730.e9–733.e9

Karamitopoulou E et al (2010) Loss of the CBX7 protein expression correlates with a more aggressive phenotype in pancreatic cancer. Eur J Cancer 46(8):1438–1444

Kauffman EC et al (2011) Role of androgen receptor and associated lysine-demethylase coregulators, LSD1 and JMJD2A, in loc.lized and advanced human bladder cancer. Mol Carcinog 50(12):931–944

Kelly TK et al (2010) H2A.Z maintenance during mitosis reveals nucleosome shifting on mitotically silenced genes. Mol Cell 39(6):901–911

Klymkowsky MW, Savagner P (2009) Epithelial-mesenchymal transition: a cancer researcher's conceptual friend and foe. Am J Pathol 174(5):1588–1593

Kouzarides T (2007) Chromatin modifications and their function. Cell 128(4):693–705

Kozomara A, Griffiths-Jones S (2011) miRBase: integrating microRNA annotation and deep-sequencing data. Nucleic Acids Res 39(Database issue):D152–D157

Ku M et al (2008) Genomewide analysis of PRC1 and PRC2 occupancy identifies two classes of bivalent domains. PLoS Genet 4(10):e1000242

Kumagai T et al (2007) Histone deacetylase inhibitor, suberoylanilide hydroxamic acid (Vorinostat, SAHA) profoundly inhibits the growth of human pancreatic cancer cells. Int J Cancer 121(3):656–665

Kumari A et al (2009) Positive regulation of human telomerase reverse transcriptase gene expression and telomerase activity by DNA methylation in pancreatic cancer. Ann Surg Oncol 16(4):1051–1059

Lachner M, Jenuwein T (2002) The many faces of histone lysine methylation. Curr Opin Cell Biol 14(3):286–298
Lander ES et al (2001) Initial sequencing and analysis of the human genome. Nature 409(6822):860–921
Lee KH et al (2009) Epigenetic silencing of MicroRNA miR-107 regulates cyclin-dependent kinase 6 expression in pancreatic cancer. Pancreatology 9(3):293–301
Li A et al (2010a) Pancreatic cancer DNMT1 expression and sensitivity to DNMT1 inhibitors. Cancer Biol Ther 9(4):5226–5237
Li A et al (2010b) Pancreatic cancers epigenetically silence SIP1 and hypomethylate and overexpress miR-200a/200b in association with elevated circulating miR-200a and miR-200b levels. Cancer Res 70(13):5226–5237
Lim S et al (2010) Lysine-specific demethylase 1 (LSD1) is highly expressed in ER-negative breast cancers and a biomarker predicting aggressive biology. Carcinogenesis 31(3):512–520
Lister R et al (2009) Human DNA methylomes at base resolution show widespread epigenomic differences. Nature 462(7271):315–322
Lohse B et al (2011) Inhibitors of histone demethylases. Bioorg Med Chem 19(12):3625–3636
Lunter G, Hein J (2004) A nucleotide substitution model with nearest-neighbour interactions. Bioinformatics 20(Suppl 1):i216–i223
Matsubayashi H et al (2003) Methylation of cyclin D2 is observed frequently in pancreatic cancer but is also an age-related phenomenon in gastrointestinal tissues. Clin Cancer Res 9(4): 1446–1452
Matsubayashi H et al (2006) DNA methylation alterations in the pancreatic juice of patients with suspected pancreatic disease. Cancer Res 66(2):1208–1217
Maunakea AK et al (2010) Conserved role of intragenic DNA methylation in regulating alternative promoters. Nature 466(7303):253–257
Md Zin R, Murch A, Charles A (2011) Pathology, genetics and cytogenetics of Wilms' tumour. Pathology 43(4):302–312
Mikkelsen TS et al (2007) Genome-wide maps of chromatin state in pluripotent and lineage-committed cells. Nature 448(7153):553–560
Ms K et al (2008) Gonadotrophin releasing hormone antagonist in IVF/ICSI. J Hum Reprod Sci 1(1):29–32
Nakamura H (2005) Microarray analysis shows that some microRNAs downregulate large numbers of target mRNAs. Scientist 19(6):25–25
Nakata B et al (2002) Prognostic value of microsatellite instability in resectable pancreatic cancer. Clin Cancer Res 8(8):2536–2540
Nalls D et al (2011) Targeting epigenetic regulation of miR-34a for treatment of pancreatic cancer by inhibition of pancreatic cancer stem cells. PLoS One 6(8):e24099
Nan X et al (1998) Transcriptional repression by the methyl-CpG-binding protein MeCP2 involves a histone deacetylase complex. Nature 393(6683):386–389
O'Hagan HM et al (2011) Oxidative damage targets complexes containing DNA methyltransferases, SIRT1, and polycomb members to promoter CpG Islands. Cancer Cell 20(5):606–619
Ohike N, Maass N et al (2003) Clinicopathological significance and molecular regulation of maspin expression in ductal adenocarcinoma of the pancreas. Cancer Lett 199(2):193–200
Ohm JE et al (2007) A stem cell-like chromatin pattern may predispose tumor suppressor genes to DNA hypermethylation and heritable silencing. Nat Genet 39(2):237–242
Okami J, Simeone DM, Logsdon CD (2004) Silencing of the hypoxia-inducible cell death protein BNIP3 in pancreatic cancer. Cancer Res 64(15):5338–5346
Okano M, Xie SP, Li E (1998) Cloning and characterization of a family of novel mammalian DNA (cytosine-5) methyltransferases. Nat Genet 19(3):219–220
Okano M et al (1999) DNA methyltransferases Dnmt3a and Dnmt3b are essential for de novo methylation and mammalian development. Cell 99(3):247–257
Olson P et al (2009) MicroRNA dynamics in the stages of tumorigenesis correlate with hallmark capabilities of cancer. Genes Dev 23(18):2152–2165

Omura N et al (2008) Genome-wide profiling of methylated promoters in pancreatic adenocarcinoma. Cancer Biol Ther 7(7):1146–1156
Ooi SK et al (2007) DNMT3L connects unmethylated lysine 4 of histone H3 to de novo methylation of DNA. Nature 448(7154):714–717
Ougolkov AV, Bilim VN, Billadeau DD (2008) Regulation of pancreatic tumor cell proliferation and chemoresistance by the histone methyltransferase enhancer of zeste homologue 2. Clin Cancer Res 14(21):6790–6796
Parsi MA et al (2008) DNA methylation alterations in endoscopic retrograde cholangiopancreatography brush samples of patients with suspected pancreaticobiliary disease. Clin Gastroenterol Hepatol 6(11):1270–1278
Peng L, Seto E (2011) Deacetylation of nonhistone proteins by HDACs and the implications in cancer. Handb Exp Pharmacol 206:39–56
Pili R et al (2012) Phase I study of the histone deacetylase inhibitor entinostat in combination with 13-cis retinoic acid in patients with solid tumours. Br J Cancer 106(1):77–84
Rauch TA et al (2009) A human B cell methylome at 100-base pair resolution. Proc Natl Acad Sci USA 106(3):671–678
Renaud S et al (2007) Dual role of DNA methylation inside and outside of CTCF-binding regions in the transcriptional regulation of the telomerase hTERT gene. Nucleic Acids Res 35(4):1245–1256
Robert MF et al (2003) DNMT1 is required to maintain CpG methylation and aberrant gene silencing in human cancer cells. Nat Genet 33(1):61–65
Ropero S, Esteller M (2007) The role of histone deacetylases (HDACs) in human cancer. Mol Oncol 1(1):19–25
Rosty C et al (2002) Overexpression of S100A4 in pancreatic ductal adenocarcinomas is associated with poor differentiation and DNA hypomethylation. Am J Pathol 160(1):45–50
Rosty C et al (2003) p16 Inactivation in pancreatic intraepithelial neoplasias (PanINs) arising in patients with chronic pancreatitis. Am J Surg Pathol 27(12):1495–1501
Ryu JK et al (2006) SK-7041, a new histone deacetylase inhibitor, induces G2-M cell cycle arrest and apoptosis in pancreatic cancer cell lines. Cancer Lett 237(1):143–154
Ryu JK et al (2011) Elevated microRNA miR-21 levels in pancreatic cyst fluid are predictive of mucinous precursor lesions of ductal adenocarcinoma. Pancreatology 11(3):343–350
Saito Y et al (2006) Specific activation of microRNA-127 with downregulation of the proto-oncogene BCL6 by chromatin-modifying drugs in human cancer cells. Cancer Cell 9(6):435–443
Sassen S, Miska EA, Caldas C (2008) MicroRNA: implications for cancer. Virchows Archiv 452(1):1–10
Sato N, Fukushima N et al (2003a) Discovery of novel targets for aberrant methylation in pancreatic carcinoma using high-throughput microarrays. Cancer Res 63(13):3735–3742
Sato N, Fukushima N et al (2003b) SPARC/osteonectin is a frequent target for aberrant methylation in pancreaticadenocarcinoma and a mediator of tumor-stromal interactions. Oncogene 22(32):5021–5030
Sato N et al (2004) Identification of maspin and S100P as novel hypomethylation targets in pancreatic cancer using global gene expression profiling. Oncogene 23(8):1531–1538
Sato N et al (2005) Epigenetic inactivation of TFPI-2 as a common mechanism associated with growth and invasion of pancreatic ductal adenocarcinoma. Oncogene 24(5):850–858
Sato N et al (2008) CpG island methylation profile of pancreatic intraepithelial neoplasia. Mod Pathol 21(3):238–244
Sato F et al (2011) MicroRNAs and epigenetics. FEBS J 278(10):1598–1609
Schildhaus HU et al (2011) Lysine-specific demethylase 1 is highly expressed in solitary fibrous tumors, synovial sarcomas, rhabdomyosarcomas, desmoplastic small round cell tumors, and malignant peripheral nerve sheath tumors. Hum Pathol 42(11):1667–1675
Schoeftner S et al (2006) Recruitment of PRC1 function at the initiation of X inactivation independent of PRC2 and silencing. EMBO J 25(13):3110–3122

Schuebel KE et al (2007) Comparing the DNA hypermethylome with gene mutations in human colorectal cancer. PLoS Genet 3(9):1709–1723

Schutte M et al (1997) Abrogation of the Rb/p16 tumor-suppressive pathway in virtually all pancreatic carcinomas. Cancer Res 57(15):3126–3130

Shogren-Knaak M et al (2006) Histone H4-K16 acetylation controls chromatin structure and protein interactions. Science 311(5762):844–847

Siegel R et al (2011) Cancer statistics, 2011: the impact of eliminating socioeconomic and racial disparities on premature cancer deaths. CA Cancer J Clin 61(4):212–236

Siegel R, Naishadham D, Jemal A (2012) Cancer statistics. CA Cancer J Clin 62(1):10–29

Song W et al (2010) Bmi-1 is related to proliferation, survival and poor prognosis in pancreatic cancer. Cancer Sci 101(7):1754–1760

Strahl BD, Allis CD (2000) The language of covalent histone modifications. Nature 403(6765):41–45

Takai D, Jones PA (2002) Comprehensive analysis of CpG islands in human chromosomes 21 and 22. Proc Natl Acad Sci USA 99(6):3740–3745

Thurman RE et al (2012) The accessible chromatin landscape of the human genome. Nature 489(7414):75–82

Tonini T et al (2004) Ezh2 reduces the ability of HDAC1-dependent pRb2/p130 transcriptional repression of cyclin A. Oncogene 23(28):4930–4937

Toyota M et al (2008) Epigenetic silencing of microRNA-34b/c and B-cell translocation gene 4 is associated with CpG island methylation in colorectal cancer. Cancer Res 68(11):4123–4132

Trojer P, Reinberg D (2006) Histone lysine demethylases and their impact on epigenetics. Cell 125(2):213–217

Tsai H-C et al (2012) Transient low doses of DNA-demethylating agents exert durable antitumor effects on hematological and epithelial tumor cells. Cancer Cell 21(3):430–446

Ueki T et al (2000) Hypermethylation of multiple genes in pancreatic adenocarcinoma. Cancer Res 60(7):1835–1839

Vincent A et al (2011) Genome-wide analysis of promoter methylation associated with gene expression profile in pancreatic adenocarcinoma. Clin Cancer Res 17(13):4341–4354

Vincenz C, Kerppola TK (2008) Different polycomb group CBX family proteins associate with distinct regions of chromatin using nonhomologous protein sequences. Proc Natl Acad Sci USA 105(43):16572–16577

Vire E et al (2006) The Polycomb group protein EZH2 directly controls DNA methylation. Nature 439(7078):871–874

Wang L et al (2004) Hierarchical recruitment of polycomb group silencing complexes. Mol Cell 14(5):637–646

Wang W et al (2009a) Significance of DNA methyltransferase-1 and histone deacetylase-1 in pancreatic cancer. Oncol Rep 21(6):1439–1447

Wang Y et al (2009b) LSD1 is a subunit of the NuRD complex and targets the metastasis programs in breast cancer. Cell 138(4):660–672

Winter JM et al (2008) Absence of E-cadherin expression distinguishes noncohesive from cohesive pancreatic cancer. Clin Cancer Res 14(2):412–418

Yamamoto H et al (2001) Genetic and clinical features of human pancreatic ductal adenocarcinomas with widespread microsatellite instability. Cancer Res 61(7):3139–3144

Yang XJ, Seto E (2007) HATs and HDACs: from structure, function and regulation to novel strategies for therapy and prevention. Oncogene 26(37):5310–5318

Yang H, Liu Y et al (2013) Tumor development is associated with decrease of TET gene expression and 5- methylcytosine hydroxylation. Oncogene 32(5):663–669

Yegnasubramanian S et al (2011) Chromosome-wide mapping of DNA methylation patterns in normal and malignant prostate cells reveals pervasive methylation of gene-associated and conserved intergenic sequences. BMC Genomics 12:313

Yen RW et al (1992) Isolation and characterization of the cDNA encoding human DNA methyltransferase. Nucleic Acids Res 20(9):2287–2291

Yu J, Li A et al (2012a) MicroRNA alterations of pancreatic intraepithelial neoplasias. Clin Cancer Res 18(4):981–992

Yu J et al (2012b) MicroRNA alterations of pancreatic intraepithelial neoplasias. Clin Cancer Res 18(4):981–992

Zhang S et al (2011) Downregulation of miR-132 by promoter methylation contributes to pancreatic cancer development. Carcinogenesis 32(8):1183–1189

Zhao XD et al (2007) Whole-genome mapping of histone H3 Lys4 and 27 trimethylations reveals distinct genomic compartments in human embryonic stem cells. Cell Stem Cell 1(3):286–298

Zinn RL et al (2007) hTERT is expressed in cancer cell lines despite promoter DNA methylation by preservation of unmethylated DNA and active chromatin around the transcription start site. Cancer Res 67(1):194–201

Innovative Technologies in the Molecular Characterization of Pancreatic Cancer

Iris H. Wei and Chandan Kumar-Sinha

Abstract The molecular characterization of pancreatic adenocarcinoma is beset with several inherent challenges. At the outset, even though pancreatic cancer is the fourth leading cause of cancer-related deaths, it is relatively rare among the most morbid malignancies, and only 15–20 % of patients are surgically resectable at presentation. As a result, there are small cohorts of tumor tissues available for research. In addition, access to the pancreas, located deep in the retroperitoneum, requires highly specialized expertise and infrastructure available only at select centers, which further limits the availability of pancreatic tissues and biopsy samples. Furthermore, pancreatic adenocarcinoma is uniquely characterized by a dense desmoplastic stroma, which typically results in no more than 20–30 % of cancer cells in grossly dissected tumor tissues. The sample-related constraints are further compounded by the abundance of proteolytic and nucleolytic enzymes in the pancreas that diminish the quality of the biomolecules used for molecular analyses. In this context, the advent of highly sensitive, high-throughput genomics platforms, *ex vivo* cultures of primary tumors, and innovative transgenic mouse models of the disease over the past decade have helped overcome many of the practical bottlenecks leading to important breakthroughs in the molecular characterization of pancreatic cancer with potential clinical significance. Here we appraise some of the most salient high-throughput technologies in genomics, proteomics, and metabolomics currently utilized in the study of cancers and review their specific applications in pancreatic cancer research.

I.H. Wei
Department of Surgery, University of Michigan Medical Center, Ann Arbor, MI 48109, USA

C. Kumar-Sinha, Ph.D. (✉)
Department of Pathology, Michigan Center for Translational Pathology, University of Michigan Medical School, 1400 E. Medical Center Dr. 5316 CCGC, Ann Arbor, MI 48109-0602, USA
e-mail: chakumar@med.umich.edu

D.M. Simeone and A. Maitra (eds.), *Molecular Genetics of Pancreatic Cancer*,
DOI 10.1007/978-1-4614-6549-2_10,

Microarrays

Schena et al. first described the use of fluorescently labeled cDNA microarrays for measuring differential gene expression patterns, by comparing mRNA extracted from root tissues versus leaf tissues in *Arabidopsis thaliana* (Schena et al. 1995). The use of fluorescence imaging and detection allowed for rapid and efficient analysis of multiple samples at a time. Since then, thousands of large-scale, microarray experiments have been published. Growing databases of publically available microarray data, such as Oncomine (www.oncomine.org), provide a powerful tool for the top–down analysis of related genes, pathways, and networks involved in different cancers (Rhodes et al. 2007).

In cancer research, microarray technology has been typically utilized to compare mRNA expression in cancer samples against normal controls, to identify differentially expressed genes that could offer diagnostic or therapeutic targets of these diseases. Microarrays also offer an efficient technique for discovering disease classifications, which can provide important prognostic information and response to therapies (Golub et al. 1999). Clinically relevant gene expression patterns and relationships between sample cohorts and gene signatures can be identified using cluster analysis algorithms (Eisen et al. 1998). Well-known examples include diffuse large B-cell lymphoma (Alizadeh et al. 2000) and breast cancer (Sørlie et al. 2001; Van't Veer et al. 2002), in which clinically significant molecular classification systems were created using gene expression profiling.

Another application of microarray technology is comparative genomic hybridization (CGH), to measure DNA sequence copy numbers by comparing differences between test DNA and normal reference DNA hybridized to normal chromosome spreads; these DNA gains/losses often serve as key events in cancer development and can offer unique therapeutic targets in cancer treatment (Kallioniemi et al. 1992; Pollack et al. 1999). Pollack et al. demonstrated this phenomenon in breast cancer, in which 12 % of all the gene expression variation in primary breast tumors was due directly to changes in gene copy number, highlighting the importance of DNA copy number variations in the dysregulation of gene expression and subsequent breast cancer development and progression (Pollack et al. 2002).

The first comparative gene expression study using microarrays in pancreatic cancer was published in 1996 (Gress et al. 1996). Since that time, there have been a few dozen pancreatic cancer microarray studies of varying size and scope. An early study using cDNA microarray analysis of nine pancreatic cancer cell lines compared to normal pancreas revealed a panel of 30 overexpressed genes, two of which, c-Myc and Rad51, were validated in patient samples by RT-PCR of frozen tissues and tissue microarray-based immunohistochemistry (Han et al. 2002). To distinguish the pancreatic cancer expression profile from the inflammatory changes that typically accompany pancreatic cancer, another group compared primary pancreatic adenocarcinoma tissues against both normal and chronic pancreatitis samples; they found 808 genes that were differentially expressed in pancreatic cancer (Friess et al. 2003). To identify genes that may correlate with metastatic spread of the disease,

Tanaka et al. performed a microarray study comparing gene expression patterns between established, parental pancreatic cancer cell lines and metastatic sublines, which yielded a panel of differentially expressed genes that may contribute to an aggressive phenotype (Tanaka et al. 2003). Using Affymetrix gene expression profiling, the ataxia-telangiectasia group D complementing gene (ATDC) was found to be uniquely overexpressed in both pancreatic cancer and pancreatic cancer precursor lesions as compared to chronic pancreatitis and normal pancreas samples, promoting tumor growth and metastasis through the Wnt/β-catenin signaling pathway (Logsdon et al. 2003; Wang et al. 2009). Limiting their focus to cell-surface genes, another group analyzed mRNA expression of pancreatic cancer specimens and normal pancreas tissue samples compared to a known database; they identified 170 targets uniquely overexpressed in cancer samples, two of which were confirmed by immunostaining of tissue microarrays (Morse et al. 2010).

By CGH, nonrandom losses and gains have been identified in pancreatic cancer cell lines (Bashyam et al. 2005) and primary patient tumor samples (Loukopoulos et al. 2007), which could offer potential therapeutic targets. More recently, Shain et al. have performed a high resolution CGH study and found that one-third of pancreatic cancers possessed a loss-of-function genetic aberration (deletion, mutation, or rearrangement) in one of the genes encoding a component of the SWI/SNF chromatin remodeling complex, highlighting the importance of this pathway in normal tumor suppression (Shain et al. 2012).

These various microarray studies have revealed pancreatic cancer to be a complex, heterogeneous disease without a globally distinct profile as is present in some other cancers. Only recently have subtypes of pancreatic adenocarcinoma been proposed by analyzing pooled gene expression microarray data sets that may correspond to different clinical behaviors (Collisson et al. 2011). The study of pancreatic cancer presents a unique, analytical challenge, which requires more sensitive, high-throughput techniques than are available with traditional microarray technology (Fig. 1).

High-Throughput Sequencing

High-throughput sequencing is a powerful tool for analyzing cancer genomics through a number of different approaches, depending on the type of sample input (RNA or DNA), the part of the genome sequenced (the entire genome, a subset of genes, the exome, or the transcriptome), and the aberrations of interest (such as point mutations, structural differences, or gene expression) (Mardis and Wilson 2009).

The original method developed for genome sequencing was Sanger sequencing, by which labeled dideoxynucleotides are used for strand termination in the synthesis of DNA from a specific primer. The corresponding lengths of the resultant fragments determine the positions of the corresponding deoxynucleotides. Despite technical advances, this method can only produce reads up to 1,000–1,200 base pairs in length (Zhang et al. 2011).

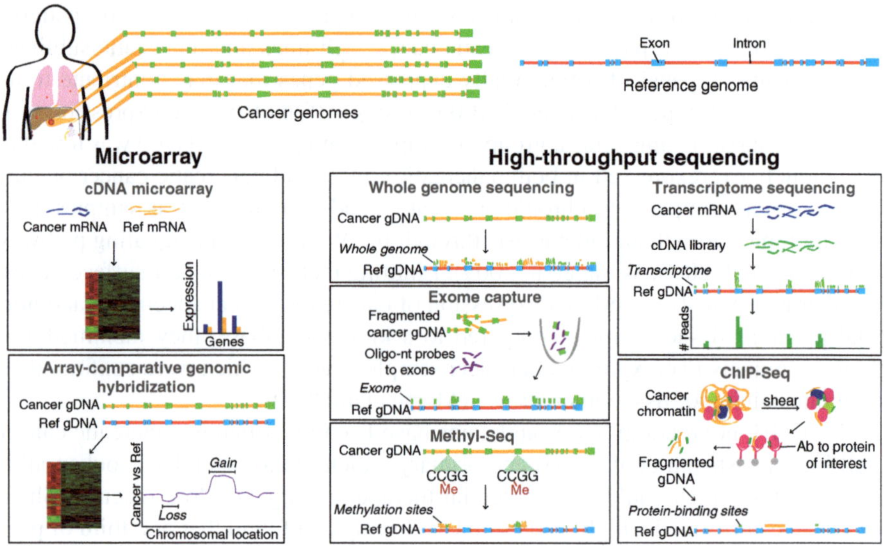

Fig. 1 Techniques for analyzing cancer genomics: microarray versus high-throughput sequencing methods. Traditional microarray measures levels of gene expression, as compared to a reference sample, or DNA copy number variation by array-comparative genomic hybridization. High-throughput sequencing methods include whole genome sequencing, exome capture (to sequence exons), Methyl-Seq (to sequence methylation sites), transcriptome sequencing (to sequence RNA transcript regions), and ChIP-seq (to sequence protein-binding sites for a specific protein of interest). *mRNA* messenger RNA, *gDNA* genomic DNA, *cDNA* complementary DNA

By the late 1990s, the international Human Genome Project (HGP) was launched with the goal to sequence the entire human genome. In order to sequence longer stretches of DNA than was possible with traditional Sanger sequencing, a new technique called "shotgun sequencing" was used, by which genomic DNA is fragmented, then aligned and reassembled based on partial sequence overlaps (Lander et al. 2001). The success of the HGP yielded complete genomes, upon which current high-throughput sequencing techniques are based.

New, high-throughput sequencing techniques utilize a similar shotgun sequencing approach, with subsequent mapping to a reference genome, as was first made possible by the HGP. Once the template DNA is divided into small fragments, adapters are ligated to the ends, from which DNA synthesis is performed in "short reads" along the entire genome. Various techniques are used to capture the identity of the individual nucleotides as they are incorporated. The reads are then reassembled by mapping them to a reference genome.

The main issues that need to be addressed with high-throughput sequencing methods are sufficient coverage and error rate. Coverage is defined as the number of overlapping reads at a site of interest. Typically, 30× coverage is needed to accurately identify individual base differences, while increased depth of coverage may be required with samples of poorer quality to decrease the error rate. In the case of genetic rearrangements, another technique for identifying these aberrations with low coverage is the use of paired end reads. Short reads are sequenced at an expected distance; interval differences when mapped to a reference genome suggest the presence of structural rearrangements without the need to sequence the intervening regions (Meyerson et al. 2010).

Applications of High-Throughput Sequencing

Whole genome sequencing offers the most comprehensive analysis of a sample genome but is resource-intensive and requires extensive coverage. To address certain questions, a targeted sequencing approach can be more efficient and effective, allowing increased coverage in the genomic area of interest (Fig. 1). In exome capture, the coding exons, which constitute only approximately 1 % of the entire genome, can be efficiently sequenced at higher coverage. The sample genomic DNA is fragmented and hybridized to oligonucleotide probes specific to exomic regions, which are then captured and sequenced. Similarly, transcriptome sequencing (also called RNA-Seq) utilizes cDNA reverse-transcribed from the RNA of interest (messenger, micro, or total) to determine levels of gene expression and possible fusions (Meyerson et al. 2010).

Epigenetic processes, such as protein-binding and DNA methylation, can also be interrogated by focused sequencing. ChIP-Seq is chromatin immunoprecipitation followed by sequencing, which has been used to identify transcription factor binding sites (Robertson et al. 2007). DNA regions bound to specific proteins of interest are enriched using antibodies to those proteins; the resultant DNA fragments are then sequenced and mapped to a reference genome to identify the protein binding sites (Park 2009). Methyl-Seq selectively sequences sites of DNA methylation, another important process in the regulation of gene expression. As CpG islands account for 99.98 % of these sites, Methyl-Seq typically involves treatment of the sample DNA with sodium bisulfite, which selectively converts unmethylated cytosines to uracils. This is then followed by hybrid selection, sequencing, and mapping to a reference genome to identify sites of methylation (Hodges et al. 2009).

High-Throughput Sequencing Platforms

High-throughput sequencing platforms may be broadly categorized by the type of template preparation, into amplification-based (Fig. 2a) versus single molecule-based (Fig. 2b) sequencing methods.

In amplification-based sequencing techniques, a library of genomic material is created by fragmentation and subsequent ligation of adapters containing priming sites, to allow for PCR amplification (Fig. 2a). Illumina, Inc. hybridizes these amplified fragments to an eight-lane glass slide, termed a "flow cell." A pool of

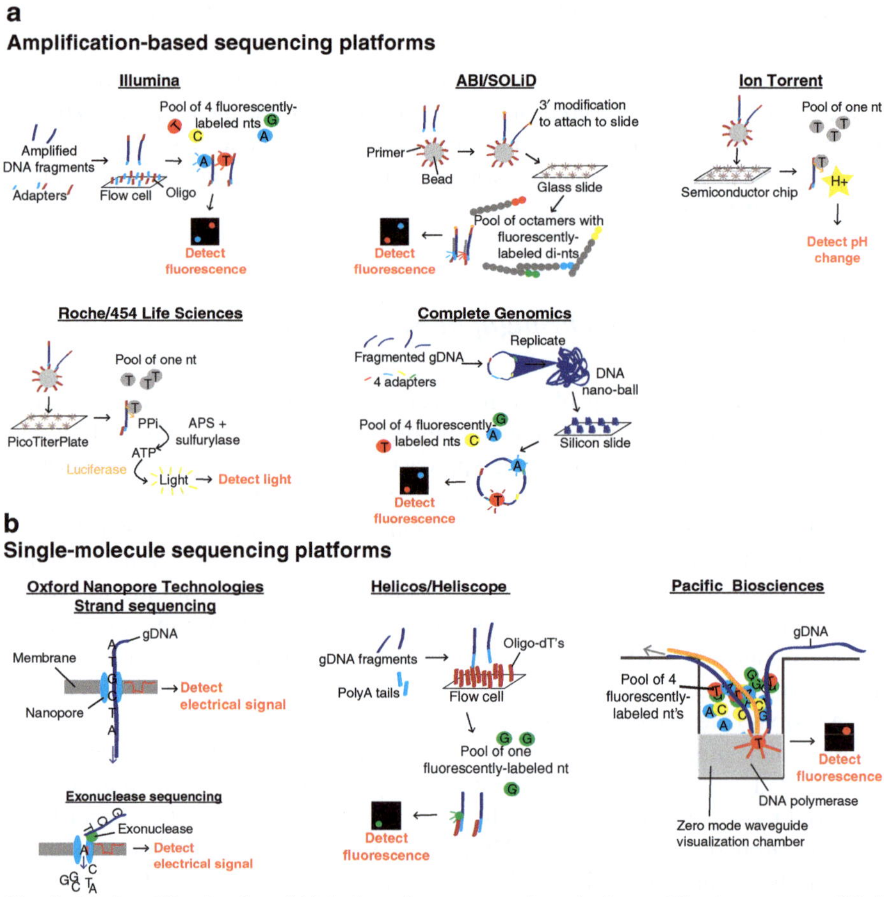

Fig. 2 (**a**) Amplification-based high-throughput sequencing platforms. Illumina uses amplified DNA fragments ligated with adapters and hybridized to a flow cell; each nucleotide has a unique fluorescent label, which is detected and recorded as it is incorporated during sequencing. With ABI/SOLiD, amplified DNA fragments are captured onto beads and attached to a glass slide. Sets of octamers containing one of four fluorescently labeled dinucleotides are ligated, and the emitted fluorescence is detected and recorded. Ion Torrent utilizes bead-hybridized DNA fragments cross-linked to a semiconductor chip. Pools of a single, unlabeled nucleotide are introduced one at a time. Incorporation of the nucleotide releases a hydrogen ion, which is detected as a local pH change and recorded. Roche/454 Life Sciences uses pyrosequencing technology. First, bead-hybridized DNA fragments are placed on a PicoTiterPlate. Pools of a single, unlabeled nucleotide are introduced one a time. Incorporation of the nucleotide releases pyrophosphate, which is converted to ATP by sulfurylase in the presence of adenosine 5′ phosphosulfate (APS). ATP catalyzes the luciferase-mediated generation of light, which is detected and recorded. With Complete Genomics technology, DNA fragments are ligated with four adapters to produce a circular plasmid that is clonally replicated to form DNA nano-balls that are placed on a silicon slide. Each nucleotide introduced has a unique fluorescent label, which is detected and recorded as it is incorporated during sequencing. *nt* nucleotide. (**b**) Single-molecule sequencing platforms. Oxford Nanopore Technologies identifies individual nucleotides by measuring changes in electrical potential across a membrane as a complete strand of DNA (strand sequencing) or a single nucleotide cleaved from a strand of DNA (exonuclease sequencing) passes through a nanopore. Helicose/Heliscope ligates polyA tails to unamplified DNA fragments, which are hybridized to oligo-dT's on a flow cell. A pool of one fluorescently labeled nucleotide is introduced at a time and incorporation detected and recorded. Pacific Biosciences uses a single DNA polymerase immobilized in each detection well. A single, unamplified strand of DNA is sequenced in a pool of four fluorescently labeled nucleotides. The zero mode waveguide visualization chamber at the bottom of each well is able to selectively detect the fluorescence emitted when a nucleotide is incorporated

modified nucleotides is introduced, with each of the four nucleotides labeled with its own unique fluorescent tag. When a single nucleotide is incorporated, DNA synthesis is terminated by the reversible terminator. The unincorporated nucleotides are then washed off, and the fluorescent probes are imaged and recorded at each site on the flow cell. The terminating 3′-OH groups and fluorescent dyes are cleaved and a new pool of fluorescently labeled, modified nucleotides is introduced (Metzker 2010).

Other companies use bead capture of the target samples, followed by amplification and enrichment. The beads are chemically cross-linked to a glass slide (ABI/SOLiD) or deposited into wells of a semiconductor chip (Ion Torrent) or PicoTiterPlate (Roche/454 Life Sciences). Similar to Illumina, ABI/SOLiD uses fluorescence emission to determine nucleotide incorporation; however, rather than single, labeled nucleotides, the latter uses octamer probes containing a terminal, fluorescently labeled dinucleotide group in a sequencing-by-ligation approach. When the probe is ligated, the emitted fluorescence is detected and recorded. The octamer is then cleaved between the fifth and sixth bases and the next pool of labeled octamers introduced. The bases are thus sequentially interrogated at overlapping intervals to improve read accuracy (Wong et al. 2011).

Ion Torrent and Roche/454 Life Sciences introduce a pool of one unlabeled nucleotide at a time. Ion Torrent recognizes the incorporation of a nucleotide by detecting a local pH change, due to the chemical release of a single hydrogen ion. This output is then converted and recorded by the semiconductor chip. Roche/454 Life Sciences uses pyrosequencing technology. Incorporation of a single nucleotide releases a pyrophosphate, which is converted to adenosine triphosphate (ATP) by sulfurylase in the presence of adenosine 5′ phosphosulfate (APS). This ATP then catalyzes the luciferase-mediated generation of light, which is detected and recorded (Metzker 2010).

Complete Genomics employs a proprietary library creation process of ligating four adapters into each DNA fragment to form stable, circular templates. These templates are then amplified into clusters called DNA "nano-balls." The nano-balls are applied to silicon slides and sequencing carried out using pools of four, labeled nucleotides, similar to the Illumina technique described above.

Single-molecule sequencing offers the major advantage of avoiding artifactual genetic errors introduced by the PCR amplification process (Fig. 2b). Oxford Nanopore Technologies utilizes a nanopore placed within a membrane, across which there is an electrical gradient. A DNA strand is then sequenced by one of two approaches. In "strand sequencing," a complete strand of DNA is passed through the nanopore; in "exonuclease sequencing," individual nucleotides are cleaved from the strand of DNA and passed through the nanopore one at a time. Characteristic changes in the electrical signal across the membrane correspond with the identity of each passing nucleotide; these signals are detected and recorded (Clarke et al. 2009).

Helicose/Heliscope ligates polyA tails to unamplified DNA fragments, which are then hybridized to oligo-dT's on a flow cell. A pool of one fluorescently labeled nucleotide is introduced at a time and its incorporation detected by capturing and recording the emitted fluorescence (Thompson and Steinmann 2010). Pacific Biosciences uses a single DNA polymerase immobilized in each detection well. A single, unamplified strand of DNA is sequenced in a pool of nucleotides labeled with one of four fluorescent probes. The zero mode waveguide visualization chamber at the bottom of each well is able in real time to selectively detect the fluorescence emitted when a nucleotide is incorporated against the background fluorescence (Eid et al. 2009).

Bioinformatics Analysis of High-Throughput Sequencing Results in Pancreatic Cancer

The subsequent sequencing readout provides a wealth of information depending on the type of input and question of interest (Fig. 3). Exome capture and whole genome sequencing allow identification of focal and global genomic amplifications and losses, rearrangements, and insertions/deletions (Fig. 3a). Using these techniques, Jones et al. sequenced 20,661 protein-coding genes in 24 pancreatic cancer samples and found an average of 63 genetic aberrations per sample, most of which were point mutations. These alterations comprised twelve cellular signaling pathways that were identified as the core processes involved in the development of pancreatic cancer (Jones et al. 2008). Focusing specifically on hereditary pancreatic cancer, Jones et al. also identified PALB2, a BRCA2-binding partner, as the second most commonly mutated gene in familial pancreatic cancer after BRCA2 (Jones et al. 2009). More recently, Campbell et al. have used massively parallel, paired-end exome sequencing to study chromosomal rearrangements and metastatic clonal

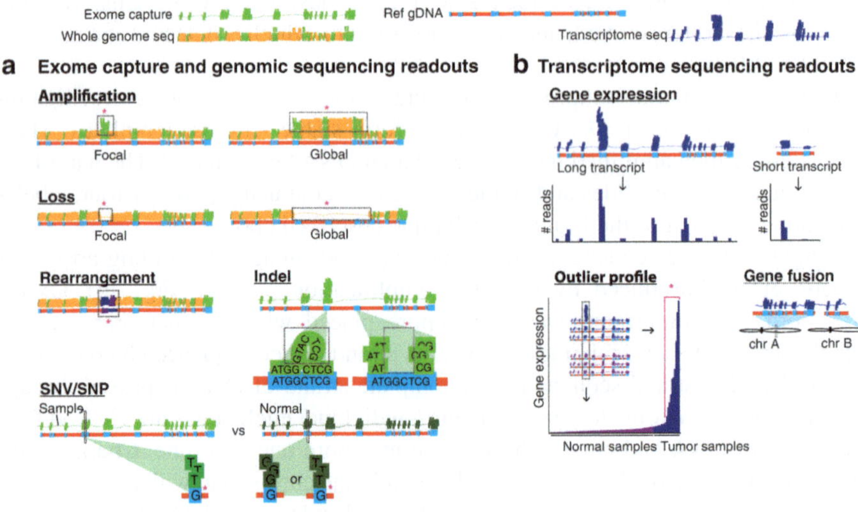

Fig. 3 Bioinformatics interpretation of high-throughput sequencing readouts. (**a**) Exome capture and whole genome sequencing allow identification of focal and global genomic amplifications and losses, rearrangements, and insertions/deletions (indels). Single nucleotide variants (SNVs) and single nucleotide polymorphisms (SNPs) can be determined by comparison of nucleotide differences against a normal sample. (**b**) Transcriptome sequencing can be used to analyze gene expression, outlier profiles (gene expression level relative to other genes and samples), and gene fusions

relationships in thirteen patients with Stage IV pancreatic adenocarcinoma. By comparing different metastases to the primary tumor of individual patients, they found that particular chromosomal rearrangements called "fold back inversions" occurred early in cancer development. In addition, they found that such genetic instability persists after metastatic spread, with metastases continuing to acquire genetic aberrations beyond those needed for primary tumor growth (Campbell et al. 2010). Subsequent analysis to quantify the genetic evolution of these metastases revealed that the time from the tumor-initiating mutation to the development of the parental clone was over 10 years, followed by another 5 years prior to the development of metastatic capabilities. According to their analysis, patients then died an average of 2 years later. This finding suggests that there may be an opportunistic window in which the development of earlier detection methods would have significant clinical impact on this lethal disease (Yachida et al. 2010).

Single nucleotide variants (SNVs) and single nucleotide polymorphisms (SNPs) can also be determined using exome and whole genome sequencing by comparison of nucleotide differences in a cancer sample against a matched normal sample. Base differences shared with the normal sample are termed SNPs, while a SNV is unique to the cancer sample. The presence and location of such genetic variations can confer clinical significance, including response to therapeutics and overall survival (Li et al. 2006; Okazaki et al. 2010).

Transcriptome sequencing offers a number of tools for the analysis of cancer genomics (Fig. 3b). Gene expression can be quantified based on the depth of coverage. During analysis, this must be normalized to the total read number mapped as well as the length of the transcript, as longer transcripts require more reads for adequate coverage. This is typically done using the expression measure RPKM or "reads per kilobase transcript per million total reads" (Mortazavi et al. 2008). Using this technique, gene expression may be efficiently measured and compared across a number of samples. One such application of this technique is the identification of outlier gene expression (i.e., a gene with high expression relative to other genes within that sample and as compared to expression levels of that gene across other samples), which may indicate a potential driver of the cancer. Transcriptome sequencing is also a powerful tool for discovering gene fusions, which are genetic aberrations characteristic of certain cancers that can serve as potent therapeutic targets (Maher et al. 2009).

An important resource in the bioinformatics analysis of pancreatic cancer is the Pancreatic Expression Database (PED, http://www.pancreasexpression.org/), a publicly available, comprehensive database of pancreatic transcriptomic, proteomic, genomic, and miRNA profiles culled from the literature. The samples include tissue and bodily fluid specimens from healthy and diseased individuals, cell lines, and mouse models, including those that have received various therapies. Currently, there are over 60,000 measurements stored, providing a powerful reference for pancreatic research (Cutts et al. 2011).

Other publically available, online resources invaluable in the study of pancreatic diseases include the Pancreapedia (http://www.pancreapedia.org/), a rigorously maintained resource for pancreatic researchers, containing high quality references and research protocols. Also, the Pancreatic Cancer Collaborative Registry Project (http://pccr.unmc.edu/pccr_project_about.html) is a growing, multicenter outcomes database to store clinical information on patients with and at high risk for developing pancreatic cancer.

Proteomics of Pancreatic Cancer

Proteomics comprehensively refers to the analysis of the identity, characterization, quantification, and interactions of proteins within a sample and is a powerful tool in cancer research. As compared to DNA and RNA analysis, which are indirect measures of gene activity, studying global protein expression patterns may provide more functionally relevant differences between cancer and normal tissues. In addition, proteins can be useful biomarkers, as they remain stable in body fluids and can be efficiently detected and measured using antibody-based methods (Ludwig and Weinstein 2005).

The most common technique used for protein separation and identification is liquid chromatography to fractionate samples, followed by mass spectrometry. In mass spectrometry, the mass-to-charge ratios of the protein fragments are calculated and compared to a known database. Protein expression levels may also be measured using fluorescently labeled antibodies on a forward or reverse phase protein microarray (Fig. 4a).

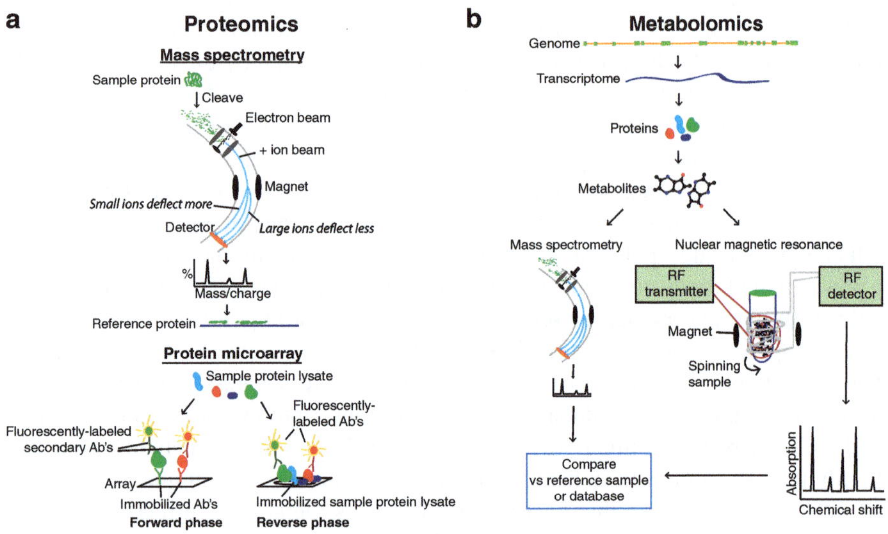

Fig. 4 Techniques for the analysis of proteomics and metabolomics. (**a**) Proteins are identified and quantitated by mass spectrometry, comparing each sample's mass-to-charge ratio to a database of known proteins. Expression levels may also be measured using fluorescently labeled antibodies on a forward- or reverse-phase protein microarray. (**b**) Metabolites can also be identified and quantitated by mass spectrometry or by structural interrogation using nuclear magnetic resonance. Results are compared against a reference sample or a known database

Over the last 5 years, proteomics has proved to be a powerful tool in elucidating molecular mechanisms and novel biomarkers of pancreatic cancer, by the analysis of cancer tissues, body fluids, and cell lines (Cecconi et al. 2011). By proteomics analysis, genes involved in glycolysis have been implicated in the development of pancreatic cancer (Mikuriya et al. 2007). Analyzing the proteomics profile of pancreatic cancer stem cells, proteins involved in the signaling pathways for apoptosis, cell proliferation, inflammation, and metastasis, were also found to be differentially expressed (Dai et al. 2010). More recently, Shi et al. have used an antibody microarray comparing a metastatic pancreatic cancer cell line to its parental line and discovered upregulation of proteins involved in tumor signal transduction and downregulation of proteins involved in cell differentiation (Shi et al. 2011).

A major focus in the field of pancreatic cancer research is the identification of highly sensitive and specific biomarkers for the early detection and surveillance of the disease. A number of such candidate biomarkers have been identified by proteomics analysis of tumor tissue, including calgranulin (Sheikh et al. 2007), synuclein-γ (Hibi et al. 2009), radixin, and moesin (Cui et al. 2009); of serum, including phosphoglycerate kinase 1 (Hwang et al. 2006; Patwa et al. 2009), Rab GDP dissociation inhibitor β, serotransferrin (Sun et al. 2007), and platelet factor 4 (Fiedler et al. 2009); and of pancreatic juice, including insulin-like growth factor binding protein-2 (Chen et al. 2006) and matrix metalloproteinase-9 (Tian et al. 2008).

Metabolomics of Pancreatic Cancer

Metabolomics profiling is also used to analyze the functional differences between cancer and benign samples. Metabolites are the small molecular end-products released by cells during metabolism. Therefore, analysis of a sample's metabolome offers a different type of functional analysis as compared to studying precursor genes and proteins, which can undergo significant epigenetic regulatory processes and posttranslational modifications (Patti et al. 2012). In addition, protein biomarkers may be difficult to detect in low concentration or against the background of higher abundance proteins. Metabolites can be identified and quantitated by mass spectrometry, or individual structures can be interrogated using nuclear magnetic resonance. Results are then compared against a reference sample or known database (Fig. 4b).

While this technique has been more extensively used to characterize other solid organ cancers, there have been few studies published in the field of pancreatic cancer. Sugimoto et al. identified a salivary metabolomics profile to distinguish pancreatic cancer patients from healthy cohorts (Sugimoto et al. 2010). More recently, metabolomics analysis of serum samples have been used to distinguish patients with pancreatic cancer from those with benign hepatobiliary disease to provide a potential diagnostic signature (Bathe et al. 2011).

Specific Issues Concerning the Molecular Characterization of Pancreatic Cancer

Tumors of pancreatic adenocarcinoma are characterized by dense desmoplastic stroma and abundant ribonucleases, which result in the reduced quantity and quality of available genomic material (Fig. 5a). The stroma can occupy up to 90 % of a tumor sample and consists primarily of fibroblasts, as well as cancer stem cells, extracellular matrix (ECM), immune cells, and scant blood vessels (Mahadevan and Von Hoff 2007). This dense microenvironment has been implicated in the development of pancreatic cancer, as well as invasion, metastasis, and chemotherapy resistance (Li et al. 2012). Recent techniques to target this complex network by Hedgehog pathway inhibition to deplete the tumor-associated stromal desmoplasia (Olive et al. 2009) and enzymatic degradation of the dominant ECM component hyaluronic acid (Jacobetz et al. 2012) have been found to successfully improve tumor perfusion and drug delivery. In addition, CD40 activation of the abundant macrophages present in

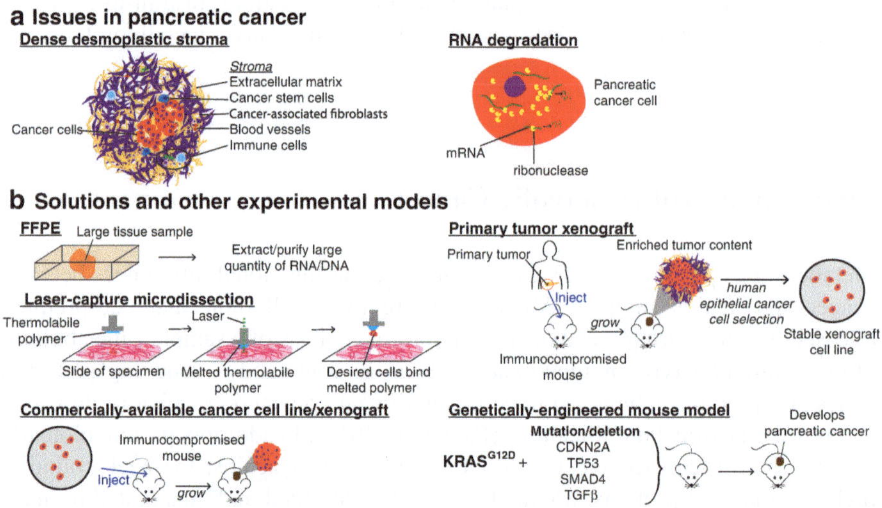

Fig. 5 Issues in the molecular characterization of pancreatic cancer; solutions and other experimental models. (**a**) Pancreatic tumors are characterized by dense desmoplastic stroma (comprised of cancer stem cells, cancer-associated fibroblasts, extracellular matrix, immune cells, and scant blood vessels) and abundant ribonucleases, which result in the reduced quantity and quality of genomic material. (**b**) New techniques to recover RNA and DNA from archived samples of formalin-fixed, paraffin-embedded (FFPE) patient samples provide increased amounts of input genomic material. Individual tumor cells can be selectively procured from the dense stromal background using laser-capture microdissection. Tumor content can also be enriched using a primary tumor xenograft, by injecting patient cancer cells into an immunocompromised mouse; the epithelial cells may be further selected by passaging to create a stable xenograft cell line. Other experimental models include commercially available cancer cell lines and genetically engineered mouse models, whose genomic material may be similarly extracted and analyzed

the stroma have also been shown to induce stromal degradation as well as promote an antitumor response (Beatty et al. 2011).

With regards to utilizing these molecular techniques to study pancreatic cancer and its interactions with this complex microenvironment, the stroma significantly contaminates the purity of the input tumor sample. In addition, the pancreas also contains the highest concentration of enzymes, including ribonucleases, in the body, which results in significant degradation of the RNA used for sequencing (Anderson et al. 2010). This can increase the error rate of the subsequent sequencing readout.

Experimental Models of Pancreatic Cancer

There are numerous methods available to overcome the above limitations characteristic of pancreatic cancer, to improve the quality and quantity of the input sample (Fig. 5b).

Primary Pancreatic Cancer Tumor Samples

New techniques to recover RNA and DNA from archived samples of formalin-fixed, paraffin-embedded (FFPE) patient samples provide increased amounts of starting genomic material. While there has been concern that the fixation process and long-term storage at room temperature can introduce DNA mutations, these errors may be successfully overcome by targeted enrichment and increased coverage to 80× (Kerick et al. 2011).

Second, laser-capture microdissection can be used to specifically analyze tumor cells by selectively procuring the cells from the dense stromal background under direct microscopy. A laser is used to melt a thermoplastic polymer, which then adheres to the cells of interest. The polymer–cell composite is carefully lifted from the slide, completely preserving the cell morphology, DNA, RNA, and protein (Espina et al. 2006).

Commercially Available Pancreatic Cancer Cell Lines

As extensively outlined elsewhere in this text, a number of important experimental models are used to study the biology of pancreatic cancer and test potential therapeutics. A valuable in vitro model is the commercially available, immortalized pancreatic cancer cell line. While there may be potential issues with regards to long-term sub-culturing and possible cross-contamination, pancreatic cancer cell lines offer a powerful tool for early stage discovery and proof-of-concept experiments. Such findings may then be further validated in vivo with the creation of xenografts by heterotopic or orthotopic injection of the cancer cell line into an immune-compromised mouse. The resultant tumor growth mimics the progression of human pancreatic cancer and its response to therapeutics, though there are biologic limitations to this technique (Deer et al. 2010).

Primary Pancreatic Cancer Cell Lines and Tumor Xenografts

Low-passage primary pancreatic cancer cell lines derived directly from primary human tumor samples can overcome some of the limitations of these long-established commercial cell lines though at the expense of high stromal contamination. Alternatively, primary tumor cells may be first enriched by creating primary tumor xenografts, by transplanting human cancer cells into immune-compromised mice, without loss of genotypic features (Rubio-Viqueira et al. 2006). Engraftment occurs at a rate of about 70 %, with successful engraftment correlating with aggressive phenotype and poor patient prognosis (Andren-Sandberg 2011; Garrido-Laguna et al. 2011). The result is an enrichment of cancer cells two-fold to five-fold, with the tumor associated stroma being gradually replaced by infiltrating murine cells (Hahn et al. 1995). Subsequent passaging of the xenograft cells further enriches the epithelial cancer cells, to provide high-tumor content experimental cells (Feldmann et al. 2009).

Genetically Engineered Mouse Models of Pancreatic Cancer

The transgenic mouse model is used as an alternative in vivo model for studying the development of pancreatic cancer, by creating genetic alterations of known cancer drivers. Most commonly, an activating KRAS mutation is combined with inactivation of a tumor suppressor, such as CDKN2A, TP53, SMAD4, or TGFβ, through mutation or deletion (Herreros-Villanueva et al. 2012). Recently, an inducible, mutant KRAS transgenic mouse model has been developed, which produces pancreas-specific, reversible, oncogenic $KRAS^{G12D}$ expression, capable of producing both pancreatic cancer development and regression (Collins et al. 2012; Ying et al. 2012).

In Hanahan and Weinberg's recent comprehensive review on the "hallmarks of cancer," the important and complex contributions of the tumor microenvironment to cancer development were highlighted (Hanahan and Weinberg 2011). In no disease is this more apparent than in the case of pancreatic cancer, with its tumor cells encased in a dense, desmoplastic stroma comprised of components that contribute to the cancer's aggressive phenotype, inhibit its response to therapies, and severely limit the relative quantity of tumor content available for study. Furthermore, as the pancreas serves as one of the major exocrine organs of the body, the high concentration of enzymes degrade what little biomolecules are available for analysis of this disease. Fortunately, with these new available techniques, many of these issues may be overcome. In addition to primary tumor samples, the genomic material from both in vitro and in vivo pancreatic cancer models may be extracted and analyzed by the various techniques discussed in this chapter, to further understand the biology of the models and how they relate to the primary disease. Although pancreatic adenocarcinoma is a lethal, genetically complex process, new advances in genomics, proteomics, and metabolomics and the techniques outlined here offer highly sensitive, efficient methods for studying and treating this disease and its complicated interactions with the tumor microenvironment.

References

Alizadeh AA, Elsen MB, Davis RE, Ma CL, Lossos IS, Rosenwald A, Boldrick JC, Sabet H, Tran T, Yu X, Powell JI, Yang L, Marü GE, Moore T, Hudson J Jr, Lu L, Lewis DB, Tibshirani R, Sherlock G, Chan WC, Greiner TC, Weisenburger DD, Armitage JO, Warnke R, Levy R, Wilson W, Grever MR, Byrd JC, Botstein D, Brown PO, Staudt LM (2000) Distinct types of diffuse large B-cell lymphoma identified by gene expression profiling. Nature 403(6769):503–511

Anderson MA, Brenner DE, Scheiman JM, Simeone DM, Singh N, Sikora MJ, Zhao L, Mertens AN, Rae JM (2010) Reliable gene expression measurements from fine needle aspirates of pancreatic tumors: effect of amplicon length and quality assessment. J Mol Diagn 12(5):566–575

Andren-Sandberg A (2011) Pancreatic cancer: animal model and molecular biology. N Am J Med Sci 3(10):441–450

Bashyam MD, Bair R, Kim YH, Wang P, Hernandez-Boussard T, Karikari CA, Tibshirani R, Maitra A, Pollack JR (2005) Array-based comparative genomic hybridization identifies localized DNA amplifications and homozygous deletions in pancreatic cancer. Neoplasia 7(6):556–562

Bathe OF, Shaykhutdinov R, Kopciuk K, Weljie AM, McKay A, Sutherland FR, Dixon E, Dunse N, Sotiropoulos D, Vogel HJ (2011) Feasibility of identifying pancreatic cancer based on serum metabolomics. Cancer Epidemiol Biomarkers Prev 20(1):140–147

Beatty GL, Chiorean EG, Fishman MP, Saboury B, Teitelbaum UR, Sun W, Huhn RD, Song W, Li D, Sharp LL, Torigian DA, O'Dwyer PJ, Vonderheide RH (2011) CD40 agonists alter tumor stroma and show efficacy against pancreatic carcinoma in mice and humans. Science 331(6024):1612–1616

Campbell PJ, Yachida S, Mudie LJ, Stephens PJ, Pleasance ED, Stebbings LA, Morsberger LA, Latimer C, McLaren S, Lin ML, McBride DJ, Varela I, Nik-Zainal SA, Leroy C, Jia M, Menzies A, Butler AP, Teague JW, Griffin CA, Burton J, Swerdlow H, Quail MA, Stratton MR, Iacobuzio-Donahue C, Futreal PA (2010) The patterns and dynamics of genomic instability in metastatic pancreatic cancer. Nature 467(7319):1109–1113

Cecconi D, Palmieri M, Donadelli M (2011) Proteomics in pancreatic cancer research. Proteomics 11(4):816–828

Chen R, Pan S, Yi EC, Donohoe S, Bronner MP, Potter JD, Goodlett DR, Aebersold R, Brentnall TA (2006) Quantitative proteomic profiling of pancreatic cancer juice. Proteomics 6(13): 3871–3879

Clarke J, Wu HC, Jayasinghe L, Patel A, Reid S, Bayley H (2009) Continuous base identification for single-molecule nanopore DNA sequencing. Nat Nanotechnol 4(4):265–270

Collins MA, Bednar F, Zhang Y, Brisset JC, Galban S, Galban CJ, Rakshit S, Flannagan KS, Adsay NV, Pasca di Magliano M (2012) Oncogenic Kras is required for both the initiation and maintenance of pancreatic cancer in mice. J Clin Invest 122(2):639–653

Collisson EA, Sadanandam A, Olson P, Gibb WJ, Truitt M, Gu S, Cooc J, Weinkle J, Kim GE, Jakkula L, Feiler HS, Ko AH, Olshen AB, Danenberg KL, Tempero MA, Spellman PT, Hanahan D, Gray JW (2011) Subtypes of pancreatic ductal adenocarcinoma and their differing responses to therapy. Nat Med 17(4):500–503

Cui Y, Wu J, Zong M, Song G, Jia Q, Jiang J, Han J (2009) Proteomic profiling in pancreatic cancer with and without lymph node metastasis. Int J Cancer 124(7):1614–1621

Cutts RJ, Gadaleta E, Hahn SA, Crnogorac-Jurcevic T, Lemoine NR, Chelala C (2011) The pancreatic expression database: 2011 update. Nucleic Acids Res 39(Database issue):D1023–D1028

Dai L, Li C, Shedden KA, Lee CJ, Quoc H, Simeone DM, Lubman DM (2010) Quantitative proteomic profiling studies of pancreatic cancer stem cells. J Proteome Res 9(7):3394–3402

Deer EL, Gonzalez-Hernandez J, Coursen JD, Shea JE, Ngatia J, Scaife CL, Firpo MA, Mulvihill SJ (2010) Phenotype and genotype of pancreatic cancer cell lines. Pancreas 39(4):425–435

Eid J, Fehr A, Gray J, Luong K, Lyle J, Otto G, Peluso P, Rank D, Baybayan P, Bettman B, Bibillo A, Bjornson K, Chaudhuri B, Christians F, Cicero R, Clark S, Dalal R, Dewinter A, Dixon J, Foquet M, Gaertner A, Hardenbol P, Heiner C, Hester K, Holden D, Kearns G, Kong X, Kuse R, Lacroix Y, Lin S, Lundquist P, Ma C, Marks P, Maxham M, Murphy D, Park I, Pham T, Phillips M, Roy J, Sebra R, Shen G, Sorenson J, Tomaney A, Travers K, Trulson M, Vieceli J, Wegener J, Wu D, Yang A, Zaccarin D, Zhao P, Zhong F, Korlach J, Turner S (2009) Real-time DNA sequencing from single polymerase molecules. Science 323(5910):133–138

Eisen MB, Spellman PT, Brown PO, Botstein D (1998) Cluster analysis and display of genome-wide expression patterns. Proc Natl Acad Sci USA 95(25):14863–14868

Espina V, Wulfkuhle JD, Calvert VS, VanMeter A, Zhou W, Coukos G, Geho DH, Petricoin EF 3rd, Liotta LA (2006) Laser-capture microdissection. Nat Protoc 1(2):586–603

Feldmann G, Rauenzahn S, Maitra A (2009) In vitro models of pancreatic cancer for translational oncology research. Expert Opin Drug Discov 4(4):429–443

Fiedler GM, Leichtle AB, Kase J, Baumann S, Ceglarek U, Felix K, Conrad T, Witzigmann H, Weimann A, Schutte C, Hauss J, Buchler M, Thiery J (2009) Serum peptidome profiling revealed platelet factor 4 as a potential discriminating peptide associated with pancreatic cancer. Clin Cancer Res 15(11):3812–3819

Friess H, Ding J, Kleeff J, Fenkell L, Rosinski JA, Guweidhi A, Reidhaar-Olson JF, Korc M, Hammer J, Buchler MW (2003) Microarray-based identification of differentially expressed growth- and metastasis-associated genes in pancreatic cancer. Cell Mol Life Sci 60(6):1180–1199

Garrido-Laguna I, Uson M, Rajeshkumar NV, Tan AC, de Oliveira E, Karikari C, Villaroel MC, Salomon A, Taylor G, Sharma R, Hruban RH, Maitra A, Laheru D, Rubio-Viqueira B, Jimeno A, Hidalgo M (2011) Tumor engraftment in nude mice and enrichment in stroma-related gene pathways predict poor survival and resistance to gemcitabine in patients with pancreatic cancer. Clin Cancer Res 17(17):5793–5800

Golub TR, Slonim DK, Tamayo P, Huard C, Gaasenbeek M, Mesirov JP, Coller H, Loh ML, Downing JR, Caligiuri MA, Bloomfield CD, Lander ES (1999) Molecular classification of cancer: class discovery and class prediction by gene expression monitoring. Science 286(5439):527–531

Gress TM, Muller-Pillasch F, Geng M, Zimmerhackl F, Zehetner G, Friess H, Buchler M, Adler G, Lehrach H (1996) A pancreatic cancer-specific expression profile. Oncogene 13(8): 1819–1830

Hahn SA, Seymour AB, Hoque AT, Schutte M, da Costa LT, Redston MS, Caldas C, Weinstein CL, Fischer A, Yeo CJ et al (1995) Allelotype of pancreatic adenocarcinoma using xenograft enrichment. Cancer Res 55(20):4670–4675

Han H, Bearss DJ, Browne LW, Calaluce R, Nagle RB, Von Hoff DD (2002) Identification of differentially expressed genes in pancreatic cancer cells using cDNA microarray. Cancer Res 62(10):2890–2896

Hanahan D, Weinberg RA (2011) Hallmarks of cancer: the next generation. Cell 144(5):646–674

Herreros-Villanueva M, Hijona E, Cosme A, Bujanda L (2012) Mouse models of pancreatic cancer. World J Gastroenterol 18(12):1286–1294

Hibi T, Mori T, Fukuma M, Yamazaki K, Hashiguchi A, Yamada T, Tanabe M, Aiura K, Kawakami T, Ogiwara A, Kosuge T, Kitajima M, Kitagawa Y, Sakamoto M (2009) Synuclein-gamma is closely involved in perineural invasion and distant metastasis in mouse models and is a novel prognostic factor in pancreatic cancer. Clin Cancer Res 15(8):2864–2871

Hodges E, Smith AD, Kendall J, Xuan Z, Ravi K, Rooks M, Zhang MQ, Ye K, Bhattacharjee A, Brizuela L, McCombie WR, Wigler M, Hannon GJ, Hicks JB (2009) High definition profiling of mammalian DNA methylation by array capture and single molecule bisulfite sequencing. Genome Res 19(9):1593–1605

Hwang TL, Liang Y, Chien KY, Yu JS (2006) Overexpression and elevated serum levels of phosphoglycerate kinase 1 in pancreatic ductal adenocarcinoma. Proteomics 6(7):2259–2272

Jacobetz, M. A., D. S. Chan, A. Neesse, T. E. Bapiro, N. Cook, K. K. Frese, C. Feig, T. Nakagawa, M. E. Caldwell, H. I. Zecchini, M. P. Lolkema, P. Jiang, A. Kultti, C. B. Thompson, D. C. Maneval, D. I. Jodrell, G. I. Frost, H. M. Shepard, J. N. Skepper, and D. A. Tuveson. 2012. Hyaluronan impairs vascular function and drug delivery in a mouse model of pancreatic cancer. *Gut*.

Jones S, Zhang X, Parsons DW, Lin JC, Leary RJ, Angenendt P, Mankoo P, Carter H, Kamiyama H, Jimeno A, Hong SM, Fu B, Lin MT, Calhoun ES, Kamiyama M, Walter K, Nikolskaya T, Nikolsky Y, Hartigan J, Smith DR, Hidalgo M, Leach SD, Klein AP, Jaffee EM, Goggins M, Maitra A, Iacobuzio-Donahue C, Eshleman JR, Kern SE, Hruban RH, Karchin R, Papadopoulos N, Parmigiani G, Vogelstein B, Velculescu VE, Kinzler KW (2008) Core signaling pathways in human pancreatic cancers revealed by global genomic analyses. Science 321(5897):1801–1806

Jones S, Hruban RH, Kamiyama M, Borges M, Zhang X, Parsons DW, Lin JC, Palmisano E, Brune K, Jaffee EM, Iacobuzio-Donahue CA, Maitra A, Parmigiani G, Kern SE, Velculescu VE, Kinzler KW, Vogelstein B, Eshleman JR, Goggins M, Klein AP (2009) Exomic sequencing identifies PALB2 as a pancreatic cancer susceptibility gene. Science 324(5924):217

Kallioniemi A, Kallioniemi OP, Sudar D, Rutovitz D, Gray JW, Waldman F, Pinkel D (1992) Comparative genomic hybridization for molecular cytogenetic analysis of solid tumors. Science 258(5083):818–821

Kerick M, Isau M, Timmermann B, Sultmann H, Herwig R, Krobitsch S, Schaefer G, Verdorfer I, Bartsch G, Klocker H, Lehrach H, Schweiger MR (2011) Targeted high throughput sequencing in clinical cancer settings: formaldehyde fixed-paraffin embedded (FFPE) tumor tissues, input amount and tumor heterogeneity. BMC Med Genomics 4:68

Lander ES, Linton LM, Birren B, Nusbaum C, Zody MC, Baldwin J, Devon K, Dewar K, Doyle M, FitzHugh W, Funke R, Gage D, Harris K, Heaford A, Howland J, Kann L, Lehoczky J, LeVine R, McEwan P, Nusbaum C, Zody MC, Baldwin J, Devon K, Dewar K, Doyle M, FitzHugh W, Funke R, Gage D, Harris K, Heaford A, Howland J, Kann L, Lehoczky J, LeVine R, McEwan P, McKernan K, Meldrim J, Mesirov JP, Miranda C, Morris W, Naylor J, Raymond C, Rosetti M, Santos R, Sheridan A, Sougnez C, Stange-Thomann N, Stojanovic N, Subramanian A, Wyman D, Rogers J, Sulston J, Ainscough R, Beck S, Bentley D, Burton J, Clee C, Carter N, Coulson A, Deadman R, Deloukas P, Dunham A, Dunham I, Durbin R, French L, Grafham D, Gregory S, Hubbard T, Humphray S, Hunt A, Jones M, Lloyd C, McMurray A, Matthews L, Mercer S, Milne S, Mullikin JC, Mungall A, Plumb R, Ross M, Shownkeen R, Sims S, Waterston RH, Wilson RK, Hillier LW, McPherson JD, Marra MA, Mardis ER, Fulton LA, Chinwalla AT, Pepin KH, Gish WR, Chissoe SL, Wendl MC, Delehaunty KD, Miner TL, Delehaunty A, Kramer JB, Cook LL, Fulton RS, Johnson DL, Minx PJ, Clifton SW, Hawkins T, Branscomb E, Predki P, Richardson P, Wenning S, Slezak T, Doggett N, Cheng JF, Olsen A, Lucas S, Elkin C, Uberbacher E, Frazier M, Gibbs RA, Muzny DM, Scherer SE, Bouck JB, Sodergren EJ, Worley KC, Rives CM, Gorrell JH, Metzker ML, Naylor SL, Kucherlapati RS, Nelson DL, Weinstock GM, Sakaki Y, Fujiyama A, Hattori M, Yada T, Toyoda A, Itoh T, Kawagoe C, Watanabe H, Totoki Y, Taylor T, Weissenbach J, Heilig R, Saurin W, Artiguenave F, Brottier P, Bruls T, Pelletier E, Robert C, Wincker P, Smith DR, Doucette-Stamm L, Rubenfield M, Weinstock K, Lee HM, Dubois J, Rosenthal A, Platzer M, Nyakatura G, Taudien S, Rump A, Yang H, Yu J, Wang J, Huang G, Gu J, Hood L, Rowen L, Madan A, Qin S, Davis RW, Federspiel NA, Abola AP, Proctor MJ, Myers RM, Schmutz J, Dickson M, Grimwood J, Cox DR, Olson MV, Kaul R, Shimizu N, Kawasaki K, Minoshima S, Evans GA, Athanasiou M,

Schultz R, Roe BA, Chen F, Pan H, Ramser J, Lehrach H, Reinhardt R, McCombie WR, de la Bastide M, Dedhia N, Blocker H, Hornischer K, Nordsiek G, Agarwala R, Aravind L, Bailey JA, Bateman A, Batzoglou S, Birney E, Bork P, Brown DG, Burge CB, Cerutti L, Chen HC, Church D, Clamp M, Copley RR, Doerks T, Eddy SR, Eichler EE, Furey TS, Galagan J, Gilbert JG, Harmon C, Hayashizaki Y, Haussler D, Hermjakob H, Hokamp K, Jang W, Johnson LS, Jones TA, Kasif S, Kaspryzk A, Kennedy S, Kent WJ, Kitts P, Koonin EV, Korf I, Kulp D, Lancet D, Lowe TM, McLysaght A, Mikkelsen T, Moran JV, Mulder N, Pollara VJ, Ponting CP, Schuler G, Schultz J, Slater G, Smit AF, Stupka E, Szustakowski J, Thierry-Mieg D, Thierry-Mieg J, Wagner L, Wallis J, Wheeler R, Williams A, Wolf YI, Wolfe KH, Yang SP, Yeh RF, Collins F, Guyer MS, Peterson J, Bork P, Brown DG, Burge CB, Cerutti L, Chen HC, Church D, Clamp M, Copley RR, Doerks T, Eddy SR, Eichler EE, Furey TS, Galagan J, Gilbert JG, Harmon C, Hayashizaki Y, Haussler D, Hermjakob H, Hokamp K, Jang W, Johnson LS, Jones TA, Kasif S, Kaspryzk A, Kennedy S, Kent WJ, Kitts P, Koonin EV, Korf I, Kulp D, Lancet D, Lowe TM, McLysaght A, Mikkelsen T, Moran JV, Mulder N, Pollara VJ, Ponting CP, Schuler G, Schultz J, Slater G, Smit AF, Stupka E, Szustakowski J, Thierry-Mieg D, Thierry-Mieg J, Wagner L, Wallis J, Wheeler R, Williams A, Wolf YI, Wolfe KH, Yang SP, Yeh RF, Collins F, Guyer MS, Peterson J, Felsenfeld A, Wetterstrand KA, Patrinos A, Morgan MJ, de Jong P, Catanese JJ, Osoegawa K, Shizuya H, Choi S, Chen YJ (2001) Initial sequencing and analysis of the human genome. Nature 409(6822):860–921

Li D, Frazier M, Evans DB, Hess KR, Crane CH, Jiao L, Abbruzzese JL (2006) Single nucleotide polymorphisms of RecQ1, RAD54L, and ATM genes are associated with reduced survival of pancreatic cancer. J Clin Oncol 24(11):1720–1728

Li X, Ma Q, Xu Q, Duan W, Lei J, Wu E (2012) Targeting the cancer-stroma interaction: a potential approach for pancreatic cancer treatment. Curr Pharm Des 18(17):2404–2415

Logsdon CD, Simeone DM, Binkley C, Arumugam T, Greenson JK, Giordano TJ, Misek DE, Kuick R, Hanash S (2003) Molecular profiling of pancreatic adenocarcinoma and chronic pancreatitis identifies multiple genes differentially regulated in pancreatic cancer. Cancer Res 63(10):2649–2657

Loukopoulos P, Shibata T, Katoh H, Kokubu A, Sakamoto M, Yamazaki K, Kosuge T, Kanai Y, Hosoda F, Imoto I, Ohki M, Inazawa J, Hirohashi S (2007) Genome-wide array-based comparative genomic hybridization analysis of pancreatic adenocarcinoma: identification of genetic indicators that predict patient outcome. Cancer Sci 98(3):392–400

Ludwig JA, Weinstein JN (2005) Biomarkers in cancer staging, prognosis and treatment selection. Nat Rev Cancer 5(11):845–856

Mahadevan D, Von Hoff DD (2007) Tumor-stroma interactions in pancreatic ductal adenocarcinoma. Mol Cancer Ther 6(4):1186–1197

Maher CA, Kumar-Sinha C, Cao X, Kalyana-Sundaram S, Han B, Jing X, Sam L, Barrette T, Palanisamy N, Chinnaiyan AM (2009) Transcriptome sequencing to detect gene fusions in cancer. Nature 458(7234):97–101

Mardis ER, Wilson RK (2009) Cancer genome sequencing: a review. Hum Mol Genet 18(R2): R163–R168

Metzker ML (2010) Sequencing technologies—the next generation. Nat Rev Genet 11(1):31–46

Meyerson M, Gabriel S, Getz G (2010) Advances in understanding cancer genomes through second-generation sequencing. Nat Rev Genet 11(10):685–696

Mikuriya K, Kuramitsu Y, Ryozawa S, Fujimoto M, Mori S, Oka M, Hamano K, Okita K, Sakaida I, Nakamura K (2007) Expression of glycolytic enzymes is increased in pancreatic cancerous tissues as evidenced by proteomic profiling by two-dimensional electrophoresis and liquid chromatography-mass spectrometry/mass spectrometry. Int J Oncol 30(4):849–855

Morse DL, Balagurunathan Y, Hostetter G, Trissal M, Tafreshi NK, Burke N, Lloyd M, Enkemann S, Coppola D, Hruby VJ, Gillies RJ, Han H (2010) Identification of novel pancreatic adenocarcinoma cell-surface targets by gene expression profiling and tissue microarray. Biochem Pharmacol 80(5):748–754

Mortazavi A, Williams BA, McCue K, Schaeffer L, Wold B (2008) Mapping and quantifying mammalian transcriptomes by RNA-Seq. Nat Methods 5(7):621–628

Okazaki T, Javle M, Tanaka M, Abbruzzese JL, Li D (2010) Single nucleotide polymorphisms of gemcitabine metabolic genes and pancreatic cancer survival and drug toxicity. Clin Cancer Res 16(1):320–329

Olive KP, Jacobetz MA, Davidson CJ, Gopinathan A, McIntyre D, Honess D, Madhu B, Goldgraben MA, Caldwell ME, Allard D, Frese KK, Denicola G, Feig C, Combs C, Winter SP, Ireland-Zecchini H, Reichelt S, Howat WJ, Chang A, Dhara M, Wang L, Ruckert F, Grutzmann R, Pilarsky C, Izeradjene K, Hingorani SR, Huang P, Davies SE, Plunkett W, Egorin M, Hruban RH, Whitebread N, McGovern K, Adams J, Iacobuzio-Donahue C, Griffiths J, Tuveson DA (2009) Inhibition of Hedgehog signaling enhances delivery of chemotherapy in a mouse model of pancreatic cancer. Science 324(5933):1457–1461

Park PJ (2009) ChIP-seq: advantages and challenges of a maturing technology. Nat Rev Genet 10(10):669–680

Patti GJ, Yanes O, Siuzdak G (2012) Innovation: metabolomics: the apogee of the omics trilogy. Nat Rev Mol Cell Biol 13(4):263–269

Patwa TH, Li C, Poisson LM, Kim HY, Pal M, Ghosh D, Simeone DM, Lubman DM (2009) The identification of phosphoglycerate kinase-1 and histone H4 autoantibodies in pancreatic cancer patient serum using a natural protein microarray. Electrophoresis 30(12):2215–2226

Pollack JR, Perou CM, Alizadeh AA, Eisen MB, Pergamenschikov A, Williams CF, Jeffrey SS, Botstein D, Brown PO (1999) Genome-wide analysis of DNA copy-number changes using cDNA microarrays. Nat Genet 23(1):41–46

Pollack JR, Sorlie T, Perou CM, Rees CA, Jeffrey SS, Lonning PE, Tibshirani R, Botstein D, Borresen-Dale AL, Brown PO (2002) Microarray analysis reveals a major direct role of DNA copy number alteration in the transcriptional program of human breast tumors. Proc Natl Acad Sci USA 99(20):12963–12968

Rhodes DR, Kalyana-Sundaram S, Mahavisno V, Varambally R, Yu J, Briggs BB, Barrette TR, Anstet MJ, Kincead-Beal C, Kulkarni P, Varambally S, Ghosh D, Chinnaiyan AM (2007) Oncomine 3.0: genes, pathways, and networks in a collection of 18,000 cancer gene expression profiles. Neoplasia 9(2):166–180

Robertson G, Hirst M, Bainbridge M, Bilenky M, Zhao Y, Zeng T, Euskirchen G, Bernier B, Varhol R, Delaney A, Thiessen N, Griffith OL, He A, Marra M, Snyder M, Jones S (2007) Genome-wide profiles of STAT1 DNA association using chromatin immunoprecipitation and massively parallel sequencing. Nat Methods 4(8):651–657

Rubio-Viqueira B, Jimeno A, Cusatis G, Zhang X, Iacobuzio-Donahue C, Karikari C, Shi C, Danenberg K, Danenberg PV, Kuramochi H, Tanaka K, Singh S, Salimi-Moosavi H, Bouraoud N, Amador ML, Altiok S, Kulesza P, Yeo C, Messersmith W, Eshleman J, Hruban RH, Maitra A, Hidalgo M (2006) An in vivo platform for translational drug development in pancreatic cancer. Clin Cancer Res 12(15):4652–4661

Schena M, Shalon D, Davis RW, Brown PO (1995) Quantitative monitoring of gene expression patterns with a complementary DNA microarray. Science 270(5235):467–470

Shain AH, Giacomini CP, Matsukuma K, Karikari CA, Bashyam MD, Hidalgo M, Maitra A, Pollack JR (2012) Convergent structural alterations define SWItch/Sucrose NonFermentable (SWI/SNF) chromatin remodeler as a central tumor suppressive complex in pancreatic cancer. Proc Natl Acad Sci USA 109(5):E252–E259

Sheikh AA, Vimalachandran D, Thompson CC, Jenkins RE, Nedjadi T, Shekouh A, Campbell F, Dodson A, Prime W, Crnogorac-Jurcevic T, Lemoine NR, Costello E (2007) The expression of S100A8 in pancreatic cancer-associated monocytes is associated with the Smad4 status of pancreatic cancer cells. Proteomics 7(11):1929–1940

Shi W, Meng Z, Chen Z, Luo J, Liu L (2011) Proteome analysis of human pancreatic cancer cell lines with highly liver metastatic potential by antibody microarray. Mol Cell Biochem 347(1–2):117–125

Sørlie T, Perou CM, Tibshirani R, Aas T, Geisler S, Johnsen H, Hastie T, Eisen MB, Van De Rijn M, Jeffrey SS, Thorsen T, Quist H, Matese JC, Brown PO, Botstein D, Lønning PE, Børresen-Dale AL (2001) Gene expression patterns of breast carcinomas distinguish tumor subclasses with clinical implications. Proc Natl Acad Sci USA 98(19):10869–10874

Sugimoto M, Wong DT, Hirayama A, Soga T, Tomita M (2010) Capillary electrophoresis mass spectrometry-based saliva metabolomics identified oral, breast and pancreatic cancer-specific profiles. Metabolomics 6(1):78–95

Sun ZL, Zhu Y, Wang FQ, Chen R, Peng T, Fan ZN, Xu ZK, Miao Y (2007) Serum proteomic-based analysis of pancreatic carcinoma for the identification of potential cancer biomarkers. Biochim Biophys Acta 1774(6):764–771

Tanaka H, Hata F, Nishimori H, Honmou O, Yasoshima T, Nomura H, Ohno K, Hirai I, Kamiguchi K, Isomura H, Hirohashi Y, Denno R, Sato N, Hirata K (2003) Differential gene expression screening between parental and highly metastatic pancreatic cancer variants using a DNA microarray. J Exp Clin Cancer Res 22(2):307–313

Thompson JF, Steinmann KE (2010) Single molecule sequencing with a HeliScope genetic analysis system. Curr Protoc Mol Biol Chapter 7:Unit7 10.

Tian M, Cui YZ, Song GH, Zong MJ, Zhou XY, Chen Y, Han JX (2008) Proteomic analysis identifies MMP-9, DJ-1 and A1BG as overexpressed proteins in pancreatic juice from pancreatic ductal adenocarcinoma patients. BMC Cancer 8:241

Van't Veer LJ, Dai H, Van de Vijver MJ, He YD, Hart AAM, Mao M, Peterse HL, Van Der Kooy K, Marton MJ, Witteveen AT, Schreiber GJ, Kerkhoven RM, Roberts C, Linsley PS, Bernards R, Friend SH (2002) Gene expression profiling predicts clinical outcome of breast cancer. Nature 415(6871):530–536

Wang L, Heidt DG, Lee CJ, Yang H, Logsdon CD, Zhang L, Fearon ER, Ljungman M, Simeone DM (2009) Oncogenic function of ATDC in pancreatic cancer through Wnt pathway activation and beta-catenin stabilization. Cancer Cell 15(3):207–219

Wong KM, Hudson TJ, McPherson JD (2011) Unraveling the genetics of cancer: genome sequencing and beyond. Annu Rev Genomics Hum Genet 12:407–430

Yachida S, Jones S, Bozic I, Antal T, Leary R, Fu B, Kamiyama M, Hruban RH, Eshleman JR, Nowak MA, Velculescu VE, Kinzler KW, Vogelstein B, Iacobuzio-Donahue CA (2010) Distant metastasis occurs late during the genetic evolution of pancreatic cancer. Nature 467(7319):1114–1117

Ying H, Kimmelman AC, Lyssiotis CA, Hua S, Chu GC, Fletcher-Sananikone E, Locasale JW, Son J, Zhang H, Coloff JL, Yan H, Wang W, Chen S, Viale A, Zheng H, Paik JH, Lim C, Guimaraes AR, Martin ES, Chang J, Hezel AF, Perry SR, Hu J, Gan B, Xiao Y, Asara JM, Weissleder R, Wang YA, Chin L, Cantley LC, DePinho RA (2012) Oncogenic Kras maintains pancreatic tumors through regulation of anabolic glucose metabolism. Cell 149(3):656–670

Zhang J, Chiodini R, Badr A, Zhang G (2011) The impact of next-generation sequencing on genomics. J Genet Genomics 38(3):95–109

Index

D.M. Simeone and A. Maitra (eds.), *Molecular Genetics of Pancreatic Cancer*,
DOI 10.1007/978-1-4614-6549-2,

H

N

O

MIX
Papier aus verantwortungsvollen Quellen
Paper from responsible sources
FSC® C105338

If you have any concerns about our products,
you can contact us on
ProductSafety@springernature.com

In case Publisher is established outside the EU,
the EU authorized representative is:
Springer Nature Customer Service Center GmbH
Europaplatz 3, 69115 Heidelberg, Germany

Printed by Libri Plureos GmbH
in Hamburg, Germany